Biological Activity and Applications of Natural Compounds

Biological Activity and Applications of Natural Compounds

Editors

Ana M. L. Seca
Laila Moujir Moujir
Farukh Sharopov

MDPI • Basel • Beijing • Wuhan • Barcelona • Belgrade • Manchester • Tokyo • Cluj • Tianjin

Special Issue Editor

Ana M. L. Seca
cE3c/GBA, FCT—University of
Azores, Ponta Delgada
Portugal

Laila Moujir Moujir
Departamento de Bioquímica,
Microbiología, Genética y
Biologia Celular,
Facultad de Farmacia,
Universidad de La Laguna
Spain

Farukh Sharopov
Research Institution
"Chinese-Tajik Innovation
Center for Natural Products",
Academy of Sciences of the
Republic of Tajikistan,
Rudaki Avenue 33, Dushanbe
Tajikistan

Editorial Office
MDPI
St. Alban-Anlage 66
4052 Basel, Switzerland

This is a reprint of articles from the Special Issue published online in the open access journal *Applied Sciences* (ISSN 2076-3417) (available at: https://www.mdpi.com/journal/applsci/special_issues/activity_natural_compounds).

For citation purposes, cite each article independently as indicated on the article page online and as indicated below:

LastName, A.A.; LastName, B.B.; LastName, C.C. Article Title. *Journal Name* **Year**, *Article Number*, Page Range.

ISBN 978-3-03936-617-0 (Hbk)
ISBN 978-3-03936-618-7 (PDF)

Contents

About the Editors

Ana M. L. Seca has a degree in Chemistry and an MSc in the Science and Technology of Paper and Forest Products, both obtained at the University of Aveiro, Portugal, where she also completed her PhD in Chemistry in 2000. Since then, she has worked as an assistant professor at the University of Azores (Portugal) and, since 2016, has been a member of the Center for Ecology, Evolution and Environmental Changes—cE3c (ABG). She has published 48 SCI papers and 8 book chapters. Her current research interests range from the isolation and identification of secondary metabolites with potential pharmacological applications to the synthesis of pharmaceutically relevant natural compound analogues.

Laila Moujir Moujir has a degree in Pharmacy from the University of La Laguna, Spain, where she also completed her PhD in Microbiology in 1988. She was granted a diploma in Health and specialized in Clinical Analysis. She was a postdoctoral fellow at the University of British Columbia (Canada) and at the Molecular Biology Center (CBM) Madrid. In 2006, she started working as an associate professor at the Department of Microbiology and Biology Cell of the University of La Laguna. During her research career, she has published more than 50 papers within her field of research, "Search and selection of antimicrobials, cytostatic and antivirals of natural products".

Farukh Sharopov completed his Doctor of Sciences degree in Natural Sciences at Ruprecht Karls University in Heidelberg, Germany, in 2015. From 2018, he has been a postdoctoral researcher at the Xinjiang Technical Institute of Physics and Chemistry, Chinese Academy of Sciences (China), and at the same time, he is a researcher at the research institution Chinese-Tajik Innovation Center for Natural Products, National Academy of Sciences of the Republic of Tajikistan. He is the author of more than 100 research papers. His research interests are natural products, drug delivery and surface chemistry.

Editorial

Natural Compounds: A Dynamic Field of Applications

Ana M. L. Seca [1,2,*] and Laila Moujir [3]

[1] cE3c-Centre for Ecology, Evolution and Environmental Changes/Azorean Biodiversity Group & Faculty of Sciences and Technology, University of Azores, Rua Mãe de Deus, 9500-321 Ponta Delgada, Portugal
[2] LAQV-REQUIMTE, University of Aveiro, 3810-193 Aveiro, Portugal
[3] Department of Biochemistry, Microbiology, Genetics and Cell Biology, Facultad de Farmacia, Universidad de La Laguna, 38206 San Cristóbal de La Laguna, Spain; lmoujir@ull.es
* Correspondence: ana.ml.seca@uac.pt; Tel.: +351-29-6650174

Received: 2 June 2020; Accepted: 7 June 2020; Published: 10 June 2020

Abstract: Nature represents an amazing source of inspiration since it produces a great diversity of natural compounds selected by evolution, which exhibit multiple biological activities and applications. A large and very active research field is dedicated to identifying biosynthesized compounds, to improve/develop new methodologies to produce/reuse natural compounds and to assess their potential for pharmaceutical, cosmetic and food industries, among others, and also to understand their mechanism of action. Here, the main results presented in each work are highlighted. The applications suggested are mostly related to pharmacological uses and involve mainly pure natural compounds and essential oils. These works are significant contributions and reinforce the dynamic field of natural products applications.

Keywords: natural compounds; therapeutic applications; essential oils; antimicrobial; antitumor; SAR

---

1. Introduction

The compounds produced by nature exhibit a great diversity of chemical structures as a result of the long, selective and evolutive process of species [1], and they constitute the active principles of natural products which have always contributed, a lot, to improve human living conditions [2]. As natural products have widespread uses in traditional medicine [3–5], and a wide range of biological effects demonstrated scientifically, they possess high scientific and industrial value [6–9].

The properties demonstrated by natural compounds constantly encourage scientific research in aspects that lead to significant advances in the identification of new natural compounds, evaluation of the biological activity displayed, understanding of how they cause a biological effect, in the development of new applications and in all cases with beneficial results for humanity.

Despite very significant advances in medicine, many diseases such as cancer, infections, diabetes and cardiovascular diseases remain without affordable, effective and safe therapy. Due to this, the most frequently explored area with relevant results is the development of new drugs from natural sources [7,8,10–14], with natural product-based new drugs being 51% of small molecules approved and launched in the market between 1981 and 2014 [15].

The demand for natural antimicrobial agents and anticancer drugs is a very active research point [16–20] since cancer and infectious diseases are a significant cause of mortality worldwide [21]. Moreover, the World Health Organization predicts that the incidence of cancer will continue to rise to over 11 million in 2030 [22]. More than 1700 clinical trials involving the natural vinca alkaloids to treat cancer are registered currently on the clinicaltrials.gov platform [23]. Furthermore, the resistance and undesirable side effects of antibiotics and antineoplastic agents used in the clinic [24,25] have become pressing problems, leading to a continuous search for new inhibitors with new mechanisms of action.

The application of natural compounds as antioxidant agents is also a hot topic [26–29], once several diseases are, at least partially, a result of free radicals' imbalance [26,30,31]. Overproduction of oxygen free radicals, when associated with the deficiency in antioxidant repair or defence mechanisms, will cause oxidative damage leading to disease development like cancer, inflammation, diabetes, cardiovascular and aging-associated diseases [26,30–32].

Natural products are also applied as an ingredient in cosmetic preparations [33,34] and in the food industry since they exhibit useful preservative properties [35,36].

The limited availability of a natural product is a difficulty when it is a promising molecule for an application. Thus, recently, it has been proposed that several techniques that can improve the natural product yield. It is the case of metabolomics, enable the rapid identification of novel compounds in complex mixtures of natural products, metabolic engineering of cells endowed with the ability of overproduction of new products and metagenomics exploring novel metabolites from microorganisms present in several environments but which remain recalcitrant to culturing [37–39].

Natural compounds are privileged structures because of their structural diversity and multiple biological activities. They are undoubtedly the ideal compounds for the rational design of new drugs and the development of new chemical entities with therapeutic potential [8,14]. Thus, research on natural compounds is necessary and very welcome.

This Special Issue is dedicated to present the most recent results about the development of natural compounds applications. Ten original research works, organized by applications, and two reviews are included in this Special Issue. Each of them contributes to the knowledge advance, insofar as they present new applications for known products, new methodologies to obtain new products or the evaluation of a given application, with the applications related to health promotion being the most frequently considered.

2. Contributions

Several original papers included in this Special Issue involve the search for new antibacterial and antifungal agents, mainly from secondary metabolites, their derivatives and essential oils of different plant species. Essential oils are a mixture of volatile compounds, mainly mono-, di- and sesquiterpenes, with high commercial value and a wide range of applications [19,40].

Piper caldense (Piperaceae family) is popularly used for the treatment of snakebites, stomach problems and as a sedative [41]. The research presented by Bezerra et al. [42] shows that essential oils from *P. caldense* leaf are composed, in major proportions, by sesquiterpenes such as caryophyllene oxide, spathulenol, γ-cadinene and bicyclogermacrene. Although the essential oils do not show antifungal activity against *Candida albicans*, they exhibit a synergistic effect with the antifungal fuconazole, which increase their activity when used combined. This modulator effect is not observed at the highest concentration. The ability of essential oils to cause complete inhibition of the hyphae prolongation was also demonstrated, being an effect superior to fuconazole [42], which suggests their application as an adjuvant in antimicrobial formulations.

Helianthus annus (Asteraceae, sunflower) has been used as a traditional medicine to treat a variety of ailments, such as rattlesnake, spider, snake and scorpion bites, fevers [43,44], food poisoning [45], burning sensation in the vagina and worms in the ears [46]. Lawson et al. [47] report the chemical composition and antifungal activity of essential oils from the aerial parts of two cultivars of *H. annuus*, "Chianti" and "Mammoth", and wild-growing *Helianthus strumosus*. Analysis of these essential oils shows they are qualitatively similar and dominated by monoterpenes, in particular sabinene, α-pinene, β-pinene and limonene. The antifungal activity of essential oils and their main constituents were evaluated against three opportunistic pathogenic fungal species, *Aspergillus niger*, *Candida albicans* and *Cryptococcus neoformans*, which mainly affect immunocompromised patients. The essential oils of *H. annus* "Chianti" and *H. strumosus* are the most active against *C. neoformans* and the authors consider that the activity is due to both enantiomers of pinene.

Carvacrol is a usual essential oils constituent, that exhibits activity against *Penicillium digitatum*, a citrus crop fungal which causes significant economic losses [48], but whose mechanism of action is not completely clarified. Using an innovative approach, metabolomics based on ^{1}H-NMR analysis, Wan et al. [49] determine the changes introduced into the *P. digitatum* metabolism and its energetic balance when this fungus is treated with carvacrol. The results show this compound induces ROS production on *P. digitatum*, which mainly disturbs the energy production by a decrease in glycolysis followed by an augmentation in gluconeogenesis involving mainly alanine, aspartate and glutamate metabolism.

Phenolic nor-triterpene are described in the literature as antibacterial agents against Gram-positive bacteria [50–52]. Moujir et al. [53] report the antimicrobial activity of five natural phenolic nor-triterpenes, isolated from *Maytenus blepharodes* and *Maytenus canariensis*, and four pristimerin derivatives synthesized. The most active compound was the derivative 6α-hydroxy-2,3-diacetoxy-pristimerol which is more active than the positive control cephotaxime against spore-forming bacteria and *Staphylococcus epidermidis*. A comprehensive structure/activity relationship (SAR) study was performed with the aim of the rational design of antimicrobial agents based on the phenolic nor-triterpene scaffold.

In addition to an application as antimicrobials, natural compounds, in pure form or as a mixture like essential oils, are also investigated as therapeutic agents with a potential application on other health problems like aging-associated diseases, cancer, diabetes or wounds.

Portulaca oleracea is used to alleviate a wide range of illnesses [54] and the extracts of this plant are known to possess a strong in vivo antioxidant capacity [55,56]. Based on the potent antioxidant activity of natural compounds oleracones isolated from this plant, Yoon et al. [57] carried out the synthesis of oleracones D–F and evaluated their lifespan extension properties using the nematode *Caenorhabditis elegans* as the experimental aging model. The oleracone E was the most active to extend the lifespan of nematodes. Therefore, this compound should be considered in the development of anti-aging formulations.

Another study where the antioxidant activity of natural compounds was assessed is proposed by Tungmunnithum et al. [58]. They optimize and validate a new green and fast microextraction procedure to obtain a phenolic acids-rich fraction from almond (*Prunus dulcis*) cold-pressed oil residue. The authors identified and quantified some of the most abundant constituents, protocatechuic, *p*-hydroxybenzoic, chlorogenic and *p*-coumaric acids and evaluated the antioxidant activity of this fraction. They suggest it as a source of antioxidant natural compounds with applications in food preservation, as medicine and in cosmetic preparations. Thus, it is given a contribution to value a very abundant by-product of the almond oil industry.

Salvia miltiorrhizae is one of the well-known traditional herbal medicines that has increased its scientific interest for its remarkable bioactivity against cardiovascular disease, renal damage, tumor angiogenesis and tumor cell invasion [59,60]. Kang et al. [61] evaluate a series of compounds isolated from *S. miltiorrhizae* Radix such as tanshinone IIA, rosmarinic acid, salvianic acid A, salvianolic acid B and caffeic acid for their cytotoxic activity and mode of action against the HCT-116 cell line (human colorectal cancer, one of the most commonly diagnosed malignant diseases [62]). Only a terpenoid, tanshinone IIA, shows an inhibitory effect on cell survival (IC$_{50}$ 61.6 µg/mL). Morphological changes observed by Hoechst staining, an increase in cleaved caspase-7 and -8 and Bax expression as well as a decrease in Bcl-2 show this compound as an inducer of apoptosis which may be a useful chemotherapy strategy for patients with colorectal cancer.

The flavonoid-type compounds named chalcones and flavanones are natural isomeric compounds which exhibit a broad range of patented therapeutic applications [63]. Since they are chemically very labile, chalcones and flavanones are also excellent scaffolds to medicinal chemistry and development of new drugs [64–66]. In this context, the optimization of the aldol condensation synthesis route by using unexplored bases to prepare hydroxylated and methoxylated chalcones and flavanones proposed by Rosa et al. [67] is well come. The application of the synthesized compounds as cytotoxic, antioxidant, antibacterial and anticholinesterase agents were evaluated, and SAR are discussed. The SAR study showed the same substituent group can cause an opposite effect on the level of different biological activities, given as a significant contribution to better understand the medicinal chemistry of these compounds.

Plants are also considered to be sources of natural compounds with antidiabetic activity, as shown by Numonov et al. [68]. They studied the chemical composition of essential oils from *Prangos pabularia* roots and describe it as a potent protein tyrosine phosphatase 1B (PTP-1B) inhibitor. Additionally, the authors used docking studies and experimental procedures to indicate coumarin osthole, one of the most abundant constituents, as probably responsible for the potent PTP-1B inhibition, a stronger effect than the positive control, and thus support the use of *P. pabularia* roots' essential oil and osthole as antidiabetic agents.

Propolis and honey are valuable natural products with several health applications already described in the literature [69,70]. However, the properties and chemical composition of these natural products significantly depend on several factors such as the plant species used in their production [71]. Using different samples of Portuguese honey, propolis and a mixture of both, Afonso et al. [72] present an original study where the antioxidant, anti-inflammatory and wound-healing activities of ethanolic extracts of these samples were evaluated. The study demonstrates that propolis extracts have higher phenolic compounds and flavonoid contents than honey extracts. On the other hand, the mixtures of honey/propolis exhibit, in general, a significant wound-healing effect, but the propolis extract 2 (propolis from the red-honey box) is the most active sample, this activity being related to its high antioxidant and anti-inflammatory effect.

The Special Issue also includes two review articles where the applications of natural sesquiterpene lactones and rosmarinic acid as potential medicines are reviewed.

Moujir et al. [73] highlight ten natural sesquiterpene lactones, alantolactone, arglabin, costunolide, cynaropicrin, helenalin, inuviscolide, lactucin, parthenolide, thapsigargin and tomentosin, critically discussing the results of in vivo studies and clinical trials involving these compounds. The work demonstrates the enormous potential of these compounds in the development of new drugs especially for the treatment of oncological and inflammatory diseases. Their application as antifungal, antitrypanosomal and analgesic agents is also discussed, although the most surprising results are related to their sensitizing effect to the action of some clinical drugs, which means their application in combination therapy. Some synthetic derivatives of these sesquiterpene lactones are also highlighted, mainly because they demonstrate very significant improvements in pharmacokinetics and selectivity, compared with the original compounds.

In its turn, Nadeem et al. review the therapeutic value of rosmarinic acid [74]. This natural phenolic compound, very abundant on several edible species such as *Rosmarinus officinalis*, *Ocimum tenuiflorum* and *Thymus mastichina*, exhibits a wide range of biological effects. The authors [74] present and discuss scientific evidence of its antitumor, anti-inflammatory, antimicrobial, antidiabetic, antioxidant, anti-aging, cardio and nephroprotective effects. Further, the mechanism of action is discussed, although the authors identified this aspect as the knowledge gap that needs further investigation.

This *Applied Sciences* Special Issue emphasizes natural compounds' potential for distinct applications. The authors, from all over the world, contribute with valuable manuscripts to this Special Issue, strengthening the high value of natural compounds and the dynamism of this research field.

Author Contributions: A.M.L.S. and L.M. conceived, designed and wrote the editorial. All authors have read and agreed to the published version of the manuscript.

Funding: This research was funded by FCT–Fundação para a Ciência e a Tecnologia, the European Union, QREN, FEDER, COMPETE, by funding the cE3c centre (UIDB/00329/2020) and the LAQV-REQUIMTE (UIDB/50006/2020) research units, and by the Spanish Ministry Science and Research (MINECO RTI2018-094356-B-C21).

Acknowledgments: Thanks are due to the University of Azores and University of La Laguna.

Conflicts of Interest: The authors L.M. and A.M.L.S. declare no conflict of interest.

References

1. Kroymann, J. Natural diversity and adaptation in plant secondary metabolism. *Curr. Opin. Plant Biol.* **2011**, *14*, 246–251. [CrossRef] [PubMed]
2. World Health Organization. *WHO Traditional Medicine Strategy: 2014–2023*; WHO Press: Geneva, Switzerland, 2013; ISBN 978.
3. Yuan, H.; Ma, Q.; Ye, L.; Piao, G. The traditional medicine and modern medicine from natural products. *Molecules* **2016**, *21*, 559. [CrossRef] [PubMed]
4. Mukherjee, P.K.; Harwansh, R.K.; Bahadur, S.; Banerjee, S.; Kar, A.; Chanda, J.; Biswas, S.; Ahmmed, S.M.; Katiyar, C.K. Development of Ayurveda—Tradition to trend. *J. Ethnopharmacol.* **2017**, *197*, 10–24. [CrossRef] [PubMed]
5. World Health Organization. *WHO Global Report on Traditional and Complementary Medicine 2019*; WHO Press: Geneva, Switzerland, 2019; Available online: https://www.who.int/traditional-complementary-integrative-medicine/WhoGlobalReportOnTraditionalAndComplementaryMedicine2019.pdf?ua=1 (accessed on 2 June 2020).
6. Martins, A.; Vieira, H.; Gaspar, H.; Santos, S. Marketed marine natural products in the pharmaceutical and cosmeceutical industries: Tips for success. *Mar. Drugs.* **2014**, *12*, 1066–1101. [CrossRef] [PubMed]
7. Mushtaq, S.; Abbasi, B.H.; Uzair, B.; Abbasi, R. Natural products as reservoirs of novel therapeutic agents. *EXCLI J.* **2018**, *17*, 420–451. [CrossRef] [PubMed]
8. Lautié, E.; Russo, O.; Ducrot, P.; Boutin, J.A. Unraveling plant natural chemical diversity for drug discovery purposes. *Front. Pharmacol.* **2020**, *11*, 397. [CrossRef] [PubMed]
9. Sorokina, M.; Steinbeck, C. Review on natural products databases: Where to find data in 2020. *J. Cheminform* **2020**, *12*, 20. [CrossRef]
10. Dias, D.A.; Urban, S.; Roessner, U. A historical overview of natural products in drug discovery. *Metabolites* **2012**, *2*, 303–336. [CrossRef] [PubMed]
11. Newman, D.J.; Cragg, G.M. Natural products as sources of new drugs over the 30 years from 1981 to 2010. *J. Nat. Prod.* **2012**, *75*, 311–335. [CrossRef] [PubMed]
12. Butler, M.S.; Robertson, A.A.B.; Cooper, M.A. Natural product and natural product derived drugs in clinical trials. *Nat. Prod. Rep.* **2014**, *31*, 1612–1661. [CrossRef] [PubMed]
13. Thomford, N.E.; Senthebane, D.A.; Rowe, A.; Munro, D.; Seele, P.; Maroyi, A.; Dzobo, K. Natural products for drug discovery in the 21st century: Innovations for novel drug discovery. *Int. J. Mol. Sci.* **2018**, *19*, 1578. [CrossRef] [PubMed]
14. Dutta, S.; Mahalanobish, S.; Saha, S.; Ghosh, S.; Sil, P.C. Natural products: An upcoming therapeutic approach to cancer. *Food Chem. Toxicol.* **2019**, *128*, 240–255. [CrossRef] [PubMed]
15. Newman, D.J.; Cragg, G.M. Natural products as sources of new drugs from 1981 to 2014. *J. Nat. Prod.* **2016**, *79*, 629–661. [CrossRef] [PubMed]
16. Rocha, D.H.A.; Seca, A.M.L.; Pinto, D.C.G.A. Seaweed secondary metabolites in vitro and in vivo anticancer activity. *Mar. Drugs* **2018**, *16*, 410. [CrossRef] [PubMed]
17. Seca, A.M.L.; Pinto, D.C.G.A. Plant secondary metabolites as anticancer agents: Successes in clinical trials and therapeutic application. *Int. J. Mol. Sci.* **2018**, *19*, 263. [CrossRef]
18. Vengurlekar, S.; Sharma, R.; Trivedi, P. Efficacy of some natural compounds as antifungal agents. *Pharmacogn. Rev.* **2012**, *6*, 91–99. [CrossRef] [PubMed]
19. D'agostino, M.; Tesse, N.; Frippiat, J.P.; Machouart, M.; Debourgogne, A. Essential oils and their natural active compounds presenting antifungal properties. *Molecules* **2019**, *24*, 3713. [CrossRef] [PubMed]
20. Ribeiro da Cunha, B.; Fonseca, L.P.; Calado, C.R.C. Antibiotic discovery: Where have we come from, where do we go? *Antibiotics* **2019**, *8*, 45. [CrossRef] [PubMed]
21. Ritchie, H.; Roser, M. Causes of Death. Published online at OurWorldInData.org. Available online: https://ourworldindata.org/causes-of-death (accessed on 2 June 2020).
22. WHO Website. Available online: https://www.who.int/cancer/resources/keyfacts/en/ (accessed on 29 May 2020).
23. ClinicalTrials.gov Website. Search Terms: "Vincristine OR Vinblastine OR Vinorelbine OR Vindesine". Available online: https://clinicaltrials.gov/ct2/results?cond=&term=vincristine+OR+vinblastine+OR+vinorelbine+OR+vindesine&cntry=&state=&city=&dist= (accessed on 29 May 2020).

24. Laws, M.; Shaaban, A.; Rahman, K.M. Antibiotic resistance breakers: Current approaches and future directions. *FEMS Microbiol. Rev.* **2019**, *43*, 490–516. [CrossRef]

25. Chatterjee, N.; Bivona, T.G. Polytherapy and targeted cancer drug resistance. *Trends Cancer* **2019**, *5*, 170–182. [CrossRef] [PubMed]

26. Forni, C.; Facchiano, F.; Bartoli, M.; Pieretti, S.; Facchiano, A.; D'Arcangelo, D.; Norelli, S.; Valle, G.; Nisini, R.; Beninati, S.; et al. Beneficial role of phytochemicals on oxidative stress and age-related diseases. *Biomed. Res. Int.* **2019**, *2019*, 8748253. [CrossRef] [PubMed]

27. Pohl, F.; Lin, P.K.T. The potential use of plant natural products and plant extracts with antioxidant properties for the prevention/treatment of neurodegenerative diseases: In vitro, in vivo and clinical trials. *Molecules* **2018**, *23*, 3283. [CrossRef] [PubMed]

28. Tavares, W.R.; Seca, A.M.L. *Inula* L. secondary metabolites against oxidative stress-related human diseases. *Antioxidants* **2019**, *8*, e122. [CrossRef]

29. Cardoso, S.M. Special Issue: The antioxidant capacities of natural products. *Molecules* **2019**, *24*, 492. [CrossRef] [PubMed]

30. Liguori, I.; Russo, G.; Curcio, F.; Bulli, G.; Aran, L.; Della-Morte, D.; Gargiulo, G.; Testa, G.; Cacciatore, F.; Bonaduce, D.; et al. Oxidative stress, aging, and diseases. *Clin. Interv. Aging* **2018**, *13*, 757–772. [CrossRef] [PubMed]

31. Nandita, H.; Manohar, M.; Gowda, D.V. Recent review on oxidative stress, cellular senescence and age-associated diseases. *Int. J. Res. Pharma. Sci.* **2020**, *11*, 1331–1342. [CrossRef]

32. Khanna, R.D.; Karki, K.; Pande, D.; Negi, R.; Khanna, R.S. Inflammation, free radical damage, oxidative stress and cancer. *Interdiscip. J. Microinflamm.* **2014**, *1*, 109. [CrossRef]

33. Mahesh, S.K.; Fathima, J.; Veena, V.G. Cosmetic potential of natural products: Industrial applications. In *Natural bio-Active Compounds*; Vol. 2: Chemistry, Pharmacology and Health Care Practices; Swamy, M.K., Akhtar, M.S., Eds.; Springer: Singapore, 2019; pp. 215–250. [CrossRef]

34. Salehi, B.; Sharifi-Rad, J.; Seca, A.M.L.; Pinto, D.C.G.A.; Michalak, I.; Trincone, A.; Mishra, A.P.; Nigam, M.; Zam, W.; Martins, N. Current trends on seaweeds: Looking at chemical composition, phytopharmacology, and cosmetic applications. *Molecules* **2019**, *24*, 4182. [CrossRef] [PubMed]

35. Faustino, M.; Veiga, M.; Sousa, P.; Costa, E.M.; Silva, S.; Pintado, M. Agro-food byproducts as a new source of natural food additives. *Molecules* **2019**, *24*, 1056. [CrossRef] [PubMed]

36. Baptista, R.C.; Horita, C.N.; Sant'Ana, A.S. Natural products with preservative properties for enhancing the microbiological safety and extending the shelf-life of seafood: A review. *Food Res. Int.* **2020**, *127*, 108762. [CrossRef] [PubMed]

37. Trindade, M.; van Zyl, L.J.; Navarro-Fernández, J.; Abd Elrazak, A. Targeted metagenomics as a tool to tap into marine natural product diversity for the discovery and production of drug candidates. *Front Microbiol.* **2015**, *6*, 890. [CrossRef] [PubMed]

38. Zhao, Q.; Zhang, J.; Li, F. Application of metabolomics in the study of natural products. *Nat. Prod. Bioprospect.* **2018**, *8*, 321–334. [CrossRef] [PubMed]

39. Chen, R.; Yang, S.; Zhang, L.; Zhou, Y.J. Advanced strategies for production of natural products in yeast. *iScience* **2020**, *23*, 100879. [CrossRef] [PubMed]

40. Aguilar-Veloz, L.M.; Calderón-Santoyo, M.; González, Y.V.; Ragazzo-Sánchez, J.A. Application of essential oils and polyphenols as natural antimicrobial agents in postharvest treatments: Advances and challenges. *Food Sci. Nutr.* **2020**. [CrossRef]

41. Cardozo-Junior, E.L.; Chaves, M.C.O. Caldensin, a new natural n-methylaristolactam from *Piper Caldense*. *Pharm. Biol.* **2003**, *41*, 216–218. [CrossRef]

42. Bezerra, J.W.A.; Rodrigues, F.C.; Pereira da Cruz, R.; Silva, L.E.; do Amaral, W.; Andrade Rebelo, R.; Begnini, I.M.; Fonseca Bezerra, C.; Iriti, M.; Varoni, E.M.; et al. Antibiotic potential and chemical composition of the essential oil of *Piper caldense* C. DC. (Piperaceae). *Appl. Sci.* **2020**, *10*, 631. [CrossRef]

43. Moerman, D.E. *Native American Ethnobotany*; Timber Press, Inc.: Portland, OR, USA, 1998.

44. Camazine, S.; Bye, R.A. A study of the medical ethnobotany of the Zuni Indians of New Mexico. *J. Ethnopharmacol.* **1980**, *2*, 365–388. [CrossRef]

45. Mesfin, F.; Demissew, S.; Teklehaymanot, T. An ethnobotanical study of medicinal plants in Wonago Woreda, SNNPR, Ethiopia. *J. Ethnobiol. Ethnomed.* **2009**, *5*, 28. [CrossRef] [PubMed]

46. Rahman, A.H.M.M. Medico-ethnobotany: A study on the tribal people of Rajshahi division, Bangladesh. *Peak J. Med. Plant Res.* **2016**, *1*, 1–8.

47. Lawson, S.K.; Sharp, L.G.; Powers, C.N.; McFeeters, R.L.; Satyal, P.; Setzer, W.N. Essential oil compositions and antifungal activity of sunflower (*Helianthus*) species growing in North Alabama. *Appl. Sci.* **2019**, *9*, 3179. [CrossRef]

48. Chen, J.; Shen, Y.; Chen, C.; Wan, C. Inhibition of key citrus postharvest fungal strains by plant extracts in vitro and in vivo: A review. *Plants* **2019**, *8*, 26. [CrossRef] [PubMed]

49. Wan, C.; Shen, Y.; Nisar, M.F.; Qi, W.; Chen, C.; Chen, J. The antifungal potential of carvacrol against *Penicillium digitatum* through ^{1}H-NMR based metabolomics approach. *Appl. Sci.* **2019**, *9*, 2240. [CrossRef]

50. De León, L.; Beltrán, B.; Moujir, L. Antimicrobial activity of 6-oxophenolic triterpenoids. Mode of action against *Bacillus subtilis*. *Planta Med.* **2005**, *71*, 313–319. [CrossRef] [PubMed]

51. De León, L.; Moujir, L. Activity and mechanism of the action of zeylasterone against *Bacillus subtilis*. *J. Appl. Microbiol.* **2008**, *104*, 1266–1274. [CrossRef] [PubMed]

52. López, M.R.; de León, L.; Moujir, L. Antibacterial properties of phenolic triterpenoids against *Staphylococcus epidermidis*. *Planta Med.* **2011**, *77*, 726–729. [CrossRef] [PubMed]

53. Moujir, L.; López, M.R.; Reyes, C.P.; Jiménez, I.A.; Bazzocchi, I.L. Structural requirements for antimicrobial activity of phenolic nor-triterpenes from Celastraceae species. *Appl. Sci.* **2019**, *9*, 2957. [CrossRef]

54. Zhou, Y.X.; Xin, H.L.; Rahman, K.; Wang, S.J.; Peng, C.; Zhang, H. *Portulaca oleracea* L.: A review of phytochemistry and pharmacological effects. *BioMed. Res. Int.* **2015**, *2015*, 925631. [CrossRef] [PubMed]

55. Silva, R.; Carvalho, I.S. In vitro antioxidant activity, phenolic compounds and protective effect against DNA damage provided by leaves, stems and flowers of *Portulaca oleracea* (Purslane). *Nat. Prod. Commun.* **2014**, *9*, 45–50. [CrossRef]

56. Ahangarpour, A.; Lamoochi, Z.; Moghaddam, H.F.; Mansouri, S.M.T. Effects of *Portulaca oleracea* ethanolic extract on reproductive system of aging female mice. *Int. J. Reprod. Biomed.* **2016**, *14*, 205–212. Available online: http://journals.ssu.ac.ir/ijrmnew/article-1-730-en.html (accessed on 2 June 2020). [CrossRef]

57. Yoon, J.A.; Lim, C.; Cha, D.S.; Han, Y.T. Synthesis and evaluation of the lifespan-extension properties of oleracones D–F, antioxidative flavonoids from *Portulaca oleracea* L. *Appl. Sci.* **2019**, *9*, 4014. [CrossRef]

58. Tungmunnithum, D.; Elamrani, A.; Abid, M.; Drouet, S.; Kiani, R.; Garros, L.; Kabra, A.; Addi, M.; Hanno, C. A quick, green and simple ultrasound-assisted extraction for the valorization of antioxidant phenolic acids from Moroccan almond cold-pressed oil residues. *Appl. Sci.* **2020**, *10*, 3313. [CrossRef]

59. Zhao, W.; Yuan, Y.; Zhao, H.; Han, Y.; Chen, X. Aqueous extract of *Salvia miltiorrhiza* Bunge-Radix Puerariae herb pair ameliorates diabetic vascular injury by inhibiting oxidative stress in streptozotocin-induced diabetic rats. *Food Chem. Toxicol.* **2019**, *129*, 97–107. [CrossRef] [PubMed]

60. Zhang, L.J.; Chen, L.; Lu, Y.; Wu, J.M.; Xu, B.; Sun, Z.G.; Zheng, S.Z.; Wang, A.Y. Danshensu has anti-tumor activity in B16F10 melanoma by inhibiting angiogenesis and tumor cell invasion. *Eur. J. Pharmacol.* **2010**, *643*, 195–201. [CrossRef] [PubMed]

61. Kang, B.; Lee, S.; Seo, C.-S.; Kang, K.S.; Choi, Y.-K. Analysis and identification of active compounds from *Salviae miltiorrhizae* radix toxic to HCT-116 human colon cancer cells. *Appl. Sci.* **2020**, *10*, 1304. [CrossRef]

62. Haggar, F.A.; Boushey, R.P. Colorectal cancer epidemiology: Incidence, mortality, survival, and risk factors. *Clin. Colon Rectal Surger.* **2009**, *22*, 191–197. [CrossRef] [PubMed]

63. Mahapatra, D.K.; Asati, V.; Bharti, S.K. An updated patent review of therapeutic applications of chalcone derivatives (2014-present). *Expert Opin. Ther. Pat.* **2019**, *29*, 385–406. [CrossRef] [PubMed]

64. Zhuang, C.; Zhang, W.; Sheng, C.; Zhang, W.; Xing, C.; Miao, Z. Chalcone: A privileged structure in medicinal chemistry. *Chem. Rev.* **2017**, *117*, 7762–7810. [CrossRef] [PubMed]

65. Gomes, M.N.; Muratov, E.N.; Pereira, M.; Peixoto, J.C.; Rosseto, L.P.; Cravo, P.V.L.; Andrade, C.H.; Neves, B.J. Chalcone derivatives: Promising starting points for drug design. *Molecules* **2017**, *22*, 1210. [CrossRef] [PubMed]

66. Farooq, S.; Ngaini, Z. Recent synthetic methodologies for chalcone synthesis (2013–2018). *Curr. Organocatalysis* **2019**, *6*, 184–192. [CrossRef]

67. Rosa, G.P.; Seca, A.M.L.; Barreto, M.C.; Silva, A.M.S.; Pinto, D.C.G.A. Chalcones and flavanones bearing hydroxyl and/or methoxyl groups: Synthesis and biological assessments. *Appl. Sci.* **2019**, *9*, 2846. [CrossRef]

68. Numonov, S.; Sharopov, F.S.; Atolikhshoeva, S.; Safomuddin, A.; Bakri, M.; Setzer, W.N.; Musoev, A.; Sharofova, M.; Habasi, M.; Aisa, H.A. Volatile secondary metabolites with potent antidiabetic activity from

the roots of *Prangos pabularia* Lindl.—Computational and experimental investigations. *Appl. Sci.* **2019**, *9*, 2362. [CrossRef]

69. Pasupuleti, V.R.; Sammugam, L.; Ramesh, N.; Gan, S.H. Honey, propolis, and royal jelly: A comprehensive review of their biological actions and health benefits. *Oxid. Med. Cell Longev.* **2017**, *2017*, 1259510. [CrossRef] [PubMed]

70. Braakhuis, A. Evidence on the health benefits of supplemental propolis. *Nutrients* **2019**, *11*, 2705. [CrossRef] [PubMed]

71. Liu, J.R.; Ye, Y.L.; Lin, T.Y.; Wang, Y.W.; Peng, C.C. Effect of floral sources on the antioxidant, antimicrobial, and anti-inflammatory activities of honeys in Taiwan. *Food Chem.* **2013**, *139*, 938–943. [CrossRef] [PubMed]

72. Afonso, A.M.; Gonçalves, J.; Luís, Â.; Gallardo, E.; Duarte, A.P. Evaluation of the *in vitro* wound-healing activity and phytochemical characterization of propolis and honey. *Appl. Sci.* **2020**, *10*, 1845. [CrossRef]

73. Moujir, L.; Callies, O.; Sousa, P.M.C.; Sharopov, F.; Seca, A.M.L. Applications of sesquiterpene lactones: A review of some potential success cases. *Appl. Sci.* **2020**, *10*, 3001. [CrossRef]

74. Nadeem, M.; Imran, M.; Aslam Gondal, T.; Imran, A.; Shahbaz, M.; Muhammad Amir, R.; Wasim Sajid, M.; Batool Qaisrani, T.; Atif, M.; Hussain, G.; et al. Therapeutic potential of rosmarinic acid: A comprehensive review. *Appl. Sci.* **2019**, *9*, 3139. [CrossRef]

Article

Antibiotic Potential and Chemical Composition of the Essential Oil of *Piper caldense* C. DC. (Piperaceae)

José Weverton Almeida Bezerra [1], Felicidade Caroline Rodrigues [1], Rafael Pereira da Cruz [2], Luiz Everson da Silva [3], Wanderlei do Amaral [3], Ricardo Andrade Rebelo [3], Ieda Maria Begnini [3], Camila Fonseca Bezerra [2], Marcello Iriti [4,*], Elena Maria Varoni [5], Henrique Douglas Melo Coutinho [6] and Maria Flaviana Bezerra Morais-Braga [2]

[1] Postgraduate Program in Plant Biology, Federal University of Pernambuco—UFPE, Recife 50670-901, PE, Brazil; weverton.almeida@urca.br (J.W.A.B.); rodriguescaroline26@gmail.com (F.C.R.)

[2] Laboratory of Applied Mycology of Cariri, Regional University of Cariri—URCA, Crato 63105-000, CE, Brazil; rafaelcruz284@gmail.com (R.P.d.C.); camilawasidi@gmail.com (C.F.B.); flavianamoraisb@yahoo.com.br (M.F.B.M.-B.)

[3] Research Laboratory of Natural Products, Federal University of Paraná—UFPR, Matinhos 83260-000, PR, Brazil; luiz_everson@yahoo.de (L.E.d.S.); wdoamaral@hotmail.com (W.d.A.); ricardorebelo@furb.br (R.A.R.); ieda@furb.br (I.M.B.)

[4] Department of Agricultural and Environmental Sciences, Milan State University, 20133 Milan, Italy

[5] Department of Biomedical, Surgical and Dental Sciences, Milan State University, 20142 Milan, Italy; elena.varoni@unimi.it

[6] Laboratory of Microbiology and Molecular Biology—LMBM, Regional University of Cariri—URCA, Crato 63105-000, CE, Brazil; hdmcoutinho@gmail.com

[*] Correspondence: marcello.iriti@unimi.it

Received: 17 December 2019; Accepted: 10 January 2020; Published: 15 January 2020

Abstract: Infections by multiresistant microorganisms have led to a continuous investigation of substances acting as modifiers of this resistance. By following this approach, the chemical composition of the essential oil from *Piper caldense* leaf and its antimicrobial potential were investigated. The antimicrobial activity was determined by broth microdilution method providing values for minimum inhibitory concentration (MIC), IC_{50}, and minimum fungicidal concentration (MFC). The essential oil was tested as a modulator for several antibiotics, and its effect on the morphology of *Candida albicans* (CA) strains was also investigated. The chemical characterization revealed an oil composed mainly of sesquiterpenes. Among them are caryophyllene oxide (13.9%), spathulenol (9.1%), δ-cadinene (7.6%) and bicyclogermacrene (6.7%) with the highest concentrations. The essential oil showed very low activity against the strains of CA with the lowest values for IC_{50} and MFC of 1790 µg/mL and 8192 µg/mL, respectively. The essential oil modulated the activity of fluconazole against CA URM 4387 strain, which was demonstrated by the lower IC_{50} obtained, 2.7 µg/mL, whereas fluconazole itself presented an IC_{50} of 7.76 µg/mL. No modulating effect was observed in the MFC bioassays. The effect on fungal morphology was observed for both CA INCQS 40006 and URM 4387 strains. The hyphae projection was completely inhibited at 4096 µg/mL and 2048 µg/mL, respectively. Thus, the oil has potential as an adjuvant in antimicrobial formulations.

Keywords: pimenta d'água; *Candida*; fungistatic effect; inhibition of dimorphism; GC/MS

1. Introduction

Infections caused by fungi, are the major problem of hospital infections, mainly due to the emergence of new infections and the generalized resistance to antibiotics [1]. Due to the rapid resistance acquired by microorganisms the search for natural products with antimicrobial activity has been constant in recent years and drugs derived from plants have contributed largely to human health, historically representing a source for the creation of new drugs [2].

One way of combating microbial resistance is the investigation of substances that can act as resistance modifiers by combining common antimicrobial drugs with some phytochemicals, and in some cases observed synergism [3,4].

Yeasts of the genus *Candida* are found as commensals in the human organism, however, because of factors that disturb the dynamics of the host can favor the growth of these fungi, from superficial infections to systemic infections [5]. Among the yeasts of this genus, *Candida albicans* stands out as the main cause of infections, being a species with great ability to change its morphology, a necessary factor for its virulence and pathogenicity [6,7].

For the treatment of infectious diseases, populations without access to medicines, especially those from underdeveloped and developing countries, use medicinal plants to combat disease. Such use of these vegetables is due to a number of advantages, such as their availability, low purchasing power and being in popular knowledge for several generations [2,8]. Among the species of the Brazilian flora, the genus *Piper* L. is one of the largest in the Piperaceae family, with 290 species and 45 varieties occurring in Brazil [9]. Some species of the genus are used as a flavoring of food and treatment of diseases [10]. A scientific study evidenced the antimicrobial potential of four species of the genus against *Staphylococcus aureus* and three strains of *Candida*, including *C. albicans* [11]. Among the species of the genus, *Piper caldense* C. DC., is popularly known in Brazil as "pimenta d'água" or "pimenta d'arda", being commonly used for the treatment of snake bites, sedative and stomach problems [10]. Research using *P. caldense* essential oil has revealed promising antimicrobial potential [12,13].

In view of the above problem, this study aimed to evaluate the antibiotic and modulator potential of volatile terpenes of *P. caldense* against *Candida albicans*, as well as to determine if the natural product is capable of reducing pleomorphism. Finally, it was evaluated by gas chromatography which terpenes were present in the essential oil.

2. Results

2.1. Chemical Composition

The chromatographic analysis of the essential oil of *P. caldense* identified 25 phytochemical constituents, corresponding to 94.3% of the total composition. Sesquiterpenes are the main class observed, reaching a percentage of 89.7% of the sample. Monoterpenes were found in very low concentration (4.6%). There were no major constituents (>20%) and no constituent in trace concentration (<1%). All the identified phytochemicals are secondary metabolites, with caryophyllene oxide (13.9%), spathulenol (9.1%), δ-cadinene (7.6%), and bicyclogermacrene (6.7%) as constituents with the highest concentrations (Table 1).

2.2. Antifungal Activity

2.2.1. Cell Viability Curve and IC$_{50}$

Regarding the antifungal activity of the essential oil of *P. caldense*, it was found that it has a low antifungal effect against strains of *C. albicans* because the IC$_{50}$ values were high, 2256.24 and 1790.24 µg/mL for the CA INCQS 40006 and CA URM 4387, respectively (Table 2) (Figures 1 and 2). However, it was observed that for the *C. albicans* URM 4387 strains, the oil potentiated the effect of the drug used, fluconazole, since it had an IC$_{50}$ of 7.73 µg/mL alone, and when associated with the essential oil, the value of evaluated parameter dropped to 2.7 µg/mL, i.e., a significant reduction. It is noteworthy that for *C. albicans* 40006 strains, no potentiating effect was found for fluconazole but an antagonistic effect, since there was an increase in the IC$_{50}$ of the oil associated with fluconazole. Thus, based on the behavior of the curve through non-linear regression, the IC$_{50}$ values of *P. caldense* oil were clinically irrelevant when evaluated alone.

Table 1. Chemical composition of the essential oil of leaves from *Piper caldense.*

Compounds	%	AI_{lit}	MF
α-Pinene	2.5	935	$C_{10}H_{16}$
Camphene	2.1	952	$C_{10}H_{16}$
α-copaene	2	1381	$C_{15}H_{24}$
(E)-Caryophyllene	2.6	1426	$C_{15}H_{24}$
Aromadendrene	2.7	1445	$C_{15}H_{24}$
γ-muurolene	4	1478	$C_{15}H_{24}$
β-selinene	3.2	1490	$C_{15}H_{24}$
Bicyclogermacrene	6.7	1492	$C_{15}H_{24}$
Germacrene D	5.3	1493	$C_{15}H_{24}$
α-muurolene	4.2	1498	$C_{15}H_{24}$
δ-Cadinene	7.6	1520	$C_{15}H_{24}$
γ-Cadinene	3.3	1521	$C_{15}H_{24}$
α-Calacorene	2.2	1551	$C_{15}H_{20}$
β-Calacorene	1.9	1572	$C_{15}H_{20}$
Spathulenol	9.1	1576	$C_{15}H_{24}O$
Caryophyllene oxide	13.9	1580	$C_{15}H_{24}O$
Globulol	2.3	1594	$C_{15}H_{26}O$
Rosifoliol	1.3	1597	$C_{15}H_{26}O$
Humulene epoxide II	1	1619	$C_{15}H_{24}O$
1.10-di-epi-Cubenol	1.6	1624	$C_{15}H_{26}O$
1-epi-Cubenol	3.4	1638	$C_{15}H_{26}O$
epi-alfa-muurolol	3.1	1640	$C_{15}H_{26}O$
α-cadinol	4.2	1650	$C_{15}H_{26}O$
α-Muurolol	2.1	1651	$C_{15}H_{26}O$
Cadalene	2	1667	$C_{15}H_{18}$
Total sesquiterpenes	89.7		
Total monoterpenes	4.6		
Total identified (%)	**94.3**		

AI_{lit}—Arithmetic Retention Indices from literature; MF: Molecular Formula.

Table 2. IC_{50} of the essential oil of *Piper caldense* (EOPc) against *Candida albicans.*

Products Tested	Yeast	
	Candida albicans **INCQS 40006**	*Candida albicans* **URM 4387**
Fluconazole (FCZ)	7.76 µg/mL	7.73 µg/mL
EOPc	2256.24 µg/mL	1790.24 µg/mL
EOPc + FCZ	12.37 µg/mL	2.7 µg/mL

INCQS: National Institute for Health Quality Control. URM: University Recife Mycology.

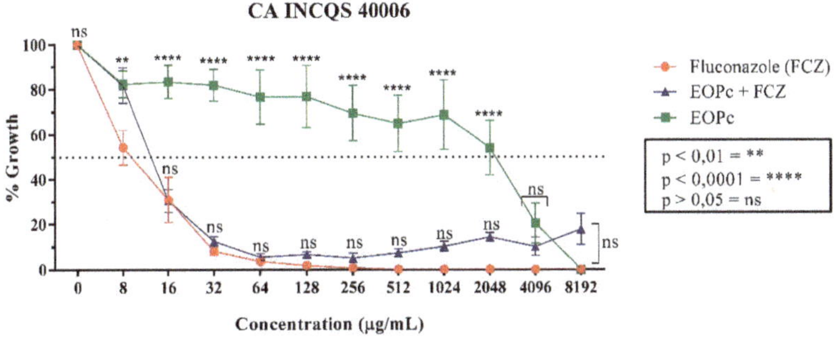

Figure 1. Anti-*Candida* potential of essential oil of *Piper caldense* (EOPc) against strains of *Candida albicans* 40006 INCQS.

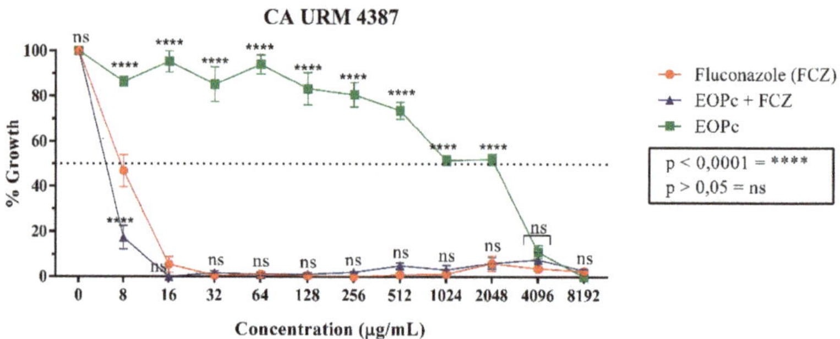

Figure 2. Anti-*Candida* potential of essential oil of *Piper caldense* (EOPc) against strains of *Candida albicans* URM 4387.

2.2.2. Cell Viability Curve and Minimum Fungicidal Concentration (MFC)

In determining MFC it was considered those samples capable of inhibiting the growth of the fungal colonies. Thus, for the CA INCQS 40006 line, none of the products alone or in combination were able to totally inhibit colony growth, since MFC was $\geq$ 16,384 µg/mL (Table 3). However, there was an MFC for the oil against CA URM 4387, in which fluconazole alone and combined with *P. caldense* oil showed a much lower MFC, 16 µg/mL, so that the oil did not modulate the fungicidal effect of fluconazole for this lineage (Table 3).

Table 3. Minimum fungicidal concentration (MFC) of essential oil of *Piper caldense* (EOPc) and fluconazole associated and isolated against strains of *Candida albicans*.

Products Tested	Yeast	
	Candida albicans **INCQS 40006**	*Candida albicans* **INCQS 40006**
Fluconazole (FCZ)	$\geq$16,384 µg/mL	16 µg/mL
EOPc	$\geq$16,384 µg/mL	8192 µg/mL
EOPc + FCZ	$\geq$16,384 µg/mL	16 µg/mL

INCQS: National Institute for Health Quality Control. URM: University Recife Mycology.

2.2.3. Activity of the *Piper caldense* in the Control of Virulence of *Candida albicans*

In the evaluation of the activity of the oil in the morphology of the yeasts of *C. albicans*, the effect was caused by impoverishment of the culture medium, so that the yeasts project the hyphae and pseudohyphae in search of nutrients. For the growth control of CA INCQS 40006 (Figure 3, Slides 1–7), it is possible to observe the formation of several hyphae (S1), whereas in the treatments with fluconazole, there is a significant decrease in the hyphae projection (S2–4), resulting in a complete inhibition of hyphae at concentration of 4096 µg/mL (S2). The EOPc showed a similar result, being as effective as fluconazole.

For the CA URM 4387 strain, fluconazole at a concentration as low as 8 µg/mL could diminish the virulence (Figure 4, S2). The oil at higher concentrations, such as 2048 µg/mL (S3), was able to inhibit the hyphae projection; on the other hand, at concentration of 512 µg/mL (S5), there was a significant decrease in hyphae projections.

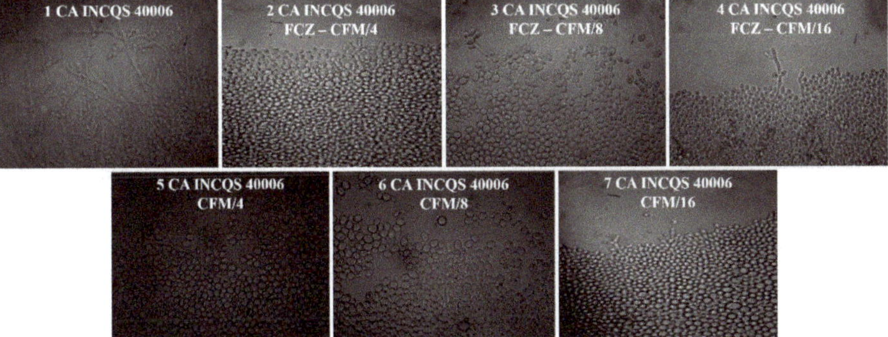

Figure 3. Effects of essential oil of *Piper caldense* on the dimorphism of *Candida albicans* INCQS 40006. Slide (S1): Growth control; S2–4: Effect of fluconazole at concentrations 4096 µg/mL (S2), 2048 µg/mL (S3), 1024 µg/mL (S4). S5–7: Effect of essential oil at concentrations of 4096 µg/mL (S5), 2048 µg/mL (S6), 1024 µg/mL (S7). Display 400× increased.

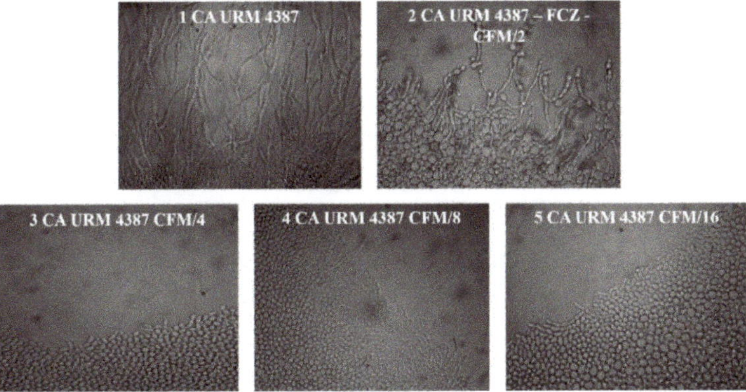

Figure 4. Effects of essential oil of *Piper caldense* on the dimorphism of *Candida albicans* URM 4387. Slide (S1): Growth control; S2: Effect of fluconazole at the concentration of 8 µg/mL. S3–5: Effect of the essential oil at 2048 µg/mL (S3), 1024 µg/mL (S4), 512 µg/mL (S5).

3. Discussion

The genus *Piper* L. presents a high number of species with medicinal, insecticidal and condiment applications, since the representatives are sources of volatile oils produced by the secondary metabolism [14]. Some species of this genus present antifungal activities, such as *Piper amalago* L. [15], *Piper aduncum* L. and *Peperomia pelúcida* (L.) Kunth [16], and antibacterial activity, among them *Piper betle* L. [17]. Thus, the selection strategy of *P. caldense* for the investigation of the antimicrobial activities of this study was based on chemotaxonomy, since there is a phylogenetic relationship between *P. caldense* and other species of the genus and possibly in the evolutionary historical branch the biosynthetic routes are similar [18].

The phytochemicals (monoterpenes and sesquiterpenes) of the essential oil of the species under study were elucidated by Rocha et al. [19], however, there were marked differences in the constitution, the first is that in our study, the caryophyllene oxide was the major constituent (13.9%), whereas in the results of the aforementioned study this sesquiterpene is not present in the oil of leaves, but in the stem essential oil (6.2%). In addition, Rocha et al. [19] states that the constituent in higher percentage is the α-cadinol, reaching compose 19% of the total composition, and in our study this sesquiterpeno is in low percentages (4.2%). This variation is justified by several factors, both intrinsic and genetic, as well as extrinsic factors such as geographic origin of the plant, cultivation, collection form, and especially the period of the year that was collected [20–22].

Although the essential oil of *P. caldense* did not present antifungal activity in low concentrations (≤500 µg/mL), it presented a modulating effect for fluconazole against strains of *C. albicans* URM 4387. So that this finding is relevant, since the introduction of azole class antibiotics (miconazole, econazole, ketoconazole, fluconazole and triazonazole) for the treatment of infections caused by *Candida* species, a growing emergence of resistant *Candida* species has been observed [23,24].

The activity of caryophyllene oxide, an oxygenated terpenoid, was tested in the laboratory against dermatophyte fungi, showing significant results, and their activity has been compared with antifungals such as cyclopiroxolamine and sulconazole [12].

Silva [25], evaluating the antifungal activity of 2-geranyl-3,4-dihydroxybenzoic acid and 3-geranyl-4-hydroxyxidozoic acid, both substances isolated from the fruits of *P. caldense*, demonstrated, respectively, a moderate and high activity against *C. albicans* strains (LM-86 and LM-111), so that the first substance showed a MIC of 512 µg/mL for both strains and the second one 32 µg/mL, also for the two strains. It is important to highlight that in our study the essential oil was used as the product, and this is a mixture of mono and sesquiterpenes, whereas in the study mentioned above the substances were derived from the benzoic acid prenylate, so that the chemical structures are quite different.

Inhibition of virulence of *C. albicans* strains by natural products was also shown by other scientists, among them Santos et al. [26], who evaluated the essential oil of *Eugenia uniflora* L. (Myrtaceae) and demonstrated that at concentrations of 8192 µg/mL there is inhibition of hyphal projection. Two other species also from the same family that have the ability to inhibit the virulence of *C. albicans* are *Psidium brownianum* Mart. ex DC. And *Psidium guajava* L., the former having medicinal properties and is used to combat infections caused by fungi of the genus *Candida* [6,27].

The activities of the natural products, concerning the antimicrobial agents act by diverse mechanisms of action such as the disintegration of the cytoplasmic membranes, destabilization of the motor proton force (MPF), altering the polarization of the membrane, and the coagulation of the cellular content [28–30].

4. Materials and Methods

4.1. Botanical Material

The collection of leaves of *Piper caldense* was performed in the municipality of Piraquara in the state of Paraná, Brazil, under the coordinates 25°29.693′ S and 49°00.844′ W at 528 m elevation (Figure 5). An exsicata was identified and deposited in the Herbarium of the Faculdades Integradas Espírita under voucher 9.103.

4.2. Extraction of Volatile Terpenes and Determination of the Chemical Composition

Healthy *P. caldense* leaves were selected and dried in a incubator at 40 °C. After dehydration, the leaves were crushed to increase their contact surface and maximize the extraction of their volatile components. For such extraction, the hydrodistillation system was used, in which 50 g of the plant material was placed in a volumetric flask with 1000 mL of distilled water, being constantly heated for 4.5 h, until the oil extraction [21].

For the chemical characterization of *P. caldense* essential oil, it was made by gas chromatography-mass spectrometry (GC/MS). Initially, the essential oil was diluted in dichloromethane to 1% concentration, then 1 µL of this solution was injected (1:20) into Agilent 6890 chromatograph, coupled to Agilent 5973N mass selective detector, wherein the injector temperature was 250 °C. For the separation of the constituents, helium gas was used as a carrier (1 mL/min) and an HP-5MS capillary column with the following specifications: 5% phenyl-95%-dimethylpolysiloxane, 30 m × 0.25 mm × 0.25 µm. For the heating ramp, the temperature started at 60 °C with a heating rate of 3 °C/min to

240 °C, totaling 60 min. The mass detector was operated in the electronic ionization mode (70 eV), at a rate of 3.15 s^{-1} sweeps and a mass range of 40 to 450 u. The transfer line was maintained at 260 °C, the ion source at 230 °C and the analyzer (quadrupole) at 150 °C.

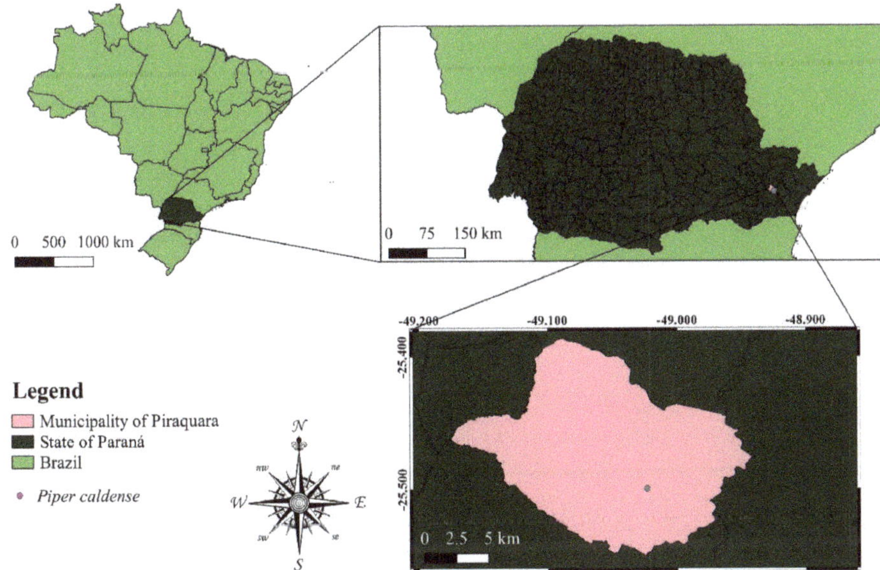

Figure 5. Location of the species *Piper caldense* in the municipality of Piraquara in the state of Paraná, Brazil.

For quantification, the diluted samples were injected into an Agilent 7890A chromatograph equipped with a flame ionization detector (FID), operated at 280 °C. We employed the same column and analytical conditions described above except for the carrier gas used, which was hydrogen at a flow rate of 1.5 mL/min. The percentage composition was obtained by the electronic integration of the FID signal by dividing the area of each component by the total area (area%).

For the determination of chemical constituents, the mass of the constituents was compared with the library (NIST and Wiley) e and also by their linear retention indexes, calculated from the injection of a homologous series of hydrocarbons (C_7-C_{26}) and compared with data from the literature [31].

4.3. Drugs, Reagents, Solution Preparation and Fungal Strains

For the antifungal test, the essential oil stock solution was prepared from 0.15 g and diluted in 1 mL of dimethyl sulfoxide (DMSO). To obtain the initial concentration of 16,384 µg/mL, the stock solution was diluted in sterile distilled water so that the DMSO concentration in the natural product had no activity in the cells tested. The reference antifungal was fluconazole (Capsule-FLUCOMED), diluted in sterile water at the same oil concentration [27]. For microbiological assays, two *Candida albicans* strains were used: CA INCQS 40006 (standard strain), obtained from the Oswaldo Cruz Culture Collection of the National Institute for Quality Control in Health (INCQS) and CA URM 4387 (clinical isolate), provided by mycology collection of the Federal University of Recife (URM - University Recife Mycology). For the antifungal activity test the culture media were used: Sabouraud Dextrose Agar (SDA) and Sabouraud Dextrose Broth (SDB). For the fungal micromorphology evaluation test the nutrient-poor Potato Dextrose Agar (PDA) culture medium prepared with solidification Agar was used. The media were prepared according to the supplier's guidelines (Difco®) and autoclaved at 121 °C within 15 min.

4.4. Determination of Minimum Inhibitory Concentration (MIC)

To perform this experiment the broth microdilution method was used according to Javadpour. First the yeasts were inoculated in ASD and kept incubated at 37 °C for 24 h. Subsequently, the inoculum were prepared by transferring small aliquots of the strains to tubes containing sterile saline, the inoculum were compared to McFarland scale resulting in a concentration of 1×10^5 cells/mL. 96-well plates were filled containing in each well 100 µL of SDB and 10% fungal inoculum. The plates were then microdiluted with 100 µL of the essential oil of P. caldense, where the well concentrations were from 8192 µg/mL to 64 µg/mL. The last well was not diluted as a growth control. Media sterility controls and substance dilution controls were also used, using only saline without fungal inoculum. The plates were incubated at 37 °C for 24 h and subsequently read on a 630 nm wavelength ELISA spectrophotometer (Thermoplate®). All assays were performed in triplicate and the results obtained were used to construct the cell viability curve and the IC_{50} of essential oil and fluconazole [6,32]. MIC was defined as the lowest concentration able to reduce the fungal growth curve.

4.5. Evaluation of Modulating Activity of Natural Product

The essential oil was tested at sub-inhibitory concentration (MIC/16) according to the method proposed by Coutinho et al., 2008 [33]. The plates were filled with a solution containing SDB, fungal inoculum and the essential oil, then 100 µL of fluconazole was mixed into the first well and serially microdiluted at a ratio of 1:1 to the penultimate well, at 1 µg/mL. Control of culture media sterility and antifungal dilution control were performed, and the MIC of fluconazole was also determined. The tests were performed in triplicate and the plates were incubated at 37 °C for 24 h. The reading was performed in an ELISA spectrophotometer (Termoplate®). The modulatory activity is defined when used in combination. The natural product enhances the action of the antifungal, showing synergism. If the opposite occurs and the natural product interferes with the action of the drug, the effect is considered antagonistic.

4.6. Determination of Minimum Fungicidal Concentration (MFC)

After the MIC test, a sterile stem was placed in each well of the plates, first the stem was used to mix the solutions contained in the wells, then small aliquots with medium, inoculum and essential oil were transferred to Petri dishes containing solid medium SDA, for yeast subculture and verification of cell viability. After 24 h of incubation, the plates were analyzed and the concentration at which no fungal colony growth was observed is considered the Minimum Fungicidal Concentration of the essential oil [34].

4.7. Effect of Natural Product on Fungal Morphology

To verify whether P. caldense volatile terpenes cause any change in fungal morphology by inhibiting hyphae emission, micromorphological sterile chambers were mounted for yeast observation. On the blade chamber (sterile) were poured 3 mL of medium PDA, poor nutrient for dilution, containing the natural product in CFM/4 concentrations CFM/8 and CFM/16. Aliquots of fungal inoculum were taken from SDA-containing Petri dishes to make two parallel strips in the solidified solid medium (PDA) and then covered by a sterile coverslip. The chamberwere taken to the incubator and after 24 h (37 °C) the culture was visualized under optical microscopy using a 400 X objective. A camera was attached to the microscope for image capture. A control for yeast growth (with hyphal emission stimulated by depletion of the medium) was performed, as well as a control with the reference antifungal fluconazole was also used for comparative purposes [35].

4.8. Statistical Analysis

The IC_{50} was calculated by means of linear regression. Subsequently, data from *P. caldense* antifungal assays were investigated by one-way analysis of variance (ANOVA), using the Bonferroni test and considered significant when $p < 0.05$. All analyzes were performed on the GraphPad Prism 6.0 software.

5. Conclusions

The essential oil of *P. caldense* presented mono- and sesquiterpene components, though not presented a major constituent. In addition, it showed low antifungal activity for *C. albicans* strains but was able to modulate the effect of standard drug (fluconazole) and decreased the virulence of these strains. In this way, the oil has in its composition constituents that are promising in the formulation of drugs used in the treatment of infectious diseases.

Author Contributions: Conceptualization, J.W.A.B., M.I., H.D.M.C. and M.F.B.M.-B.; Data curation, J.W.A.B., F.C.R., R.P.d.C., L.E.d.S., E.M.V., H.D.M.C. and M.F.B.M.-B.; Formal analysis, J.W.A.B., F.C.R., R.P.d.C. and L.E.d.S.; Methodology J.W.A.B., F.C.R., R.P.d.C., L.E.d.S., W.d.A., R.A.R., I.M.B., and E.M.V.; Supervision, M.I., H.D.M.C. and M.F.B.M.-B.; Methodology, C.F.B.; Writing—original draft, J.W.A.B., F.C.R., R.P.d.C. and L.E.d.S.; Writing—review & editing, M.I., E.M.V., H.D.M.C. and M.F.B.M.-B. All authors have read and agreed to the published version of the manuscript.

Funding: The authors thank the Conselho Nacional de Desenvolvimento Científico e Tecnológico (CNPq) by research grants, Fundação Cearense de Apoio ao Desenvolvimento Científico e Tecnológico (FUNCAP) by the concession of equipment and projects, and the Coordenação de Aperfeiçoamento de Pessoal de Nível Superior (CAPES) for supporting research.

Conflicts of Interest: The authors declare no conflict of interest.

References

1. Barbosa-Filho, V.M.; Waczuk, E.P.; Leite, N.P.; Menezes, I.R.A.; Costa, J.G.M.; Lacerda, S.R.; Adedara, I.A.; Coutinho, H.D.M.; Posser, T.; Kamdem, J.P. Phytocompounds and modulatory effects of *Anacardium microcarpum* (cajui) on antibiotic drugs used in clinical infections. *Drug. Des. Dev. Ther.* **2015**, *9*, 5965–5972.

2. Rodrigues, F.C.; Santos, A.T.L.; Machado, A.J.T.; Bezerra, C.F.; Freitas, T.S.; Coutinho, H.D.M.; Morais-Braga, M.F.B.; Bezerra, J.W.A.; Duarte, A.E.; Kandem, J.P.; et al. Chemical composition and anti-*Candida* potencial of the extracts of *Tarenaya spinosa* (Jacq.) Raf. (Cleomaceae). *Comp. Immunol. Microb.* **2019**, *64*, 14–19. [CrossRef] [PubMed]

3. Chaves, T.P.; Fernandes, F.H.A.; Santana, C.P.; Santos, J.S.; Medeiros, F.D.; Felismino, D.C.; Santos, V.L.; Catão, R.M.R.; Coutinho, H.D.M.; Medeiros, A.C.D. Evaluation of the interaction between the *Poincianella pyramidalis* (Tul.) LP queiroz extract and antimicrobials using biological and analytical models. *PLoS ONE* **2016**, *11*, 1–23. [CrossRef]

4. Chaves, T.P.; Pinheiro, R.E.; Melo, E.S.; Soares, M.J.S.; Souza, J.S.M.; Andrade, T.B.; Lemos, T.L.G.; Coutinho, H.D.M. Essential oil of *Eucalyptus camaldulensis* Dehn potentiates β-lactam activity against *Staphylococcus aureus* and *Escherichia coli* resistant strains. *Ind. Crop. Prod.* **2018**, *112*, 70–74. [CrossRef]

5. Karkowska-Kuleta, J.; Rapala-Kozik, M.; Kozik, A. Fungi pathogenic to humans: molecular bases of virulence of *Candida albicans*, *Cryptococcus neoformans* and *Aspergillus fumigatus*. *Acta Biochim. Pol.* **2009**, *56*, 211–224. [CrossRef] [PubMed]

6. Morais-Braga, M.F.B.; Sales, D.L.; Carneiro, J.N.P.; Machado, A.J.T.; Santos, A.T.L.; Freitas, M.A.; Martins, G.M.A.B.; Leite, N.F.; Matos, Y.M.L.S.; Tintino, S.R.; et al. *Psidium guajava* L. and *Psidium brownianum* Mart ex DC.: Chemical composition and anti–*Candida* effect in association with fluconazole. *Microb. Pathogenesis.* **2016**, *95*, 200–207. [CrossRef]

7. Sudbery, P.; Gow, N.; Berman, J. The distinct morphogenic states of *Candida albicans*. *Trends. Microbiol.* **2004**, *12*, 317–324. [CrossRef]

8. Albuquerque, U.P.; Medeiros, P.M.D.; Ramos, M.A.; Júnior, F.; Soares, W.; Nascimento, A.L.B.; Avilez, W.M.T.; Melo, J.G.D. Are ethnopharmacological surveys useful for the discovery and development of drugs from medicinal plants? *Rev. Bras. Farmacogn.* **2014**, *24*, 110–115. [CrossRef]

9. Guimarães, E.F.; Carvalho-Silva, M.; Monteiro, D.; Medeiros, E.S.; Queiroz, G.A. Piperaceae in Lista de Espécies da Flora do Brasil. Jardim Botânico do Rio de Janeiro, 2015. Available online: http://floradobrasil. jbrj.gov.br/jabot/floradobrasil/FB12756 (accessed on 7 June 2018).

10. Cardozo-Junior, E.L.; Chaves, M.C.O. Caldensin, a new natural n-methylaristolactam from *Piper caldense*. *Pharm. Biol.* **2003**, *41*, 216–218. [CrossRef]

11. Alves, H.S.; Rocha, W.R.V.; Fernandes, A.F.C.; Nunes, L.E.; Pinto, D.S.; Costa, J.I.V.; Chaves, M.C.O.; Catão, R.M.R. Antimicrobial activity of products obtained from *Piper* species (Piperaceae). *Rev. Cub. Plant. Med.* **2016**, *21*, 168–180.

12. Yang, D.; Michel, L.; Chaumont, J.P.; Millet-Clerc, J. Use of caryophyllene oxide as an antifungal agent in an in vitro experimental model of onychomycosis. *Mycopathol.* **2000**, *148*, 79–82. [CrossRef] [PubMed]

13. Xu, W.H.; Li, X.C. Antifungal compounds from *Piper* species. *Current Bio. Comp.* **2011**, *7*, 262–267. [CrossRef] [PubMed]

14. Ribeiro, R.V.; Bieski, I.G.C.; Balogun, S.O.; Oliveira-Martins, D.T. Ethnobotanical study of medicinal plants used by Ribeirinhos in the North Araguaia microregion, Mato Grosso, Brazil. *J. Ethnopharmacol.* **2017**, *205*, 69–102. [CrossRef] [PubMed]

15. Carrara, V.D.S.; Souza, A.D.; Dias-Filho, B.P.; Nakamura, C.V.; Paulo, L.F.D.; Young, M.; Svidzinski, T.I.E.; García-Cortez, D.A. Chemical composition and antifungal activity of the essential oil from *Piper amalago* L. *Lat. Am. J. Pharm.* **2010**, *29*, 1459–1462.

16. Hastuti, U.S.; Ummah, Y.P.I.; Khasanah, H.N. Antifungal activity of *Piper aduncum* and *Peperomia pellucida* leaf ethanol extract against *Candida albicans*. *AIP. Conf. Proc.* **2017**, *w/v*, 789–798.

17. Junairiah; Ni'matuzahroh; Zuraidassanaaz, N.I.; Sulistyorini, L. Antifungal and antibacterial activity of black betel (*Piper betle* L. var Nigra) extract. *Biosci. Res.* **2017**, *14*, 750–755.

18. Guirado, O.A.A.; Cuéllar, A.C. Estrategias en la selección de las plantas medicinales a investigar. *Rev. Cub. Plant. Med.* **2008**, *13*, 1–10.

19. Rocha, D.S.; Silva, J.M.D.; Navarro, D.M.D.A.F.; Camara, C.A.G.; Lira, C.S.D.; Ramos, C.S. Potential antimicrobial and chemical composition of essential oils from *Piper caldense* tissues. *J. Mex. Chem. Soc.* **2016**, *60*, 148–151.

20. Bezerra, J.W.A.; Costa, A.R.; Freitas, M.A.; Rodrigues, F.C.; Souza, M.A.; Silva, A.R.P.; Santos, A.T.L.; Linhares, K.V.; Coutinho, H.D.M.; Morais-Braga, M.F.B. Chemical composition, antimicrobial, modulator and antioxidant activity of essential oil of *Dysphania ambrosioides* (L.) Mosyakin & Clemants. *Comp. Immunol. Microb.* **2019**, *65*, 58–64.

21. Miguel, M.G.; Duarte, F.; Venâncio, F.; Tavares, R. Variation in themain componentes of the essential oils isolated from *Thymbra capitata* L. (Cav.) and *Origanum vulgare* L. *J. Agr. Food. Chem.* **2005**, *53*, 8162–8168.

22. Bezerra, J.W.A.; Costa, A.R.; Silva, M.A.P.; Rocha, M.I.; Boligon, A.A.; Rocha, J.B.T.; Barros, L.M.; Kamdem, J.P. Chemical composition and toxicological evaluation of *Hyptis suaveolens* (L.) Poiteau (LAMIACEAE) in *Drosophila melanogaster* and *Artemia salina*. *S. Afr. J. Bot.* **2017**, *113*, 437–442. [CrossRef]

23. Martins, N.; Barros, L.; Henriques, M.; Silva, S.; Ferreira, I.C. Activity of phenolic compounds from plant origin against *Candida* species. *Ind. Crop. Prod.* **2015**, *74*, 648–670. [CrossRef]

24. Sardi, J.C.O.; Scorzoni, L.; Bernardi, T.; Fusco-Almeida, A.M.; Giannini, M.M. *Candida* species: current epidemiology, pathogenicity, biofilm formation, natural antifungal products and new therapeutic options. *J. Med. Microbiol.* **2013**, *62*, 10–24. [CrossRef] [PubMed]

25. Silva, G.A.T.D. *Estudo fitoquímico das folhas e frutos de Piper caldense C. DC. (Piperaceae)*; Universidade Federal da Paraíba: João Pessoa, Brazil, 2013.

26. Santos, J.F.S.; Rocha, J.E.; Bezerra, C.F.; Silva, M.K.N.; Matos, Y.M.L.S.; Freitas, T.S.; Santos, A.T.L.; Cruz, R.P.; Machado, A.F.T.; Rodrigues, T.H.S.; et al. Chemical composition, antifungal activity and potential anti-virulence evaluation of the *Eugenia uniflora* essential oil against *Candida* spp. *Food. Chem.* **2018**, *261*, 233–239. [CrossRef]

27. Morais-Braga, M.F.B.; Carneiro, J.N.P.; Machado, A.J.T.; Sales, D.L.; Santos, A.T.; Boligon, A.A.; Athayde, M.L.; Menezes, I.R.A.; Souza, D.S.L.; Costa, J.G.M.; et al. Phenolic composition and medicinal usage of *Psidium guajava* L.: Antifungal activity or inhibition of virulence? *Saudi J. Biol. Sci.* **2016**, *24*, 302–313. [CrossRef]

28. Duarte, A.E.; Menezes, I.R.A.; Morais-Braga, M.F.B.; Leite, N.F.; Barros, L.M.; Waczuk, E.P.; Silva, M.A.P.; Boligon, A.; Rocha, J.B.T.; Souza, D.O.; et al. Antimicrobial activity and Modulatory effect of essential oil from the leaf of *Rhaphiodon echinus* (Nees & Mart) Schauer on some antimicrobial drugs. *Molecules* **2016**, *21*, 743–753.

29. Sikkema, J.; Bont, J.A.; Poolman, B. Interactions of cyclic hydrocarbons with biological membranes. *J. Biol. Chem.* **1994**, *269*, 8022–8028.

30. Santos, F.S.M.; Bezerra, J.W.A.; Kamdem, J.P.; Boligon, A.A.; Anraku, M.M.; Silva, A.R.P.; Fidelis, K.R.; Leite, N.F.; Pinho, A.I.; Coutinho, H.D.M.; et al. Polyphenolic composition, antibacterial, modulator and neuroprotective activity of *Tarenaya spinosa* (Jacq.) Raf.(Cleomaceae). *Asian. Pac. J. Trop. Biomed.* **2019**, *9*, 12–17.

31. Adams, R.P. *Identification of Essential Oil Components by Gas Chromatography/Mass Spectrometry*; Allured Publishing Corporation: Carol Stream, IL, USA, 2007; Volume 456.

32. Javadpour, M.M.; Juban, M.M.; Lo, W.C.; Bishop, S.M.; Alberty, J.B.; Cowell, S.M.; Becker, C.L.; Mclaughlin, M.L. De novo antimicrobial peptides with low mammalian cell toxicity. *J. Med. Chem.* **1996**, *39*, 3107–3113. [CrossRef]

33. Coutinho, H.D.M.; Costa, J.G.; Lima, E.O.; Falcão-Silva, V.S.; Siqueira-Júnior, J.P. Enhancement of the antibiotic activity against a multiresistant *Escherichia coli* by *Mentha arvensis* L. and chlorpromazine. *Chemotherapy* **2008**, *54*, 328–330. [CrossRef]

34. Ernst, E.J.; Klepser, M.E.; Ernst, M.E.; Messer, S.A.; Pfaller, M.A. *In vitro* pharmacodynamic properties of MK-0991 determined by time-kill methods. *Micology* **1999**, *33*, 75–80. [CrossRef]

35. Sidrim, J.J.C.; Rocha, M.F.G. *Micologia Médica à luz de autores contemporâneos*, 1st ed.; Guanabara Koogan: Rio de Janeiro, Brazil, 2010.

Article

Essential Oil Compositions and Antifungal Activity of Sunflower (*Helianthus*) Species Growing in North Alabama

Sims K. Lawson [1,2], Layla G. Sharp [1,2], Chelsea N. Powers [2], Robert L. McFeeters [2], Prabodh Satyal [3] and William N. Setzer [2,3,*]

[1] Department of Biological Sciences, University of Alabama in Huntsville, Huntsville, AL 35899, USA
[2] Department of Chemistry, University of Alabama in Huntsville, Huntsville, AL 35899, USA
[3] Aromatic Plant Research Center, 230 N 1200 E, Suite 100, Lehi, UT 84043, USA
*   Correspondence: wsetzer@chemistry.uah.edu; Tel.: +1-256-824-6519

Received: 1 May 2019; Accepted: 2 August 2019; Published: 5 August 2019

Abstract: *Helianthus* species are North American members of the Asteraceae, several of which have been used as traditional medicines by Native Americans. The aerial parts of two cultivars of *Helianthus annuus*, "Chianti" and "Mammoth", and wild-growing *H. strumosus*, were collected from locations in north Alabama. The essential oils were obtained by hydrodistillation and analyzed by gas chromatography—mass spectrometry. The *Helianthus* essential oils were dominated by monoterpene hydrocarbons, in particular α-pinene (50.65%, 48.91%, and 58.65%, respectively), sabinene (6.81%, 17.01%, and 1.91%, respectively), β-pinene (5.79%, 3.27%, and 4.52%, respectively), and limonene (7.2%, 7.1%, and 3.8%, respectively). The essential oils were screened against three opportunistic pathogenic fungal species, *Aspergillus niger*, *Candida albicans*, and *Cryptococcus neoformans*. The most sensitive fungus was *C. neoformans* with minimum inhibitory concentration (MIC) values of 78, 156, and 78 µg/mL, respectively.

Keywords: *Helianthus annuus*; *Helianthus strumosus*; *Aspergillus niger*; *Candida albicans*; *Cryptococcus neoformans*; α-pinene

1. Introduction

Helianthus L., the sunflowers, is a genus in the family Asteraceae, tribe Heliantheae, made up of 51 North American species [1]. *Helianthus annuus* L. (common sunflower) is native to North America and the current range of wild forms of *H. annuus* are central and western United States, southern Canada, and northern Mexico [2]. The common sunflower is one of the earliest domesticated plants in the Americas. There is evidence that the plant was domesticated in Tabasco, Mexico, around 2600 B.C. [3], and independently in the southeastern United States around 2800 B.C. [2,4]. Several Native American tribes used *H. annuus* in traditional medicine [5]. For example, the White Mountain Apache used a poultice of the crushed plants to treat snakebites; the Hopi used the plant as a spider bite medicine; the Jemez applied the juice of the plant to cuts; the Pima used a decoction of the leaves to treat fevers [5]; and the Zuni natives of New Mexico used the roots to treat rattlesnake bites [6]. In addition, *H. annuus* is used as a traditional herbal medicine in many locations where it has been introduced. Ethiopians use *H. annuus* in teas to treat food poisoning [7]. In Bangladesh the seeds and/or the flowers are crushed and used for snake bites, scorpion bites, and a variety of other ailments, such as burning sensation in the vagina and worms in the ears [8].

Helianthus strumosus L. (woodland sunflower) is a rhizomatous perennial plant, growing up to two meters tall and is native to eastern North America [9–11]. These plants are strongly aromatic. Leaves are up to 10 cm long and cuneate to subcordate in shape. The composite flower heads can be up to 9 cm at the peduncle. The ray flowers are a dark yellow color with orange-brown disc flowers in the center. These flowers are common along roadsides and in open fields and are sometimes found in forests. The Iroquois used a decoction of the roots as an anthelmintic [5].

Invasive fungal infections are becoming increasingly common in immunocompromised patients, such as those receiving cancer chemotherapy, transplant patients receiving immunosuppressant drugs, and HIV patients [12]. The predominant fungal pathogens are *Aspergillus* spp. [13,14] and *Candida* spp. [15,16] among others [12]. *Aspergillus niger* is a haploid filamentous parasitic fungus that is commonly known for the disease "black mold" on fruits, vegetables, and nuts [17]. *Aspergillus* conidia (fungal "spores") are environmentally widespread and inhalation can lead to opportunistic pulmonary aspergillosis, chiefly attributed to *A. fumigatus*, *A. flavus*, and *A. tubingensis*, as well as *A. niger* [18]. In immunocompromised individuals, however, the infection can progress to invasive systemic aspergillosis [19]. *Candida albicans* is another opportunistic pathogenic fungus that commonly colonizes the human body [20]. The organism can cause superficial infections of the mucosa, but can lead to invasive candidiasis in immunocompromised patients [21]. Cryptococcosis is a fungal infection caused by *Cryptococcus neoformans* [22]. The fungus is widespread in the environment and typically enters the body through inhalation where it can cause pulmonary infection [23]. However, the organism has the ability to cross the blood brain barrier and in immunocompromised patients, cryptococcosis can lead to cryptococcal meningoencephalitis with increased intracranial pressure [24,25]. As part of our continuing investigation of antifungal activity of essential oils [26] as well as essential oils from the Asteraceae growing in north Alabama [27], we have collected and analyzed the essential oils from the aerial parts of *H. annuus* and *H. strumosus*, and we have carried out in vitro antifungal screening of the essential oils against *A. niger*, *C. albicans*, and *C. neoformans*.

2. Materials and Methods

2.1. Plant Materials

The two cultivars of *H. annuus* ("Chianti" and "Mammoth") were cultivated, grown without fertilizer or pesticides, in a rural area near Gurley in north Alabama (34°38′29″N, 86°24′39″W, elevation 199 m) and the aerial parts were collected on 4 and 6 August 2018. Aerial parts of *H. strumosus* were collected on 10 August 2018 from wild-growing plants near Huntsville, Alabama (34°42′42″N, 86°32′35″W, elevation 354 m). The plants were identified by S.K. Lawson. Voucher specimens have been deposited in the herbarium of the University of Alabama in Huntsville (20180729-183243 and 20190402-111732). The fresh plant materials (78.14, 80.32, and 65.47 g, respectively) were hydrodistilled using a Likens–Nickerson apparatus with continuous extraction with dichloromethane for 3 h. The dichloromethane was carefully evaporated, and the residual essential oils weighed using an analytical balance to give the essential oils (82.3, 20.3, and 24.0 mg, respectively).

2.2. Gas Chromatographic—Mass Spectral Analysis

The *Helianthus* essential oils were analyzed by GC-MS with a Shimadzu GCMS-QP2010 Ultra with a ZB-5 capillary column as previously described [28]. Identification of the chemical components was carried out by comparison of the retention indices, calculated with respect to a homologous series of normal alkanes using the arithmetic index [29], and by comparison of their mass spectra with those reported in the Adams [30], NIST17 [31], FFNSC 3 [32], and our own in-house library [33]. Concentrations shown in Table 1 (average of three measurements ± standard deviations) are based on peak integration without standardization.

Table 1. Chemical compositions of *Helianthus annuus* "Chianti", *H. annuus* "Mammoth", and *H. strumosus* aerial parts essential oils.

RI [a]	RI [b]	Compound	Percent Composition [c]		
			H. annuus "Chianti"	*H. annuus* "Mammoth"	*H. strumosus*
800	797	(3Z)-Hexenal	0.06 ± 0.01	Tr [d]	tr
801	801	Hexanal	0.35 ± 0.02	0.24 ± 0.04	0.41 ± 0.03
810	796	2-Hexanol	—	—	0.07 ± 0.00
849	846	(2E)-Hexenal	1.13 ± 0.05	0.83 ± 0.05	1.96 ± 0.03
861	854	(2E)-Hexenol	—	—	tr
864	863	1-Hexanol	—	—	0.09 ± 0.00
921	921	Tricyclene	0.37 ± 0.00	0.21 ± 0.01	0.18 ± 0.00
924	924	α-Thujene	0.17 ± 0.00	0.23 ± 0.01	0.1
932	932	α-Pinene	50.65 ± 0.32	48.91 ± 0.64	58.65 ± 0.14
946	945	α-Fenchene	—	—	tr
948	946	Camphene	7.26 ± 0.03	3.72 ± 0.03	3.38 ± 0.02
952	953	Thuja-2,4(10)-diene	0.05 ± 0.01	—	tr
971	969	Sabinene	6.81 ± 0.04	17.01 ± 0.18	1.91 ± 0.00
977	974	β-Pinene	5.79 ± 0.04	3.27 ± 0.01	4.52 ± 0.02
988	988	Myrcene	0.42 ± 0.01	0.30 ± 0.03	9.79 ± 0.03
1004	1003	*p*-Mentha-1(7),8-diene	—	—	tr
1006	1002	α-Phellandrene	—	0.08 ± 0.01	0.05 ± 0.01
1008	1008	δ-3-Carene	—	—	tr
1016	1014	α-Terpinene	—	tr	—
1024	1020	*p*-Cymene	0.06 ± 0.03	0.09 + 0.01	0.07 ± 0.00
1028	1024	Limonene	7.20 ± 0.03	7.11 ± 0.11	3.79 ± 0.01
1030	1025	β-Phellandrene	0.24 ± 0.00	0.21 ± 0.14	0.29 ± 0.01
1031	1026	1,8-Cineole	0.06 ± 0.00	0.07 ± 0.02	tr
1034	1032	(Z)-β-Ocimene	—	—	tr
1044	1044	(E)-β-Ocimene	—	tr	0.41 ± 0.01
1057	1054	γ-Terpinene	0.10 ± 0.00	0.25 ± 0.01	tr
1069	1065	*cis*-Sabinene hydrate	—	—	tr
1084	1086	Terpinolene	0.10 ± 0.01	0.16 ± 0.01	tr
1099	1099	α-Pinene oxide	0.10 ± 0.01	—	tr
1105	1100	Nonanal	—	—	tr
1109	1108	*p*-Mentha-2,8-dien-1-ol	0.36 ± 0.00	0.10 ± 0.02	tr
1112	1114	(3E)-4,8-Dimethyl-1,3,7-nonatriene	0.09 ± 0.01	0.13 ± 0.02	0.18 ± 0.00
1121	1124	Chrysanthenone	0.05 ± 0.00	—	—
1127	1122	α-Campholenal	0.33 ± 0.01	0.05 ± 0.01	0.06 ± 0.01
1140	1135	*trans*-Pinocarveol	0.37 ± 0.05	tr	0.06 ± 0.02
1141	1137	*cis*-Verbenol	0.10 ± 0.01	—	tr
1145	1140	*trans*-Verbenol	1.50 ± 0.02	0.24 ± 0.02	0.26 ± 0.00
1163	1160	Pinocarvone	0.11 ± 0.01	tr	tr
1171	1165	Borneol	0.73 ± 0.01	0.07 ± 0.02	0.12 ± 0.00
1180	1174	Terpinen-4-ol	0.09 ± 0.01	0.19 ± 0.01	tr
1187	1179	*p*-Cymen-8-ol	0.05 ± 0.01	—	—
1195	1186	α-Terpineol	—	tr	—
1195	1195	Myrtenal	0.29 ± 0.02	tr	0.10 ± 0.01
1206	1201	Decanal	tr	—	—
1207	1204	Verbenone	0.28 ± 0.04	0.11 ± 0.01	0.07 ± 0.00
1219	1215	*trans*-Carveol	0.16 ± 0.00	—	tr
1283	1287	Bornyl acetate	7.13 ± 0.04	3.02 ± 0.04	4.97 ± 0.01
1294	1298	*trans*-Pinocarvyl acetate	0.08 ± 0.01	tr	tr
1382	1387	β-Bourbonene	0.21 ± 0.02	0.18 ± 0.01	tr
1386	1387	β-Cubebene	tr	tr	tr

Table 1. *Cont.*

RI [a]	RI [b]	Compound	Percent Composition [c]		
			H. annuus "Chianti"	*H. annuus* "Mammoth"	*H. strumosus*
1387	1389	β-Elemene	0.05 ± 0.01	0.17 ± 0.01	tr
1392	1392	(Z)-Jasmone	—	—	tr
1416	1419	β-Ylangene	0.07 ± 0.01	0.15 ± 0.01	tr
1417	1417	β-Caryophyllene	0.33 ± 0.03	0.54 ± 0.09	0.84 ± 0.00
1427	1434	γ-Elemene	—	—	tr
1428	1431	β-Gurjunene (=Calarene)	0.62 ± 0.01	0.86 ± 0.01	tr
1430	1432	*trans*-α-Bergamotene	0.06 ± 0.00	0.14 ± 0.03	tr
1442	1442	6,9-Guaiadiene	tr	tr	—
1446	1453	Geranyl acetone	tr	tr	—
1454	1452	α-Humulene	0.19 ± 0.02	0.29 ± 0.01	0.20 ± 0.00
1479	1484	Germacrene D	3.32 ± 0.03	6.84 ± 0.09	3.68 ± 0.02
1487	1489	β-Selinene	0.12 ± 0.01	tr	tr
1493	1493	*epi*-Cubebol	0.12 ± 0.04	—	—
1494	1500	Bicyclogermacrene	—	0.16 ± 0.01	0.07 ± 0.01
1513	1514	Cubebol	0.14 ± 0.02	tr	tr
1516	1522	δ-Cadinene	tr	0.07 ± 0.00	tr
1547	1548	Elemol	—	tr	0.07 ± 0.01
1559	1561	(E)-Nerolidol	0.10 ± 0.02	0.09 ± 0.03	0.64 ± 0.03
1575	1574	Germacrene D-4β-ol	0.37 ± 0.02	0.46 ± 0.00	0.46 ± 0.00
1581	1582	Caryophyllene oxide	0.16 ± 0.09	tr	0.37 ± 0.01
1608	1608	Humulene epoxide II	—	—	0.05 ± 0.01
1636	1639	Caryophylla-4(12),8(13)-dien-5β-ol	—	—	tr
1638	1643	Hedycaryol	—	—	0.10 ± 0.01
1641	1638	τ-Cadinol	0.18±0.01	tr	0.14 ± 0.01
1654	1649	β-Eudesmol	0.10 ± 0.02	—	0.16 ± 0.03
1655	1652	α-Cadinol	—	tr	tr
1663	1665	Intermediol	tr	0.56 ± 0.01	—
1683	1685	Germacra-4(15),5,10(14)-trien-1α-ol	0.13 ± 0.04	—	0.51 ± 0.01
1686	1687	Eudesma-4(15),7-dien-1β-ol	—	—	0.14 ± 0.00
1689	1690	(Z)-*trans*-α-Bergamotol	—	—	0.12 ± 0.01
1699	1695	6-*epi*-Shyobunol	—	—	0.05 ± 0.01
		Monoterpene hydrocarbons	79.21	81.56	83.13
		Oxygenated monoterpenoids	11.80	3.85	5.64
		Sesquiterpene hydrocarbons [e]	4.97	9.40	4.79
		Oxygenated sesquiterpenoids [e]	1.39	1.11	2.79
		Green leaf volatiles	1.54	1.07	2.53
		Others	0.091	0.13	0.18
		Total Identified	99.01	97.12	99.05

[a] RI = Retention index determined with reference to a homologous series of *n*-alkanes on a ZB-5 column. [b] RI values from the databases (NIST17 [31], FFNSC 3 [32], Adams [30], or Satyal [33]). [c] Average of three measurements ± standard deviations. [d] tr = "trace" (<0.05%). [e] Sesquiterpenoids are considered tentatively identified based on MS and RI.

2.3. Antifungal Screening Assays

The *Helianthus* essential oils were screened for antifungal activity against *Aspergillus niger* (ATCC 16888), *Candida albicans* (ATCC 18804), and *Cryptococcus neoformans* (ATCC 24607) using the broth dilution technique as previously described [26,34]. Antifungal screening was carried out in triplicate.

3. Results and Discussion

3.1. Essential Oil Compositions

Hydrodistillation of *Helianthus* aerial parts gave pale yellow essential oils in 0.105%, 0.025%, and 0.037% yield (*w*/*w*) for *H. annuus* "Chianti", *H. annuus* "Mammoth", and *H. strumosus*, respectively.

The essential oil compositions for the three essential oils are compiled in Table 1. A perusal of the table reveals that the three *Helianthus* essential oils are qualitatively similar. The major components for *H. annuus* "Chianti" were α-pinene (50.65%), camphene (7.26%), limonene (7.20%), bornyl acetate (7.13%), sabinene (6.81%), and β-pinene (5.79%). The essential oil of *H. annuus* "Mammoth" was also dominated by α-pinene (48.91%), followed by sabinene (17.01%), limonene (7.11%), and germacrene D (6.84%). *H. strumosus* essential oil was also rich in α-pinene (58.65%), as well as myrcene (9.79%) and bornyl acetate (4.97%).

The compositions of *H. annuus* essential oils cultivated in north Alabama are very similar to those reported by Adams and co-workers for populations growing in the southern plains of the United States [35]. The essential oils of *H. annuus* from Pisa, Tuscany, Italy [36]; Lagos, Nigeria [37]; or from western United States [35] had much lower concentrations of α-pinene and correspondingly higher concentrations of germacrene D. In marked contrast to the essential oils of *Helianthus*, essential oils of *Rudbeckia fulgida* Aiton and *Rudbeckia hirta* L. (Asteraceae, Heliantheae) from north Alabama were devoid of α-pinene, but rich in sesquiterpene hydrocarbons [27].

3.2. Antifungal Activity

The *Helianthus* essential oils were screened for antifungal activity against three potentially pathogenic fungi, *Aspergillus niger*, *Candida albicans*, and *Cryptococcus neoformans*, as shown in Table 2. The most susceptible fungus was *C. neoformans*. Both *H. annuus* "Chianti" and *H. strumosus* essential oils showed minimum inhibitory concentration (MIC) values of 78 µg/mL. It is tempting to suggest that the major component, α-pinene, is responsible for the observed anti-*Cryptococcus* activity; all three *Helianthus* essential oils have around 50% α-pinene. Furthermore, α-pinene has shown antifungal activity against *C. neoformans* with an MIC around 70 µg/mL [38,39]. In addition, α-pinene-rich (46.1% α-pinene) commercial *Myrtis communis* essential oil showed a similar antifungal activity against *C. neoformans* (MIC = 78 µg/mL) [26]. Conversely, commercial *Cupressus sempervirens* essential oil, with 49.7% α-pinene was less active against *C. neoformans* (MIC = 313 µg/mL) [26]. There may be synergistic or antagonistic effects of α-pinene with minor components. Limonene [39,40] and β-pinene [39], have also shown antifungal activity against *C. neoformans*; camphene, however, was inactive [41]. Although we do not know which of the enantiomers is present in the *Helianthus* essential oils, we have screened both (+)- and (−)-α-pinene, (+)- and (−)-limonene, and (−)-β-pinene against the three fungal strains, as shown in Table 2. Consistent with previous investigations, (−)-β-pinene and (+)-limonene both showed activity against *C. neoformans* with MIC values of 39 and 78 µg/mL. Furthermore, both enantiomers of α-pinene were active against *C. neoformans*; MIC = 20 and 39 µg/mL for (+)- and (−)-α-pinene, respectively.

Table 2. Antifungal activities (minimum inhibitory concentration (MIC), µg/mL) of *Helianthus* essential oils and major components [a].

Material	Fungal Species		
	Aspergillus niger	*Candida albicans*	*Cryptococcus neoformans*
H. annuus "Chianti"	625	625	78
H. annuus "Mammoth"	625	625	156
H. strumosus	625	1250	78
(+)-α-Pinene	625	313	20
(−)-α-Pinene	156	625	39
(−)-β-Pinene	156	625	39
(+)-Limonene	1250	625	78
(−)-Limonene	2500	1250	313
Amphotericin B	0.78	0.78	1.56

[a] Each MIC determination was carried out in triplicate.

4. Conclusions

Helianthus essential oils have been shown to be rich in α- and β-pinenes, sabinene, and limonene, and have demonstrated poor antifungal activities against *A. niger* and *C. albicans*, but promising activity against *C. neoformans* (although much lower activity than the reference antifungal drug amphotericin B). These and other monoterpene-rich essential oils deserve further exploration as alternative and complementary agents to combat fungal infections; further studies against more susceptible fungi are recommended.

Author Contributions: Conceptualization, S.K.L. and W.N.S.; methodology, S.K.L., L.G.S. and C.N.P.; software, P.S.; validation, W.N.S.; formal analysis, W.N.S.; investigation, S.K.L., L.G.S. and C.N.P.; resources, R.L.M.; data curation, W.N.S.; writing—original draft preparation, S.K.L., L.G.S. and W.N.S.; writing—review and editing, all authors; project administration, W.N.S.

Funding: This research received no external funding.

Acknowledgments: P.S. and W.N.S. participated in the project as part of the activities of the Aromatic Plant Research Center (APRC, https://aromaticplant.org/).

Conflicts of Interest: The authors declare no conflict of interest.

References

1. Mabberley, D.J. *Mabberley's Plant-Book*, 3rd ed.; Cambridge University Press: Cambridge, UK, 2008.
2. Smith, B.D. Eastern North America as an independent center of plant domestication. *Proc. Natl. Acad. Sci. USA* **2006**, *103*, 12223–12228. [CrossRef] [PubMed]
3. Lentz, D.L.; Pohl, M.D.; Alvarado, J.L.; Tarighat, S.; Bye, R. Sunflower (*Helianthus annuus* L.) as a pre-Columbian domesticate in Mexico. *Proc. Natl. Acad. Sci. USA* **2008**, *105*, 6232–6237. [CrossRef] [PubMed]
4. Blackman, B.K.; Scascitelli, M.; Kane, N.C.; Luton, H.H.; Rasmussen, D.A.; Bye, R.A.; Lentz, D.L.; Rieseberg, L.H. Sunflower domestication alleles support single domestication center in eastern North America. *Proc. Natl. Acad. Sci. USA* **2011**, *108*, 14360–14365. [CrossRef] [PubMed]
5. Moerman, D.E. *Native American Ethnobotany*; Timber Press, Inc.: Portland, OR, USA, 1998.
6. Camazine, S.; Bye, R.A. A study of the medical ethnobotany of the Zuni Indians of New Mexico. *J. Ethnopharmacol.* **1980**, *2*, 365–388. [CrossRef]
7. Mesfin, F.; Demissew, S.; Teklehaymanot, T. An ethnobotanical study of medicinal plants in Wonago Woreda, SNNPR, Ethiopia. *J. Ethnobiol. Ethnomed.* **2009**, *5*, 28. [CrossRef] [PubMed]
8. Rahman, A.H.M.M. Medico-ethnobotany: A study on the tribal people of Rajshahi division, Bangladesh. *Peak J. Med. Plant Res.* **2016**, *1*, 1–8.
9. Kartesz, J.T. BONAP's North American Plant Atlas. Available online: http://bonap.net/Napa/TaxonMaps/Genus/County/Helianthus (accessed on 19 April 2019).
10. Tropicos.org. Missouri Botanical Garden. Available online: www.tropicos.org (accessed on 19 April 2019).
11. eFloras.org. *Helianthus Strumosus* Linnaeus. Available online: http://www.efloras.org/florataxon.aspx?flora_id=1&taxon_id=242416641 (accessed on 24 April 2019).
12. Richardson, M.; Lass-Flörl, C. Changing epidemiology of systemic fungal infections. *Clin. Microbiol. Infect.* **2008**, *14*, 5–24. [CrossRef] [PubMed]
13. Erjavec, Z.; Kluin-Nelemans, H.; Verweij, P.E. Trends in invasive fungal infections, with emphasis on invasive aspergillosis. *Clin. Microbiol. Infect.* **2009**, *15*, 625–633. [CrossRef]
14. Galimberti, R.; Torre, A.C.; Baztán, M.C.; Rodriguez-Chiappetta, F. Emerging systemic fungal infections. *Clin. Dermatol.* **2014**, *30*, 633–650. [CrossRef]
15. Dean, D.A.; Burchard, K.W. Fungal infection in surgical patients. *Am. J. Surg.* **1996**, *171*, 374–382. [CrossRef]
16. Miceli, M.H.; Díaz, J.A.; Lee, S.A. Emerging opportunistic yeast infections. *Lancet Infect. Dis.* **2011**, *11*, 142–151. [CrossRef]
17. Hocking, A.D. *Aspergillus* and related teleomorphs. In *Food Spoilange Microorganisms*; Blackburn, C.d.W., Ed.; CRC Press: Boca Raton, FL, USA, 2006; pp. 451–487. ISBN 0-8493-9156-3.

18. Shittu, O.B.; Adelaja, O.M.; Obuotor, T.M.; Sam-Wobo, S.O.; Adenaike, A.S. PCR-internal transcribed spacer (ITS) genes sequencing and phylogenetic analysis of clinical and environmental *Aspergillus* species associated with HIV-TB co infected patients in a hospital in Abeokuta, southwestern Nigeria. *Afr. Health Sci.* **2016**, *16*, 141–148. [CrossRef] [PubMed]

19. Paulussen, C.; Hallsworth, J.E.; Álvarez-Pérez, S.; Nierman, W.C.; Hamill, P.G.; Blain, D.; Rediers, H.; Lievens, B. Ecology of aspergillosis: Insights into the pathogenic potency of *Aspergillus fumigatus* and some other *Aspergillus* species. *Microb. Biotechnol.* **2017**, *10*, 296–322. [CrossRef] [PubMed]

20. Kabir, M.A.; Hussain, M.A.; Ahmad, Z. *Candida albicans*: A model organism for studying fungal pathogens. *Int. Sch. Res. Netw. Microbiol.* **2012**, *2012*, 538694. [CrossRef] [PubMed]

21. Gow, N.A.R.; Yadav, B. Microbe profile: *Candida albicans*: A shape-changing, opportunistic pathogenic fungus of humans. *Microbiology* **2017**, *163*, 1145–1147. [CrossRef]

22. Hay, R.J. Fungal infections. *Clin. Dermatol.* **2006**, *24*, 201–212. [CrossRef]

23. Paterson, D.L.; Singh, N. *Cryptococcus neoformans* infection. *Liver Transpl.* **2002**, *8*, 846–847. [CrossRef]

24. Gaona-Flores, V.A. Central nervous system and *Cryptococcus neoformans*. *N. Am. J. Med. Sci.* **2013**, *5*, 492–493. [CrossRef]

25. Armstrong-James, D.; Meintjes, G.; Brown, G.D. A neglected epidemic: Fungal infections in HIV/AIDS. *Trends Microbiol.* **2014**, *22*, 120–127. [CrossRef]

26. Powers, C.N.; Osier, J.L.; McFeeters, R.L.; Brazell, C.B.; Olsen, E.L.; Moriarity, D.M.; Satyal, P.; Setzer, W.N. Antifungal and cytotoxic activities of sixty commercially-available essential oils. *Molecules* **2018**, *23*, 1549. [CrossRef]

27. Stewart, C.D.; Jones, C.D.; Setzer, W.N. Leaf essential oil compositions of *Rudbeckia fulgida* Aiton, *Rudbeckia hirta* L., and *Symphyotrichum novae-angliae* (L.) G.L. Nesom (Asteraceae). *Am. J. Essent. Oils Nat. Prod.* **2014**, *2*, 36–38.

28. Satyal, P.; Hieu, H.V.; Chuong, N.T.H.; Hung, N.H.; Sinh, L.H.; Van The, P.; Tai, T.A.; Hien, V.T.; Setzer, W.N. Chemical composition, *Aedes* mosquito larvicidal activity, and repellent activity against *Triatoma rubrofasciata* of *Severinia monophylla* leaf essential oil. *Parasitol. Res.* **2019**, *118*, 733–742. [CrossRef] [PubMed]

29. Van den Dool, H.; Kratz, P.D. A generalization of the retention index system including linear temperature programmed gas-liquid partition chromatography. *J. Chromatogr.* **1963**, *11*, 463–471. [CrossRef]

30. Adams, R.P. *Identification of Essential Oil Components by Gas Chromatography/Mass Spectrometry*, 4th ed.; Allured Publishing: Carol Stream, IL, USA, 2007.

31. *NIST17*; National Institute of Standards and Technology: Gaithersburg, MD, USA, 2017.

32. Mondello, L. *FFNSC 3*; Shimadzu Scientific Instruments: Columbia, MD, USA, 2016.

33. Satyal, P. Development of GC-MS Database of Essential Oil Components by the Analysis of Natural Essential Oils and Synthetic Compounds and Discovery of Biologically Active Novel Chemotypes in Essential Oils. Ph.D. Thesis, University of Alabama in Huntsville, Huntsville, AL, USA, 2015.

34. Sahm, D.H.; Washington, J.A. Antibacterial susceptibility tests: Dilution methods. In *Manual of Clinical Microbiology*; Balows, A., Hausler, W.J., Herrmann, K.L., Isenberg, H.D., Shamody, H.J., Eds.; American Society for Microbiology: Washington, DC, USA, 1991.

35. Adams, R.P.; TeBeest, A.K.; Holmes, W.; Bartel, J.A.; Corbet, M.; Parker, C.; Thornburg, D. Geographic variation in volatile leaf oils (terpenes) in natural populations of *Helianthus annuus* (Asteraceae, Sunflowers). *Phytologia* **2017**, *99*, 130–138.

36. Ceccarini, L.; Macchia, M.; Flamini, G.; Cioni, P.L.; Caponi, C.; Morelli, I. Essential oil composition of *Helianthus annuus* L. leaves and heads of two cultivated hybrids "Carlos" and "Florom 350". *Ind. Crops Prod.* **2004**, *19*, 13–17.

37. Ogunwande, I.A.; Flamini, G.; Cioni, P.L.; Omikorede, O.; Azeez, R.A.; Ayodele, A.A.; Kamil, Y.O. Aromatic plants growing in Nigeria: Essential oil constituents of *Cassia alata* (Linn.) Roxb. and *Helianthus annuus* L. *Rec. Nat. Prod.* **2010**, *4*, 211–217.

38. Lima, I.O.; Oliveira, R.d.A.G.; Lima, E.d.O.; de Souza, E.L.; Farias, N.P.; Navarro, D.d.F. Inhibitory effect of some phytochemicals in the growth of yeasts potentially causing opportunistic infections. *Rev. Bras. Ciênc. Farm.* **2006**, *41*, 199–203. [CrossRef]

39. Pinto, E.; Hrimpeng, K.; Lopes, G.; Vaz, S.; Gonçalves, M.J.; Cavaleiro, C.; Salgueiro, L. Antifungal activity of *Ferulago capillaris* essential oil against *Candida*, *Cryptococcus*, *Aspergillus* and dermatophyte species. *Eur. J. Clin. Microbiol. Infect. Dis.* **2013**, *32*, 1311–1320. [CrossRef]

40. Pinto, E.; Gonçalves, M.J.; Cavaleiro, C.; Salgueiro, L. Antifungal activity of *Thapsia villosa* essential oil against *Candida, Cryptococcus, Malassezia, Aspergillus* and dermatophyte species. *Molecules* **2017**, *22*, 1595. [CrossRef]
41. Tirillini, B.; Velasquez, E.R.; Pellegrino, R. Chemical composition and antimicrobial activity of essential oil of *Piper angustifolium*. *Planta Med.* **1996**, *62*, 372–373. [CrossRef]

 *applied sciences*

Article

The Antifungal Potential of Carvacrol against *Penicillium Digitatum* through ^{1}H-NMR Based Metabolomics Approach

Chunpeng Wan [1], Yuting Shen [1], Muhammad Farrukh Nisar [1,†], Wenwen Qi [1], Chuying Chen [1,*] and Jinyin Chen [1,2,*]

1   Collaborative Innovation Center of Post-Harvest Key Technology and Quality Safety of Fruits and Vegetables in Jiangxi Province, Jiangxi Key Laboratory for Postharvest Technology and Non-destructive Testing of Fruits & Vegetables, Jiangxi Agricultural University, Nanchang 330045, China; chunpengwan@jxau.edu.cn (C.W.); ytshenchina@126.com (Y.S.); farrukh.nisar@hotmail.com (M.F.N.); qwwbaobei@163.com (W.Q.)
2   Pingxiang University, Pingxiang, Jiangxi 337055, China
*   Correspondence: cy.chen@jxau.edu.cn (C.C.); jinyinchen@126.com (J.C.); Tel.: +86-791-8381-3158 (C.C. & J.C.)
†   Current Address: Department of Physiology and Biochemistry, Cholistan University of Veterinary and Animal Sciences (CUVAS), Bahawalpur 63100, Pakistan.

Received: 29 April 2019; Accepted: 28 May 2019; Published: 30 May 2019

Abstract: Carvacrol (5-Isopropyl-2-methylphenol), a volatile oil constituent, mainly exists in Labiaceae family plants. Carvacrol has long been studied for its natural antifungal potential and food preservative potential. However, its exact mode of action, especially against *Penicillium digitatum* (*P. digitatum*), remains unexplored. Herein, a ^{1}H-NMR-based metabolomic technique was used to investigate the antifungal mechanism of carvacrol against *P. digitatum*. The metabolomic profiling data showed that alanine, aspartate, glutamate, and glutathione metabolism were imbalanced in the fungal hyphae. A strong positive correlation was seen between aspartate, glutamate, alanine, and glutamine, with a negative correlation among glutathione and lactate. These metabolic changes revealed that carvacrol-induced oxidative stress had disturbed the energy production and amino acid metabolism of *P. digitatum*. The current study will improve the understanding of the metabolic changes posed by plant-based fungicides in order to control citrus fruit green mold caused by *P. digitatum*. Moreover, the study will provide a certain experimental and theoretical basis for the development of novel citrus fruit preservatives.

Keywords: amino acid metabolism; carvacrol; metabolomics data; oxidative stress; *Penicillium digitatum*

---

1. Introduction

Rotting of fruits and vegetables have been a frequent and serious problem for thousands of years. Even in developed countries, the citrus yield loss reaches up to 25%, with developing countries facing almost double this yield loss of citrus crop. This yield loss is causing serious annual economic losses and creating hurdles for the development of the citrus fruits-related industry [1]. Citrus fruits (Family Rutaceae, subfamily Aurantioideae) are highly susceptible to decay caused by more than 20 postharvest fungal infections leading, to approximately half the crop wasted due to fungal diseases [2]. Among various fungal strains, *Penicillium* fungi cause serious plant diseases and damage to the citrus crop during storage, accounting for about 10–30% [3]. The *P. digitatum* infects the citrus fruits, causes green mold disease, and generates huge (60–80%) yield losses under ambient conditions [4]. In order to control such fungal disease-mediated yield losses, prochloraz, imazalil, and thiabendazole are extensively used as potential antifungal agents [5]. The literature reported serious health issues aroused by excessive use of certain routine and synthetic antifungal agents. Over the past few decades, the intentions have been to ensure the hygiene and safety levels of fruit products along with promoting maximum biodegradability. In order to achieve this ambition, researchers are still trying to screen natural products with strong antifungal potential from plants. Moreover, it has also been tried to explore new preservation technologies that can replace synthetic fungicides. Compared with synthetic fungicides, natural antifungal substances have attained much attention in recent years [6–9].

Various essential oils refined from *Melaleuca alternifolia* have strong inhibitory effects on multiple fungal strains, particularly *P. italicum* Wehmer and *P. digitatum* Sacc., and hence can be used as natural preservatives in agricultural and food processing [10]. Previously we have reported that cinnamaldehyde contained in cinnamon has a good antifungal effect against *P. italicum* and clarifies the method for dynamic detection of the antifungal effect of cinnamaldehyde [11]. In addition, GC-MS and GC-FID techniques successfully separated the essential oils components *viz.* limonene (87.02%), linalyl acetate (53.76%) and linalool (39.74%) from peels, leaves and flowers of *Citrus aurantium* var amara, neroli oil appeared as the best fungistatic agent, which reduced the *P. digitatum* sporulation to 25 % at a 50 mg/mL concentration of oil [12].

Carvacrol (5-Isopropyl-2-methylphenol) is the major volatile oil constituent from Labiaceae family plants, such as oregano, mint, and thyme [13]. Carvacrol has many medicinal and edible applications, including antifungal, antimicrobial, anti-inflammatory, insecticidal, antitumor, and other effects [14]. Carvacrol inhibited the growth of *P. digitatum* and *P. italicum* in in vitro and in vivo experiments of lemon [15]. Another study in Morocco, reported that *Thymus leptobotrys* and *T. riatarum* have carvacrol contents of 76.94% and 32.24%, respectively, and the former species has the highest bioactivity among the tested plant species in that study. In addition, the study reported a complete inhibition of germination of the spore of *P. digitatum* and *P. italicum* at 500 µL/L [16]. The results of the antimicrobial activity assay showed that carvacrol, cinnamaldehyde, and *trans*-cinnamaldehyde significantly reduced the growth of *P. digitatum*; particularly, the carviniol appeared as the most effective remedy to control *P. digitatum* [17].

Metabolomics is an important part of systems biology, which is quite similar to genomics and proteomics [18]. Metabolomics is distinguished from other related sciences in terms if its ability to find the relative relationship of metabolites with physiological and pathological changes through quantitative analyses of metabolites with low molecular weight (1000) [19]. Recently, the ^{1}H-NMR based metabolomics approach was used to reveal the antimicrobial mechanism, which generally includes amino acid biosynthesis and energy-associated metabolism [20,21]. Although carvacrol already been proved to have significant antifungal properties, the underlying mechanism is not clear yet. In the current study, we applied the ^{1}H-NMR-based metabolomics approach to evaluate the antifungal potential of carvacrol on *P. digitatum* and explored the underlying antifungal mechanism. Moreover, how carvacrol-induced oxidative stress disturbs the energy production and amino acid metabolism in *P. digitatum* shall also be studied through metabolic profiles.

2. Materials and Methods

2.1. Chemicals

Carvacrol (purity 99%) purchased from Shanghai Aladdin Co., Ltd. (C804847), and because of its insolubility in water, it was previously dissolved in ethanol (50%, *w/w*) to obtain a stock solution with a proper concentration of 10 mg· mL^{-1}.

2.2. Preparation of P. Digitatum Spores

The highly pathogenic fungus *P. digitatum* was isolated from an infected orange with typical green mold symptoms and maintained on potato dextrose agar (PDA: 200 g peeled potatoes, 20 g glucose, 18 g agar powder and 1 L distilled water) medium plates at 25 °C. The preparation of *P. digitatum* spore suspension was based on a previous method [22], and the tested pathogen was incubated at 25 °C for

7 days. The seven-days-old plate was washed with sterile water and then gently dispersed by spread bacteria beads to release spores. Finally, the spore suspensions were filtered through a sterile cotton ball in a funnel to remove mycelia and PDA fragments and adjusted to the suitable concentration of 1×10^6 spores mL^{-1} with the aid of a hemocytometer.

2.3. Antifungal Effects of Carvacrol against P. Digitatum

The inhibitory effect of carvacrol on the mycelial growth of *P. digitatum* was determined as previously reported [22]. Briefly, the 0, 0.0625, 0.125, 0.25, 0.5, and 1 mL of carvacrol stock solution was diluted with 2, 1.9375, 1.875, 1.75, 1.5, 1 mL of sterile 0.5% Tween 80 and mixed with 18 mL of PDA for obtaining the final concentrations of 0 (control), 0.03125, 0.0625, 0.125, 0.25, and 0.5 mg·mL^{-1}. The mycelial disks (5 mm in diameter), cut from the periphery of a seven-days-old culture using a stainless-steel punch, was placed in the center of each Petri dish (90 mm in diameter). Then, all plates were incubated at 25 °C for seven days. Four replicates were used per treatment and the experiment was carried out at two separate times. Mycelial growth inhibition (MGI) of carvacrol treatment against control was calculated using the following equation:

$$\text{MGI}\ (\%) = \frac{Dc - Dt}{Dc - 5} \times 100$$

where *Dc* and *Dt* were the mean colony diameter of control and treated sets, respectively.

The minimum inhibitory concentration (MIC) was defined as the lowest carvacrol concentration that completely inhibited the growth of *P. digitatum* after 48 h of incubation at 25 °C. The minimum fungicidal concentration (MFC) was considered the lowest concentration of carvacrol with no visible fungal growth on a PDA plate after a following 5 days incubation at 25 °C [22].

2.4. The Effect of Carvacrol on Mycelial Weights and Water-Retention Rate

The effects of carvacrol on the wet and dry weights as well as the water-retention rate of *P. digitatum* mycelial were determined by the method described by Tian et al. with some modifications [23]. Briefly, 100 µL of *P. digitatum* containing 10^6 spores mL^{-1} was inoculated into 50 mL of potato dextrose broth (PDB) medium and then was incubated in a rotary shaker (150 rpm) at 25 °C. After shake incubating for 48 h, the carvacrol solution at final concentrations of 0 (control), 0.03125, 0.0625, 0.125, 0.25 and 0.5 mg·mL^{-1} were added into the above-mentioned PDB and then incubated for 24 h at 25 °C in a rotary shaker. The mycelia from the carvacrol treated and control PDB was collected by filtering used a Buchner funnel and washed three times with sterile water. The wet weights of the mycelia were measured, the mycelia were dried at 70 °C for 12 h, and the dry weights were then measured using an analytical balance (Ms105, Mettler Toledo, Greifensee, Switzerland). The water-retention rates of the carvacrol-treated and control groups were calculated using the following equation:

$$\text{Water-retention rate}\ (\%) = \frac{Ww - Wd}{Ww} \times 100$$

where *Ww* and *Wd* were the mean of wet and dry weights in carvacrol treated and control sets, respectively.

2.5. Sample Preparation for ^{1}H NMR Spectroscopy

The collected mycelial treated with MIC carvacrol for 4, 8 and 12 h respectively were washed three times with pre-cooled PBS buffer solution and then added with 3.8 mL pre-cooled methanol-water mixture (1/0.9, v/v). The mixture was placed on the ice for 4 min of sonication bathing, and then added with 4 mL trichloromethane. After full oscillation, the mixture was centrifuged at 4 °C and 10,000 rpm for 10 min. The upper methanol water phase was placed in the nitrogen blowing instrument (NBI,

HSC-24B, Tianjin Hengao Technology Development Co. Ltd, Tianjin, China) to blow off the methanol, and then the supernatants were freeze-dried, and then stored under $-80°C$ until NMR analysis [11].

In the NMR measurements, the samples were dissolved in 550 mL 99.8% D_2O phosphate buffer (0.2 M, pH = 7.0), which contained 0.05% (w/v) 3-(trimethylsilyl) sodium propionate -2, 2, 3, 3-d4 (TSP). After rotating for 15 s and centrifuging for 10 min at 12,000 rpm and 4 °C, the supernatant was transferred to a clean nuclear magnetic resonance tube (5 mm) for analysis. The ^{1}H-NMR spectrum was recorded in the 298 K at 500 MHz nuclear magnetic resonance spectroscopy (Bruker Avance III, Bruker Instruments, Darmstadt, Germany). Field-frequency locking with D_2O and TSP was used as a reference for chemical shift (1H, D 0.00). The carrpurcell-meiboom-gill sequence [90 (t-180-t) n-acquisition], which is edited by lateral relaxation, has a total spin-echo delay (2 n t) of 40 ms. The spectra were recorded in 64K data points with 128 scans, ranging from -5–15 ppm. By multiplying FIDS with exponential weighting function (corresponding to line broadening at 0.5Hz), Fourier transform was performed on the spectrum.

2.6. Spectral Pre-Processing and Data Analysis

All ^{1}H NMR spectra were phase and baseline corrections, and the peak was manually calibrated using Topspin software (Bruker BioSpin GmbH, version 3.5, Rheinstetten, Karlsruhe, Germany). Then the peak was exported to ASCII file using Mestrec (version 4.9.6, Mestrelab Research SL, Santiago de Compostela, Spain) and imported into R software (https://www.r-project.org/) for further analysis. The region containing residual water signal was removed. The spectrum was combined with an adaptive intelligent algorithm. Before multivariate data analysis, the remaining trash cans were the probability quotient normalization and Pareto scale.

Orthogonal signal correction partial least squares discriminant analysis (OSC-PLS-DA) was used to reveal the metabolic differences among groups. Score charts are used to show clustering between categories, and load/s charts are used to identify different metabolites between two groups. Differential metabolites are metabolites in the upper right quadrant and the lower left quadrant of the S diagram. According to the correlation coefficient from blue to red, the loading graph is color coded. The model was validated internally by repeated two cross-validations and externally by a permutation test.

2.7. Statistical Analysis

Data on the inhibitory effects of carvacrol on mycelial growth, mycelial weights, and water-retention rate were analyzed by variance analysis (ANOVA) using the SPSS version 17.0 (SPSS Inc., Chicago, IL, USA). The significant difference of means was performed with the Duncan's test at $p < 0.05$.

3. Results

3.1. Effects of Carvacrol on P. Digitatum Mycelial Growth on PDA

The inhibitory effect of carvacrol on the growth of *P. digitatum* was quite obvious and a significant growth inhibition on PDA medium was seen in a dose-dependent manner ($p < 0.05$) (Figure 1). The increasing carvacrol concentrations had higher mycelial growth inhibition (MGI). As a whole, over one-fifth of the *P. digitatum* mycelial growth was inhibited at 0.0625 mg/mL of carvacrol, but 0.125 mg/mL concentration inhibited more than half (54.84%) of the mycelial growth. The higher carvacrol concentrations (0.25 and 0.5 mg/mL) completely inhibited the mycelial growth of *P. digitatum* (Figure 1).

Based on the observation of *P. digitatum* in mycelial growth on PDA medium with carvacrol treatments at 0, 0.03125, 0.0625, 0.125, 0.25, and 0.5 mg/mL during the incubation period at 25 °C, carvacrol treatments at the concentrations of 0.125 mg/mL and 0.25 mg/mL completely inhibited *P. digitatum* in mycelial growth at the 2nd day and 7th days of incubation, respectively. Therefore, the values of MIC and MFC were 0.125 mg/mL and 0.25 mg/mL, respectively (Figure 2).

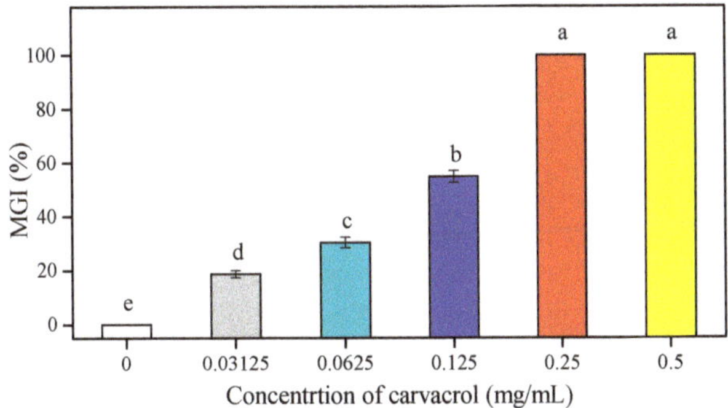

Figure 1. The antifungal efficacy of carvacrol on the in vivo mycelial growth inhibition (MGI) of *P. digitatum* on PDA. Bars indicate the mean ± standard deviation (S.D.) and those labeled with different letters (**a**, **b**, **c**, **d**, and **e**) were significantly different according to Duncan's test ($p < 0.05$).

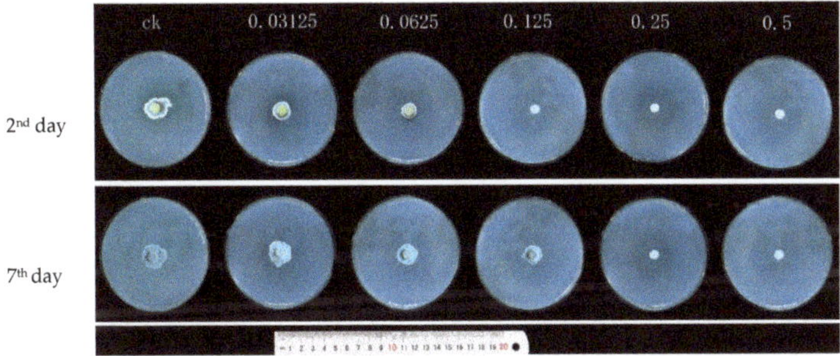

Figure 2. The effect of the growth diameter of *P. digitatum* treated with different concentrations of carvacrol (0, 0.03125, 0.0625, 0.125, 0.25, and 0.50 mg/mL) at 2 days and 7 days post-inoculation (dpi).

3.2. Effects of Carvacrol on Mycelial Weights in PDB

The mycelial weights of *P. digitatum* in carvacrol treatment and control groups are shown in Table 1. The data showed that mycelial growth biomass was strongly inhibited with increasing the carvacrol concentration. Initially, the wet and dry weights were 3.736 g and 0.393 g/100 mL at the lower carvacrol concentration of 0.0625 mg/mL, respectively. At higher carvacrol concentrations (0.125, 0.25 and 0.50 mg/mL), the effect on mycelial weights was recorded at a significant level ($p < 0.05$) in comparison with the control group.

Table 1. The mycelial weights and water-retention rate of *P. digitatum* treated with several concentrations of carvacrol.

Concentrations (mg/mL)	Mycelial Weight (g/100 mL)		Water-retention Rate (%)
	Wet Weight	**Dry Weight**	
0	4.868 ± 0.288 [a]	0.419 ± 0.011 [a]	91.38 ± 0.29 [a]
0.03125	4.487 ± 0.185 [a]	0.407 ± 0.008 [a]	90.89 ± 0.33 [a]
0.0625	3.736 ± 0.113 [b]	0.393 ± 0.015 [ab]	89.47 ± 0.48 [b]
0.125	3.065 ± 0.267 [c]	0.368 ± 0.016 [bc]	87.69 ± 0.77 [c]
0.25	2.895 ± 0.170 [c]	0.357 ± 0.012 [c]	87.57 ± 0.39 [c]
0.5	2.671 ± 0.232 [c]	0.345 ± 0.020 [c]	86.96 ± 0.65 [c]

Values are mean ± S.E. The data followed by different letters within the column are significantly different according to Duncan's test ($p < 0.05$).

3.3. Effect of Carvacrol on Water-Retention Rate of P. Digitatum

Water is the main component in the fungal cell and account for about 90% of mycelial fresh weight and plays an important role in regulating of cell osmotic pressure. The water-retention rate is used as an index of lipid peroxidation that is related to membrane damage leading to cell aging. The lower water-retention index shows higher membrane damages, and vice versa. The effect of carvacrol on the water-retention rate of *P. digitatum* mycelia is shown in Table 1. Different concentrations of carvacrol treatments were used to evaluate the mycelial membrane damage. The water-retention rate of *P. digitatum* mycelia was decreased and treated by carvacrol in a dose-dependent manner. The results demonstrate that the water-retention rate of *P. digitatum* mycelia was more significantly inhibited by the higher concentrations (0.125, 0.25, and 0.50 mg·mL^{-1}) of carvacrol.

3.4. Metabolites Identified in 1H-NMR Spectra

The representative 500 MHz ^{1}H-NMR spectra of mycelia obtained from the control group and MIC carvacrol administration group are shown in Figure 3. Chemical shifts assignments of metabolites were shown in the Supplementary Material (Supplementary Table S1). Nuclear magnetic resonance (NMR) was allocated by searching publicly-accessible metabolome databases (such as the Human Metabonomics Database and Madison Qingdao Metabonomics Joint Database) based on chemical changes reported in the literature.

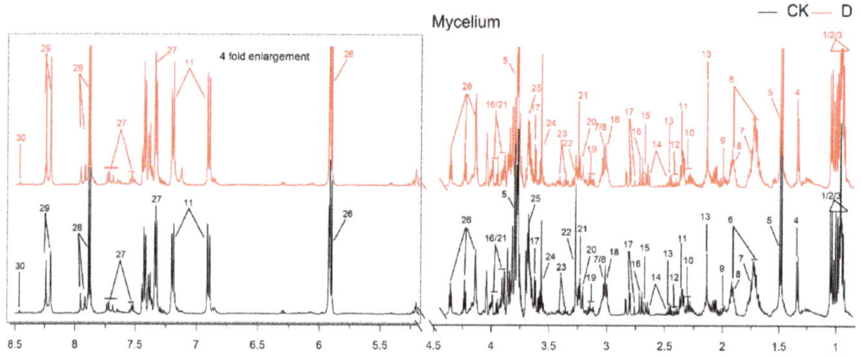

Figure 3. The typical 500 MHz CPMG ^{1}H-NMR spectra for 2 groups. Keys: 1. Isoleucine; 2. Leucine; 3. Valine; 4. Lactate; 5. Alanine; 6. Lysine; 7. Putrescine; 8. 4-Aminobutyrate; 9. Acetate; 10. Glutamate; 11. Acetaminophen; 12. Succinate; 13. Glutamine; 14. Glutathione; 15. 5,6-Dihydrouracil; 16. Aspartate; 17. Sarcosine; 18. Phenylalanine; 19. Ethanolamine; 20. Choline; 21. Betaine; 22. Arginine; 23. Methanol; 24. Glycine; 25. π-Methylhistidine; 26. Uracil; 27. Tryptophan; 28. Xanthine; 29. Adenine; 30. Formate.

3.5. Multivariate Analysis of 1H-NMR Spectral Data

To evaluate the antifungal activity of carvacrol on *P. digitatum*, the OSC-PLS-DA model was constructed and all NMR data obtained from the control group (CK) and carvacrol administration group (D) at 4, 8 and 12 h were analyzed. In Figure 4B–D, each point manifested a sample, and each clustering represented a corresponding metabolic pattern in different groups. Figure 4B shows that the two groups were not well separated at 4 h. However, with the passage of time, the control group and the drug group were separated further and further, and the CK and D groups were the furthest away in the scores plot at 12 h (Figure 4D). This result shows that the metabolomic changes in D group increased from 4 h to 12 h. At 12 h, the metabolic spectrum of group D changed fundamentally, reflecting the rapid response of the strain to carvacrol. The trajectory plot (Figure 4A) also exhibited a good separation between the CK and D groups, showing an apparent time-dependent antifungal activity of carvacrol on *P. digitatum*.

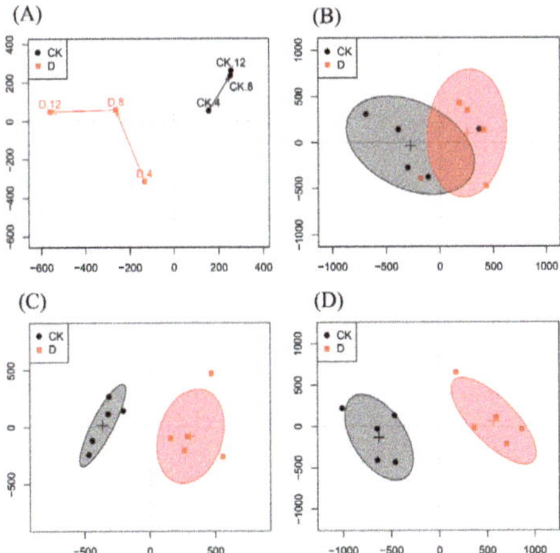

Figure 4. (**A**): Score trajectory of the OSC-PLS-DA analysis in the CK group and D group at 4, 8, 12 h. (**B–D**): Scores plots of CK and D groups at 4 h, 8 h, and 12 h, respectively.

In the 12 h S-plot (Figure 5B), different shapes and colors of the dots show different metabolites. The contribution of these metabolites to the group is related to their distance to the center; variables farther away from the center contribute more to the group separation. On the basis of the correlation coefficient, the 12-hours loading plots (Figure 5C,D) are coded with cold and warm tones, and the correlation increases gradually from blue to red. Significant decrease of lactate, alanine, glutamate, glutamine, glutathione, aspartate, sarcosine, phenylalanine, ethanolamine, choline, arginine, methanol, glycine, π-Methylhistidine and xanthine, and marked increase of leucine, uracil, tryptophan and adenine were found in the carvacrol dosed group.

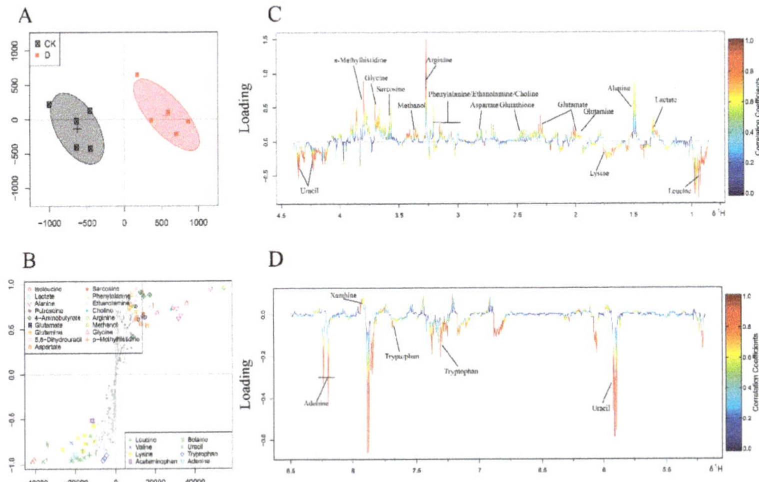

Figure 5. OSC-PLS-DA analysis of NMR data at 12 h. (**A**) The OSC-PLS-DA Score plot. (**B**) S-plot. (**C,D**) Color-coded loadings plots. The use of color bars, where red and blue represent metabolites, is statistically significant or insignificant in facilitating the separation of groups. In the CK group, the peak of the positive and negative states showed that the decrease and increase of metabolites were correlated with the score plot.

The results of the normality test of metabolites are shown in Table 2. The folding changes (FC) of metabolites and their associated P values were calculated and corrected in color tables. Collapse change values are color-coded after log conversion. Cell units were filled with red or blue to indicate the increase or decrease respectively of metabolites in the carvacrol-treated group compared with the control group.

Table 2. Differential expression of metabolites between two groups at 12 h.

No	Metabolites	D/CK	
		[a] FC	[b] P
1	Isoleucine	0.04	
2	Leucine	0.5	*
3	Valine	−0.11	
4	Lactate	−0.52	*
5	Alanine	−0.82	**
6	Lysine	0.25	
7	Putrescine	−0.18	
8	4-Aminobutyrate	−0.27	
9	Acetate	0.1	
10	Glutamate	−0.41	*
11	Acetaminophen	0.04	
12	Succinate	−0.26	
13	Glutamine	−1.17	***
14	Glutathione	−0.65	*
15	5-6-Dihydrouracil	−0.3	
16	Aspartate	−0.71	**
17	Sarcosine	−1.02	**
18	Phenylalanine	−1.65	***
19	Ethanolamine	−0.51	*
20	Choline	−1.51	***
21	Betaine	0.06	
22	Arginine	−1.86	***
23	Methanol	−0.75	**
24	Glycine	−1.09	**
25	π-Methylhistidine	−1	*
26	Uracil	1.13	**
27	Tryptophan	0.37	*
28	Xanthine	−1.42	***
29	Adenine	1.47	***
30	Formate	−0.07	

[a] FC: Color-coded according to the fold-change value; Color coded according to the $\log_2$ (FC), red represents increased and blue represents decreased concentrations of metabolites. Color bar

-1   -0.5   0   0.5   1

[b] P values corrected by Benjamini-Hochberg methods were calculated based on a parametric Student's t-test or a nonparametric Mann-Whitney test (dependent on the conformity to the normal distribution). * $p < 0.05$, ** $p < 0.01$, *** $p < 0.001$.

4. Discussion

The current study comprehensively evaluated the antifungal activity of carvacrol against *P. digitatum* through ^{1}H-NMR metabonomics. Following the important metabolites selected by OSC-PLS-DA loading/S-diagram, METPA (http://www.metaboanalyst.ca) (Figure 6A) and KEGG (http://www.genome.jp/kegg/) were used for path analysis to determine biologically significant metabolic patterns and related pathways (Figure 6B–E).

Alanine, aspartate, glutamate, and glutathione metabolisms were imbalanced in the fungal hyphae (Figure 6). These metabolisms were also found in the correlation network of D group at 12 h. A strong positive correlation between aspartate, glutamate, alanine, and glutamine was seen, and a negative correlation between glutathione and lactate was also observed. These metabolic changes revealed that carvacrol induced the generation of heavy oxidative stress to disturb the energy and amino acid metabolism of *P. digitatum* (Supplementary Figure S1).

A significant decrease in the level of glutamate, glutamine, glycine and precursors of glutathione was observed in the D group. As an ROS scavenger, the level of glutathione also markedly decreased in the D group. Glutathione being endogenously expressed antioxidant enzyme that played an important role in the protection of cells against ROS by quickly eliminating free radicals [24]. Glutathione was consumed in large quantities in fungi against carvacrol-induced ROS. To sustain a certain level of glutathione in fungus, the synthesis of glutathione from glutamate, glutamine and glycine has to be enhanced [25], which will lead to the degradation of glutamate and glycine. In addition, a heavy load of ROS could enhance peroxidation of membrane lipids, leading to damage the cellular membranes [26]. The decreased level of choline and ethanolamine were seen in the D group compared with the CK group, and both are considered as the key players in the stability and integrity of cell membranes [27]. Therefore, any decrease in choline and ethanolamine may be due to excessive utilization for repairing the damaged membranes caused by ROS which are indicators of decreased level of sarcosine in D group [28]. In the current study, the sarcosine level in group D was lower than that in group CK, and was linked with the decrease of choline and glycine levels. These results reflect the disturbance of ROS to the sarcosine pathway caused by carvacrol.

The self-protection mechanism of fungus made it consume huge amounts of energy to avoid ROS mediated stress and repair the cell membranes. In order to produce more ATPs, the aerobic respiration in fungal hypha was enhanced while the glycolysis pathway was checked to a great extent. Such reduction in glycolysis was supported by the decreased level of lactate in D group. In addition, a marked decrease in phenylalanine, alanine, aspartate, and arginine levels was observed in D group. The glycogenic amino acids underwent degradation [29] and reflected an enhancement in the gluconeogenesis pathway that compensates the deficiency of glucose. The results further suggested that carvacrol helps produce energy and amino acid metabolic disorders in *P. digitatum*.

A significant increase in the transcriptional initiators e.g., uracil and adenine were found in the D group compared with the CK group (Table 2) [30]. The increased level of uracil and adenine revealed that carvacrol caused RNA damage to *P. digitatum*.

In this pioneering study, the antifungal activity of carvacrol against *P. digitatum* was studied by using the metabolomics approach based on ^{1}H-NMR. Carvacrol can cause an abnormal metabolic state of *P. digitatum* by interfering with different metabolic pathways such as energy and amino acid metabolisms. Furthermore, carvacrol could also induce damage to the RNA molecules and ROS production. Metabolomics provide a powerful and feasible tool for evaluating antifungal activity and exploring its potential mechanism.

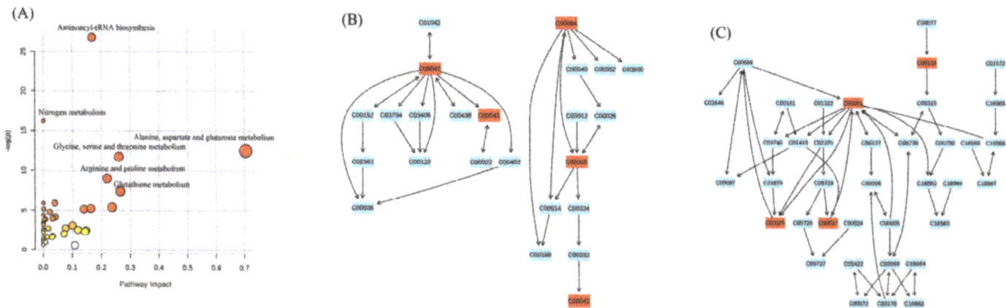

Figure 6. (**A**) In this study, metabolic analysis was used to analyze the pathway topology related to antifungal activity. The term "log p" is the conversion of the original p value calculated from enrichment analysis, and "impact" is the path impact value calculated from the path topology analysis. The bubble area is proportional to the effect of each path, and the color indicates the importance from the highest red to the lowest white. (**B**) Alanine, aspartate, glutamate metabolism; (**C**) glutathione metabolism.

Supplementary Materials: The following are available online at http://www.mdpi.com/2076-3417/9/11/2240/s1.

Author Contributions: Conceptualization, C.W. and J.C.; methodology, C.W. and C.C.; formal analysis, C.W. and C.C.; investigation, Y.S.; W.Q. and C.C.; data curation, C.C.; writing—original draft preparation, C.W.; Y.S.; W.Q. and C.C.; writing—review and editing, C.W. and M.F.N.; supervision, C.W.; project administration, C.W. and C.C.; funding acquisition, J.C.

Funding: This research was funded by National Natural Science Foundation of China (NO.31760598) and Advantage Innovation Team Project of Jiangxi Province (NO.20181BCB24005).

Conflicts of Interest: The authors declare no conflict of interest.

References

1. Nicosia, M.G.; Pangallo, S.; Raphael, G.; Romeo, F.V.; Strano, M.C.; Rapisarda, P.; Droby, S.; Schena, L. Control of postharvest fungal rots on citrus fruit and sweet cherries using a pomegranate peel extract. *Postharvest Biol. Technol.* **2016**, *114*, 54–61. [CrossRef]

2. Chen, J.; Shen, Y.; Chen, C.; Wan, C. Inhibition of key citrus postharvest fungal strains by plant extracts in vitro and in vivo: A review. *Plants* **2019**, *8*, 26. [CrossRef]

3. Chen, C.; Cai, N.; Chen, J.; Peng, X.; Wan, C. Chitosan-Based Coating Enriched with Hairy Fig (Ficus hirta Vahl.) Fruit Extract for "Newhall" Navel Orange Preservation. *Coatings* **2018**, *8*, 445. [CrossRef]

4. Zafar, I.; Zora, S.; Ravjit, K.; Saeed, A. Management of citrus blue and green moulds through application of organic elicitors. *Australas. Plant Pathol.* **2012**, *41*, 69–77.

5. Hao, W.; Li, H.; Hu, M.Y.; Liu, Y.; Rizwan-ul-Haq, M. Integrated control of citrus green and blue mold and sour rot by Bacillus amyloliquefaciens incombination with tea saponin. *Postharvest Biol. Technol.* **2011**, *59*, 316–323. [CrossRef]

6. Palou, L.; Ali, L.; Fallik, E.; Romanazzi, G. GRAS, plant- and animal-derived compounds as alternatives to conventional fungicides for the control of postharvest dieseases of fresh horticultural produce. *Postharvest Biol. Technol.* **2016**, *122*, 41–52. [CrossRef]

7. Gong, T.; Li, C.; Bian, B.; Wu, Y.; Dawuda, M.M.; Liao, W. Advances in application of small molecule compounds for extending the shelf life of perishable horticultural products: A review. *Sci. Hortic.* **2018**, *230*, 25–34. [CrossRef]

8. Zhang, M.; Xu, L.; Zhang, L.; Guo, Y.; Qi, X.; He, L. Effects of quercetin on postharvest blue mold control in kiwifruit. *Sci. Hortic.* **2018**, *228*, 18–25. [CrossRef]

9. Ncama, K.; Magwaza, L.S.; Mditshwa, A.; Tesfay, S.Z. Plant-based edible coatings for managing postharvest quality of fresh horticultural produce: A review. *Food Packag. Shelf Life* **2018**, *16*, 157–167. [CrossRef]

10. Zhang, X.F.; Guo, Y.J.; Guo, L.Y.; Jiang, H.; Ji, Q.H. In Vitro Evaluation of Antioxidant and Antimicrobial Activities of Melaleuca alternifolia Essential Oil. *BioMed Res. Int.* **2018**, *2018*, 2396109. [CrossRef]

11. Wan, C.P.; Pei, L.; Chen, C.Y.; Peng, X.; Li, M.X.; Chen, M.; Wang, J.S.; Chen, J.Y. Antifungal Activity of Ramulus cinnamomi Explored by ^{1}H-NMR Based Metabolomics Approach. *Molecules* **2017**, *22*, 2237. [CrossRef]

12. Trabelsi, D.; Hamdane, A.M.; Said, M.B.; Abdrrsbba, M. Chemical composition and antifungal activity of essential oils from flowers, leaves and peels of Tunisian Citrus aurantium against Penicillium digitatum and Penicillium italicum. *J. Essent. Oil Bear. Plants* **2016**, *19*, 1660–1674. [CrossRef]

13. Can Baser, K.H. Biological and pharmacological activities of carvacrol and carvacrol bearing essential oils. *Curr. Pharm. Des.* **2008**, *14*, 3106–3119. [CrossRef]

14. Zotti, M.; Colaianna, M.; Morgese, M.; Tucci, P.; Schiavone, S.; Avato, P.; Trabace, L. Carvacrol: From ancient flavoring to neuromodulatory agent. *Molecules* **2013**, *18*, 6161–6172. [CrossRef]

15. Pérez-Alfonso, C.O.; Martínez-Romero, D.; Zapata, P.J.; Serrano, M.; Valero, D.; Castillo, S. The effects of essential oils carvacrol and thymol on growth of Penicillium digitatum and P. italicum involved in lemon decay. *Int. J. Food Microbiol.* **2012**, *158*, 101–106.

16. Boubaker, H.; Karim, H.; Hamdaoui, A.E.; Msanda, F.; Leach, D.; Bombarda, I.; Vanloot, P.; Abbad, A.; Boudyach, E.H.; Aoumar, A.B. Chemical characterization and antifungal activities of four Thymus species essential oils against postharvest fungal pathogens of citrus. *Ind. Crop. Prod.* **2016**, *86*, 95–101. [CrossRef]

17. Sun, X.; Narciso, J.; Wang, Z.; Ference, C.; Bai, J.; Zhou, K. Effects of chitosan-essential oil coatings on safety and quality of fresh blueberries. *J. Food Sci.* **2014**, *79*, M955–M960. [CrossRef]

18. Keun, H.C.; Ebbels, T.M.; Antti, H.; Bollard, M.E.; Beckonert, O.; Schlotterbeck, G.; Senn, H.; Niederhauser, U.; Holmes, E.; Lindon, J.C.; et al. Analytical Reproducibility in ^{1}H NMR-Based Metabonomic Urinalysis. *Chem. Res. Toxicol.* **2002**, *15*, 1380–1386. [CrossRef]

19. Liu, X.; Zhang, L.; You, L.; Cong, M.; Zhao, J.; Wu, H.; Li, C.; Liu, D.; Yu, J. Toxicological responses to acute mercury exposure for three species of Manila clam Ruditapes philippinarum by NMR-based metabolomics. *Environ. Toxicol. Pharmacol.* **2011**, *31*, 323–332. [CrossRef]

20. Liu, Q.; Wu, J.E.; Lim, Z.Y.; Aggarwal, A.; Yang, H.; Wang, S. Evaluation of the metabolic response of Escherichia coli to electrolysed water by ^{1}H NMR spectroscopy. *LWT Food Sci. Technol.* **2017**, *79*, 428–436. [CrossRef]

21. Picone, G.; Laghi, L.; Gardini, F.; Lanciotti, R.; Siroli, L.; Capozzi, F. Evaluation of the effect of carvacrol on the Escherichia coli 555 metabolome by using ^{1}H-NMR spectroscopy. *Food Chem.* **2013**, *141*, 4367–4374. [CrossRef]

22. Chen, C.; Qi, W.; Peng, X.; Chen, J.; Wan, C. Inhibitory Effect of 7-Demethoxytylophorine on Penicillium italicum and its Possible Mechanism. *Microorganisms* **2019**, *7*, 36. [CrossRef]

23. Tian, J.; Wang, Y.; Zeng, H.; Li, Z.; Zhang, P.; Tessema, A.; Peng, X. Efficacy and possible mechanisms of perillaldehyde in control of Aspergillus niger causing grape decay. *Int. J. Food Microbiol.* **2015**, *202*, 27–34. [CrossRef]

24. Chen, C.; Fu, Y.H.; Li, M.H.; Li, M.H.; Ruan, L.Y.; Xu, H.; Chen, J.F.; Wang, J.S. Nuclear magnetic resonance-based metabolomics approach to evaluate preventive and therapeutic effects of Gastrodia elata Blume on chronic atrophic gastritis. *J. Pharm. Biomed. Anal.* **2019**, *164*, 231–240. [CrossRef]

25. Ballatori, N.; Jacob, R.; Boyer, J.L. Intrabiliary glutathione hydrolysis. A source of glutamate in bile. *J. Biol. Chem.* **1986**, *261*, 7860–7865.

26. Apel, K.; Hirt, H. Reactive oxygen Species: Metabolism, Oxidative Stress, and Signal Transduction. *Annu. Rev. Plant Biol.* **2004**, *55*, 373–399. [CrossRef]

27. Fu, Y.; Si, Z.; Li, P.; Li, M.; Zhao, H.; Jiang, L.; Xing, Y.; Hong, W.; Ruan, L.; Wang, J.S. Acute psychoactive and toxic effects of D. metel on mice explained by ^{1}H NMR based metabolomics approach. *Metab. Brain Dis.* **2017**, *32*, 1295–1309. [CrossRef]

28. Boncompagni, E.; Østerås, M.; Poggi, M.C.; le Rudulier, D. Occurrence of Choline and Glycine Betaine Uptake and Metabolism in the Family Rhizobiaceae and Their Roles in Osmoprotection. *Appl. Environ. Microbiol.* **1999**, *65*, 2072–2077.

29. Armstrong, C.W.; McGregor, N.R.; Sheedy, J.R.; Buttfield, I.; Butt, H.L.; Gooley, P.R. NMR metabolic profiling of serum identifies amino acid disturbances in chronic fatigue syndrome. *Clin. Chim. Acta* **2012**, *413*, 1525–1531. [CrossRef]

30. Sakurai, K.; Shinkai, S. Molecular Recognition of Adenine, Cytosine, and Uracil in a Single-Stranded RNA by a Natural Polysaccharide: Schizophyllan. *J. Am. Chem. Soc.* **2000**, *122*, 1–8. [CrossRef]

Article

Structural Requirements for Antimicrobial Activity of Phenolic Nor-Triterpenes from Celastraceae Species

Laila Moujir [1], Manuel R. López [1], Carolina P. Reyes [2], Ignacio A. Jiménez [2] and Isabel L. Bazzocchi [2,*]

[1] Departamento de Bioquímica, Microbiología, Biología Celular y Genética, Facultad de Farmacia, Universidad de La Laguna, 38206 La Laguna, Tenerife, Spain
[2] Instituto Universitario de Bio-Orgánica Antonio González, Departamento de Química Orgánica, Universidad de La Laguna, Avda. Astrofísico Francisco Sánchez 2, 38206 La Laguna, Tenerife, Spain
* Correspondence: ilopez@ull.es; Tel.: +34-922-318576

Received: 27 June 2019; Accepted: 19 July 2019; Published: 24 July 2019

Featured Application: Identifying the structural requirements of phenolic nor-triterpene framework as potential antimicrobial agents.

Abstract: The emergence of pathogenic bacteria-resistant strains is a major public health issue. In this regard, natural product scaffolds offer a promising source of new antimicrobial drugs. In the present study, we report the antimicrobial activity against Gram-positive and Gram-negative bacteria and the yeast *Candida albicans* of five phenolic nor-triterpenes (**1–5**) isolated from *Maytenus blepharodes* and *Maytenus canariensis* in addition to four derivatives (**6–9**), three of them reported for the first time. Their stereostructures have been elucidated on the basis of spectroscopic analysis, including one-dimensional (1D) and two-dimensional (2D) NMR techniques, spectrometric methods, and comparison with data reported in the literature. To understand the structural basis for the antimicrobial activity of this type of compounds, we have performed an in-depth study of the structure–activity relationship (SAR) of a series of previously reported phenolic nor-triterpenes. The SAR analysis was based on the skeleton framework, oxidation degree, functional groups, and regiosubstitution patterns, revealing that these aspects modulate the antimicrobial activity.

Keywords: *Maytenus*; celastroloids; semisynthesis; antibacterial activity; structure–activity relationship

---

1. Introduction

The resistance of common pathogens to standard antibiotic therapies is rapidly becoming a major public health problem all over the world [1], and consequently, there is a need to develop new structural and mechanistic classes of antibiotic agents. In this regard, the development of new antibiotics inspired in natural product scaffolds seems the best short-term solution to address antibiotic resistance [2].

The Celastraceae family is distributed mainly in tropical and subtropical regions of the world including North Africa, South America, and East Asia, and their species have a long history in traditional medicine [3]. The most representative genus in this family is *Maytenus*, with more than 225 species [4]. In the Amazonian region, species of this genus are well known for their use in the treatment of rheumatism, gastrointestinal diseases, and as an antitumoral for skin cancer [4]. The therapeutic potential of *Maytenus* species has been mainly attributed to celastroloids, chemotaxonomic markers of the family [5]. The term celastroloid refers to methylenequinone nor-triterpenes with a 24-nor-*D:A*-friedo-oleanane skeleton. Celastrol [6] and pristimerin [7] are the first and most frequently reported celastroloids, and later on, this term was extended to related phenolic nor-triterpenes [5,8] and their dimer and trimer congeners [9]. This particular class of natural products shows a wide range of bioactivities, including cytotoxic [10,11], anti-inflammatory [12], antioxidant [13], antiparasitic [14], and insecticidal [15] properties. Concerning their antimicrobial activity, pristimerin, tingenone, celastrol, and netzahualcoyone (Figure 1) exhibit inhibitory activity against Gram-positive bacteria [8,16,17], and the mode of action of netzahualcoyone against *Bacillus subtilis* and *Escherichia coli* has been investigated [18,19]. Regarding the phenolic nor-triterpenes, studies on their antimicrobial activity, mechanism of action against Gram-positive bacteria, and preliminary structure–activity relationship have been reported [20–25].

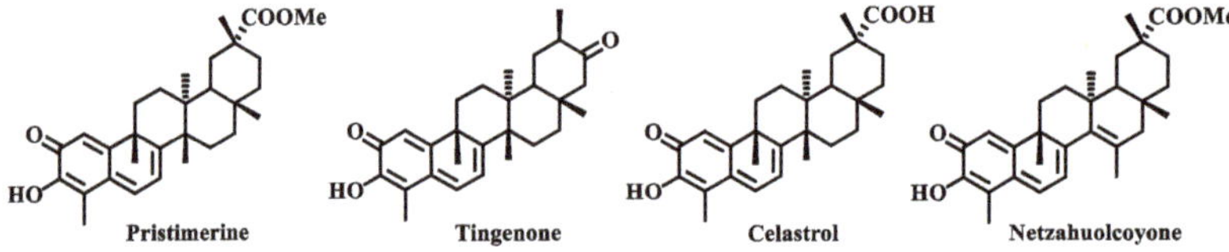

Figure 1. Most frequently reported antimicrobial celastroloids from Celastraceae species.

As part of an intensive investigation into antimicrobial metabolites from Celastraceae species, we report herein on the minimal inhibitory concentrations (MICs) of five natural phenolic nor-triterpenes (**1–5**) and four derivatives (**6–9**), three of them reported for the first time, against Gram-positive and Gram-negative bacteria, and the yeast *Candida albicans*. The structure–activity relationship study of compounds **1–9** was expanded by the known antimicrobial activity of a series of phenolic nor-triterpenes (**10–26**), previously reported by our research group [20–23,25], to deepen our knowledge of the structural requirements for their activity.

2. Materials and Methods

2.1. General Procedures

Optical rotations were measured on a Perkin Elmer 241 automatic polarimeter in $CHCl_3$ at 25 °C, and the $[\alpha]_D$ values are given in 10^{-1} deg cm^2 g^{-1}. UV spectra were obtained in absolute EtOH on a JASCO V-560 instrument. IR (film) spectra were measured in $CHCl_3$ on a Bruker IFS 55 spectrophotometer. ^1H (400 or 500 MHz) and ^{13}C (100 or 125 MHz) NMR spectra were recorded on Bruker Avance 400 or 500 spectrometers; chemical shifts are given in ppm and coupling constants in hertz. Samples were dissolved ($CDCl_3$: δ_H 7.26, δ_C 77.0). EI-MS and EI-HRMS were recorded on a Micromass Autospec spectrometer. Silica gel 60 (particle size 15–40 µm) for column chromatography and silica gel 60 F254 for analytical (TLC) and preparative thin-layer chromatography (PTLC) were purchased from Macherey-Nagel. Sephadex LH-20 was obtained from Pharmacia Biotech. Shimadzu high-performance liquid chromatography (HPLC) equipment consisted of a pump LKB 2248 solvent delivery module, SPD-6V detector set at 254 nm, using a semipreparative silica gel column (Waters µ-Porosil®, 15 cm × 1.6 mm, particle size 10 µm). The mobile phase consisted of a mixture of *n*-hexane-EtOAc (8:2) in isocratic mode with a flow rate of 9 mL/min. The degree of purity of the compounds was over 95%, as indicated by a single peak in HPLC and NMR. All solvents used were of analytical grade (Panreac), and the reagents, used without purification, were purchased from Sigma-Aldrich. Pristimerin, used as starting material, was isolated from the root bark of *M. blepharodes* and *M. canariensis*, as previously described [20,21].

2.2. Phenolic Nor-Triterpenes

The natural phenolic nor-triterpenes **1** (6-oxopristimerol) [20,26], **3** (7,8-dihydro-6-oxoiguesterol, canarol) [27], **16** (6-oxotingenol) [20], **17** (3-*O*-methyl-6-oxotingenol) [20], and **20** (6-oxoiguesterol) [20] were isolated from the root bark of *Maytenus canariensis*, and compounds **2** (7-hydroxy-6-oxopristimerol) [21,28], **4** (blepharodol) [21], **5** (7α-hydroxyblepharodol) [27], **10** (blepharotriol) [21], **12** (zeylasteral) [27,29], **13** (demethylzeylasteral) [21,29], **14** (zeylaterone) [27,29], **15** (demethylzeylasterone) [21,29], **22** (7-oxoblepharodol) [21], and **25** (6-deoxoblepharodol) [21] from *Maytenus blepharodes*. Derivatives **11** (2,3-

O-dimethoxyblepharodol) [21], **18** (2,3-*O*-dimethyl-6-oxotingenol) [20], **19** (2,3-*O*-diacetoxy-6-oxoting enol) [20], **21** (2,3-*O*-dimethyl-6-oxoiguesterol) [20], **23** (2-*O*-methoxy-7-oxoblepharodol) [21], **24** (pristimerol) [21,30], and **26** (8-epi-6-deoxoblepharodol) [21] were obtained following the methodology previously described [20,21]. The structures of these compounds are depicted in Figure 2. Moreover, the semisynthesis of derivate **6**, not previously described, was achieved by acetylation of pristimerin, a main quinone–methide triterpene isolated from the root bark of *Maytenus* species [27]. In addition, derivatives **7–9** were prepared by catalytic reduction and further acetylation of pristimerin. Derivative **7** is a known synthetic compound [31] whose ^{1}H and ^{13}C NMR data have not been previously assigned, whereas derivatives **8** and **9** are reported herein for the first time (Scheme 1, Figures S1–S4.).

2.2.1. Preparation of Compound 6

To a stirred solution of pristimerin (31.0 mg) in pyridine (0.15 mL), anhydride acetic (0.1 mL) and a catalytic amount of 4-dimethylamino-pyridine were added. The resulting orange-red solution was stirred for 14 h at room temperature, until TLC showed complete conversion. This yellow solution was concentrated under reduced pressure, and the residue was purified by preparative TLC using *n*-hexane-EtOAc (1:1) as eluent to afford compound **6** (14.0 mg, 37.0%).

6α-Hydroxy-2,3-diacetylpristimerol (**6**). Pale yellow amorphous solid; $[\alpha]_D^{20}$ − 39.7 (*c* 0.15, MeOH); UV (EtOH) λ_{max} (log ε) 340 (7.3), 204 (7.4) nm; IR ν_{max} (film) 3517, 2948, 2872, 1770, 1730, 1459, 1373, 1215, 1106, 758 cm^{-1}; ^{1}H NMR (500 MHz, CDCl$_3$) δ 0.59 (3H, s, Me-27), 1. 09 (3H, s, Me-28), 1.16 (3H, s, Me-30), 1.24 (3H, s, Me-26), 1.35 (3H, s, Me-25), 2.29 (3H, s, OAc-3), 2.32 (3H, s, Me-23), 2.33 (3H, s, OAc-2), 3.54 (3H, s, OMe-29), 5.06 (1H, dd, *J* = 3.0, 10.8 Hz, H-6), 5.76 (1H, d, *J* = 3.0 Hz, H-7), 7.03 (1H, s, H-1). ^{13}C NMR (125 MHz, CDCl$_3$) δ 12.3 (q, C-23), 18.6 (q, C-27), 20.7 (q, C-26), 21.0 (q, OAc-2), 21.6 (q, OAc-3), 29.1 (t, C-15), 30.2 (t, C-21), 30.4 (t, C-12), 30.9 (t, C-19), 31.2 (s, C-17), 31.9 (q, C-28), 33.1 (q, C-30), 34.5 (t, C-22), 35.1 (t, C-11), 36.9 (t, C-16), 37.4 (q, C-25), 38.3 (s, C-9), 38.6 (s, C-13), 40.8 (s, C-20), 44.6 (d, C-18), 44.8 (s, C-14), 51.9 (q, OMe-29), 66.4 (d, C-6), 116.9 (d, C-1), 117.5 (s, C-7), 128.7 (s, C-4), 130.8 (s, C-5), 139.1 (s, C-3), 142.9 (s, C-2), 151.8 (s, C-10), 159.4 (s, C-8), 168.4 (s, OAc-2), 168.7 (s, OAc-3), 179.2 (s, C-29); EI-MS *m/z* (%) 566 [M$^+$] (39), 551 (14), 534 (52), 506 (32), 492 (100), 450 (42), 201 (95); EI-HRMS *m/z* 556.3032 (calcd. for C$_{34}$H$_{46}$O$_7$, 556.3048).

2.2.2. Preparation of Compounds 7–9

A mixture of pristimerin (622.0 mg) and Pd/C 5% (100 mg) in acetic acid (15 mL) was stirred under 1 atmosphere of hydrogen for 3 h. The reaction mixture was filtered through a pad of celite, the solution quenched by addition of saturated aqueous sodium bicarbonate solution, and the aqueous residue extracted with dichloromethane (3 × 30 mL). Then, to the crude, pyridine (1.5 mL) dissolved in acetic anhydride (1.5 mL) was added, and the reaction mixture was stirred for 12 h at room temperature. Upon completion of the reaction, the solution was concentrated on a cold finger with liquid nitrogen. The residue was purified by HPLC using *n*-hexane-ethyl acetate (8:2) as eluent to give compound **7** (21.9 mg, 2.9%, t$_R$ = 11.4 min), reported elsewhere [31], and derivatives **8** (76.4 mg, 10.3%, t$_R$ = 12.3 min) and **9** (32.1 mg, 4.3%, t$_R$ = 12.6 min), not previously reported.

2,3-Diacetylpristimerol (**7**). Pale yellow amorphous solid; $[\alpha]_D^{20}$ − 15.5 (*c* 0.15, MeOH); UV (EtOH) λ_{max} (log ε) 278 (7.3), 203 (7.4), 201 (7.3) nm; IR ν_{max} (film) 2927, 2870, 1775, 1729, 1649, 1371, 1214, 1188, 756 cm^{-1}; ^{1}H NMR (500 MHz, CDCl$_3$) δ 0.58 (3H, s, Me-27), 1.08 (3H, s, Me-28), 1.17 (3H, s, Me-30), 1.22 (3H, s, Me-26), 1.35 (3H, s, Me-25), 2.06 (3H, s, Me-23), 2.28 (3H, s, OAc-2), 2.31 (3H, s, OAc-3), 3.06 (1H, d br, *J* = 15.7 Hz, H-6β), 3.32 (1H, dd, *J* = 5.1, 15.7 Hz, H-6α), 3.55 (3H, s, OMe-29), 5.73 (1H, d, *J* = 5.1 Hz, H-7), 7.00 (1H, s, H-1). ^{13}C NMR (100 MHz, CDCl$_3$) δ 12.5 (q, C-23), 18.2 (q, C-27), 20.4 (q, OAc-2), 20.7 (q, OAc-3), 22.7 (q, C-26), 28.1 (t, C-6), 28.9 (t, C-15), 30.2 (t, C-12), 30.2 (t, C-19), 30.5 (t, C-22), 31.6 (q, C-28), 32.8 (q, C-30), 34.3 (q, C-25), 34.4 (t, C-11), 34.8 (s, C-17), 34.8 (t, C-21), 36.8 (t, C-16), 37.2 (s, C-13), 37.5 (s, C-9), 40.4 (s, C-20), 43.7 (s, C-14), 44.4 (d, C-18), 51.5 (q, OMe-29), 116.7 (d, C-1), 116.9 (d, C-7), 127.8 (s, C-4), 131.5 (s, C-5), 138.0 (s, C-3), 140.6 (s, C-2), 147.5 (s, C-10), 149.0 (s, C-8), 168.4 (s,

OAc-2), 168.7 (s, OAc-3), 178.9 (s, C-29); EI-MS *m/z* (%): 550 [M$^+$] (1), 535 (8), 475 (16), 322 (7), 229 (17), 149 (9), 57 (100); EI-HRMS *m/z* 550.3167 (calcd. for C$_{34}$H$_{46}$O$_6$, 550.3192).

2,3-Diacetyl-6-deoxoblepharodol (**8**). Pale yellow amorphous solid; [α]$_{\text{D}}^{20}$ – 29.5 (*c* 0.20, MeOH); UV (EtOH) λ*max* (log ε) 278 (7.3), 203 (7.3), 201 (7.2) nm; IR ν*max* (film) 2937, 1773, 1730, 1462, 1371, 1213, 1040, 756 cm^{-1}; ^{1}H NMR (500 MHz, CDCl$_3$) δ 0.78 (3H, s, Me-27), 0.88 (3H, s, Me-26), 1.10 (3H, s, Me-28), 1.60 (6H, s, Me-25, Me-30), 1.98 (3H, s, Me-23), 2.26 (3H, s, OAc-3), 2.30 (3H, s, OAc-2), 2.59 (1H, dd, *J* = 11.8, 15.7 Hz, H-6α), 2.60 (1H, dd, *J* = 6.1, 15.7 Hz, H-6α), 3.59 (3H, s, OMe), 6.89 (1H, s, H-1). ^{13}C NMR (100 MHz, CDCl$_3$) δ 12.5 (q, C-23), 15.9 (q, C-26), 17.3 (q, C-27), 18.2 (t, C-7), 20.4 (q, OAc-3), 20.6 (q, OAc-2), 27.3 (q, C-25), 28.6 (t, C-6), 28.9 (t, C-21), 30.0 (t, C-12), 30.1 (s, C-17), 30.2 (t, C-19), 30.5 (t, C-15), 31.8 (q, C-28), 31.9 (q, C-30), 33.8 (t, C-11), 36.2 (t, C-22), 36.5 (t, C-16), 37.2 (s, C-9), 38.9 (s, C-13), 39.4 (s, C-14), 40.5 (s, C-20), 43.5 (d, C-8), 44.5 (d, C-18), 51.5 (q, OMe-29), 116.2 (d, C-1), 129.4 (s, C-4), 132.7 (s, C-5), 138.0 (s, C-3), 139.9 (s, C-2), 149.7 (s, C-10), 168.4 (s, OAc-3), 168.8 (s, OAc-2), 179.1 (s, C-29); EI-MS *m/z* (%) 552 [M$^+$] (10), 510 (49), 495 (5), 468 (100), 453 (15), 249 (26), 190 (9); EI-HRMS *m/z* 552.3457 (calcd. for C$_{34}$H$_{48}$O$_6$, 552.3451).

2,3-Diacetyl-8-*epi*-6-deoxoblepharodol (**9**). Pale yellow amorphous solid; [α]$_{\text{D}}^{20}$ – 29.5 (*c* 0.20, MeOH); UV (EtOH) λ$_{\text{max}}$ (log ε) 277 (9.4), 203 (9.4), 200 (9.4) nm; IR ν$_{\text{max}}$ (film) 2956, 1771, 1728, 1643, 1370, 1214, 1184, 1039, 757 cm^{-1}; ^{1}H NMR (500 MHz, CDCl$_3$) δ 0.81 (3H, s, Me-27), 0.88 (2H, m, H-22), 1.08 (3H, s, Me-28), 1.16 (3H, s, Me-30), 1.21 (3H, s, Me-26), 1.42 (1H, m, H-8) 1.46 (3H, s, Me-25), 1.49 (1H, m, H-18), 1.58 (1H, m, H-19β), 1.86 (2H, m, H-16), 2.02 (3H, s, Me-23), 2.21 (2H, m, H-22), 2.26 (3H, s, OAc-3), 2.29 (3H, s, OAc-2), 2.35 (1H, d br, H-19α), 2.72 (2H, m, *J* = 16 Hz, H-6), 3.56 (3H, s, OMe), 6.98 (1H, s, H-1). ^{13}C NMR (100 MHz, CDCl$_3$) δ 12.6 (q, C-23), 18.9 (q, C-27), 20.4 (q, OAc-2), 20.7 (q, OAc-3), 26.5 (t, C-12), 25.8 (q, C-26), 28.6 (t, C-6), 30.0 (t, C-12), 30.1 (t, C-6), 30.2 (t, C-15), 30.5 (t, C-7), 30.6 (t, C-19), 30.7 (s, C-17), 30.8 (t, C-21), 31.3 (q, C-28), 32.8 (q, C-30), 33.3 (t, C-11), 36.8 (q, C-25), 35.7 (t, C-22), 37.1 (t, C-16), 38.3 (s, C-14), 38.7 (s, C-9), 40.4 (s, C-13), 40.6 (s, C-20), 46.7 (d, C-18), 51.5 (q, OMe-29), 56.0 (d, C-8), 1193 (d, C-1), 128.0 (s, C-4), 135.4 (s, C-5), 137.5 (s, C-3), 140.2 (s, C-2), 148.5 (s, C-10), 168.4 (s, OAc-2), 168.8 (s, OAc-3), 180.7 (s, C-29); EI-MS *m/z* (%) 552 [M$^+$] (9), 510 (43), 495 (10), 468 (100), 453 (22), 435 (6), 287 (13), 249 (25), 190 (22); EI-HRMS *m/z* 552.3457 (calcd. for C$_{34}$H$_{48}$O$_6$, 552.3451).

2.3. Antimicrobial Activity

Strains used for determining antimicrobial activity included *Bacillus subtilis* ATCC 6051, *B. alvei* ATCC 6344, *B. cereus* ATCC 21772, *B. megaterium* ATCC 25848, *B. pumilus* ATCC 7061, *Staphylococcus aureus* ATCC 6538, *S. epidermidis* ATCC 14990, *S. saprophyticus* ATCC 15305, *Enterococcus faecalis* ATCC 29212, *Mycobacterium smegmatis* ATCC 19420, *Escherichia coli* ATCC 9637, *Proteus mirabilis* CECT 170, (from Type Culture Spanish Collection), *Pseudomonas aeruginosa* AK 958 (from the University of British Columbia, Department of Microbiology collection), *Salmonella sp.* CECT 456, *Klebsiella oxytoca* LMM2 (clinical isolate, University of La Laguna), and *Candida albicans* CECT 1039. The bacteria cultures were developed in nutrient broth (NB) or brain heart infusion broth (for *E. faecalis* and *M. smegmatis* containing 0.06% Tween 80), and the yeast was cultured in Sabouraud liquid medium at 37 °C. All culture media were purchased from Oxoid. The minimum inhibitory concentrations (MICs) were determined for each compound in triplicate by broth microdilution method (range 0.08–40 µg/mL) in 96 well microtitre plates, according to the M07-A9 of the CLSI (Clinical and Laboratory Institute) [32]. Wells with the same proportions of dimethyl sulfoxide (DMSO) were used as controls and never exceeded 1% (*v/v*). The starting microorganism concentration was approximately 1–5 × 10^5 CFU/mL, and growth was monitored by measuring the increase in optical density (OD) at 550 nm (OD$_{550}$) with a microplate reader (Multiskan Plus II, Tritertek, Huntsville, AL, USA) and viable count in agar plates. The MIC was defined as the lowest concentration of compound that completely inhibits growth of the organisms compared with that of the control after incubation time.

3. Results and Discussion

3.1. Chemistry

Compounds **1–5**, **10**, **12–17**, **20**, **22**, and **25** (Figure 2) were isolated, purified, and characterized in our laboratory from two *Maytenus* species, *M. blepharodes* and *M. canariensis*, as previously described [20, 21,27]. Derivatives **11**, **18**, **19**, **21**, **23**, **24**, and **26**, achieved by acetylation or methylation from natural compounds, are described by González et al. [20] and Rodríguez et al. [21]. Moreover, derivatives **6–9** were prepared from pristimerin (Scheme 1). The semisynthesis of **6** was achieved by acetylation of pristimerin, and derivatives **7–9** were obtained by catalytic reduction and further acetylation of pristimerin, as described in the experimental section. Their structures were greatly aided by comparison of their spectroscopic data with those previously reported for related compounds **24–26** [21]. Even so, a complete set of two-dimensional (2D) NMR spectra (COSY, ROESY, HSQC and HMBC) was acquired for the new derivatives to gain the complete assignment of the ^{1}H and ^{13}C NMR resonances (see Experimental Section, Figures S1–S4). Derivative **7** was identified as the previously described 2,3-diacetylpristimerol [31], however, its ^{1}H and ^{13}C NMR assignments have not been previously reported. Moreover, derivatives **6**, **8**, and **9** are described herein for the first time, and their structures were elucidated as described below.

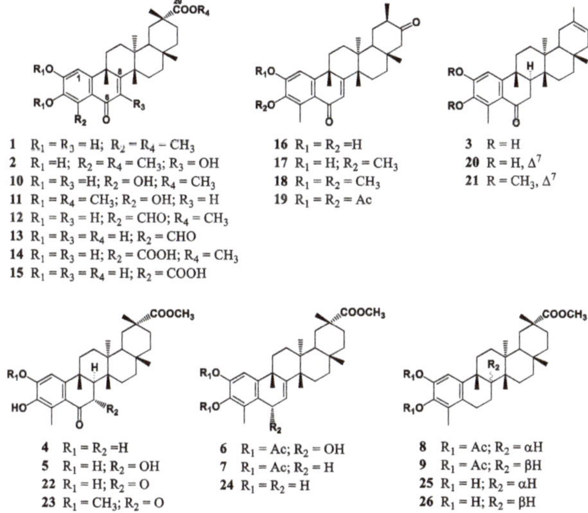

Figure 2. Natural and derivatives phenolic nor-triterpenes included in the SAR studies. SAR: structure–activity relationship.

Scheme 1. Synthesis of analogues **6–9** from pristimerin.

Derivative **6** was obtained as a pale yellow amorphous solid with $[\alpha]_D^{20}$ − 39.7 (c 0.15, MeOH). The molecular formula, $C_{34}H_{46}O_7$, was established by EI-HRMS (m/z 556.3048 [M$^+$], calcd. 556.3032). The IR absorptions revealed the presence of hydroxyl (3517 cm^{-1}) and ester carbonyl (1770 and 1730 cm^{-1}) groups and an aromatic ring (1459 and 758 cm^{-1}). Two acetoxy and a carboxylic methyl ester group in **6** were evident from the ^{1}H NMR signals at δ 2.29 (3H, s), 2.33 (3H, s), and 3.54 (3H, s), in combination with those in the ^{13}C NMR (δ 20.6, 21.0, 51.9, 168.4, and 168.7). Besides the carbons of acetoxy and methoxy groups, the ^{13}C NMR and DEPT data revealed 29 carbon resonances, including one carboxylic carbon, one aromatic ring, one trisubstituted double bond, one oxymethine carbon, five sp^3 quaternary carbons, one methine, seven methylenes, and six methyls. Analysis of 2D NMR data, including HSQC, ^{1}H-^{1}H COSY, ROESY, and HMBC experiments, and comparison with data reported for pristimerol (**24**) [30] indicated that **6** is a phenolic triterpene with a pristimerin framework. In particular, HMBC correlations of Me-23 with C-3, C-4, and C-5; H-1 with C-2, C-3, C-5, C-9, and C-10; H-6 with C-4, C-5, C-8, and C-10, and correlations of Me-25 with C-8, C-9, and C-10 established the structure of A and B rings. The regiosubstitution of the acetate groups and the relative configuration of H-6 were deduced by a ROESY experiment, showing ROEs (Rotating-frame Overhauser Effects) between H-6/OMe-25, H-1/OAc-2, and Me-23/OAc-3 (Figure 3). Thus, the structure of **6** was established as 6α-hydroxy-2,3-diacetoxy-pristimerol.

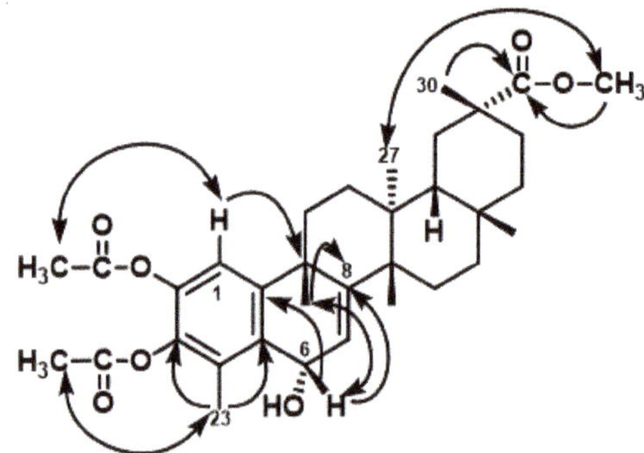

Figure 3. Selected HMBC (single arrow) and ROE (doubled arrow) correlations for compound **6**.

Derivative **7** was assigned the molecular formula $C_{34}H_{46}O_6$, determined by EI-HRMS. The mass spectrum exhibited peaks characteristic to acetate groups (M$^+$-15-60, m/z 475, CH$_3$, CH$_3$COOH). This was confirmed by the ^{1}H and ^{13}C NMR spectra, which included signals for two acetyl groups (δ$_H$ 2.28 (s), 2.31 (s), δ$_C$ 20.4 (q), 20.7 (q), 168.4 (q), 168.7 (s)). Its ^{1}H NMR spectrum showed signals for six methyl groups (δ$_H$ 0.58, 1.08, 1.17, 1.22, 1.35, and 2.06), a methoxy group at δ$_H$ 3.55, a vinylic proton at δ$_H$ 5.73 (d, J = 5.1 Hz), and an aromatic proton at δ$_H$ 7.00 (s). The ^{13}C NMR spectrum displayed 34 signals, which were assigned to one methoxy, eight methyls, eight methylenes, three methines, and fourteen quaternary carbons, including six sp^2 and three carboxylic carbons. Extensive study of the $^{2,3}J_{\text{C-H}}$ correlations (HMBC) allowed us to establish the structure of A and B rings and build the nor-triterpene skeleton for **7**. The regiosubstitution of the acetate groups was deduced by a ROESY experiment, showing ROE effects between H-1/OAc-2 and Me-23/OAc-3. This data established the structure of **7** as 2,3-diacetoxy-pristimerol.

Compounds **8** and **9**, both with the molecular formula $C_{34}H_{48}O_6$ (EI-HRMS), differ from that of **7** by the presence of two additional hydrogen atoms. The most remarkable difference in their NMR spectra, when compared to that of compound **7**, was the absence of the double bond C-7–C-8. Thus, in the ^{13}C NMR spectra of **8** and **9**, the signals assigned to C-7 and C-8 resonated in the region of aliphatic carbons (δ$_C$ 18.2 and 43.5 in **8**, and δ$_C$ 30.5 and 56.0 in **9**). Moreover, differences between **8** and **9** were also observed for the chemical shifts of Me-25 (δ$_H$ 1.60, δ$_C$ 27.3 in **8**, and δ$_H$ 1.46, δ$_C$ 36.8 in **9**) and

Me-26 (δ_H 0.88, δ_C 15.9 in **8**, and δ_H 1.21, δ_C 25.8 in **9**). Analysis of 2D COSY, HSQC, and HMBC spectra allowed us to define their core structures. The ROE effects of H-8/Me-25 and Me-26 observed in a ROESY experiment for compound **8**, and those of H-8/Me-27 observed for compound **9**, further supported the stereochemical assignment for epimers **8** and **9**. Their structures were also confirmed by chemical correlations; hydrolysis of **8** and **9** yielded compounds whose spectroscopic data were identical to those previously reported for 6-deoxoblepharodol (**25**) and 8-*epi*-6-deoxoblepharodol (**26**) [21], respectively. Consequently, the structure of compounds **8** and **9** were deduced to be 2,3-diacetyl-6-deoxoblepharodol and 2,3-diacetyl-8-*epi*-6-deoxoblepharodol, respectively.

3.2. Antimicrobial Evaluation

Compounds **1–9** were tested on Gram-positive and Gram-negative bacteria, and the yeast *Candida albicans* by the broth microdilution method. Moreover, compounds **24–26**, closely related to **6–9**, were also assayed against *B. alvei*, *B. megaterium*, and *B. pumilus* for SAR studies. The results (see Table 1) showed that spore-forming bacteria, such as the genus *Bacillus* are more sensitive to these compounds, among which compounds **6** and **24** exhibiting higher activity (MIC values of 1 and 2.5 μg/mL against *B. subtilis* and *B. cereus*, respectively) than cephotaxime, used as a positive control. Furthermore, the effectiveness against the Gram-positive cocci was restricted to compound **6** (*S. epidermidis* and *E. faecalis*, MIC 5–2.5 and 20–10 μg/mL, respectively). On the other hand, compounds **1–9** were inactive against the Gram-negative bacteria and the yeast *C. albicans* (MIC > 40 μg/mL), which is in line with previous studies on this type of compound [20–23,25].

Table 1. Minimal inhibitory concentration (MIC, μg/mL) [1] against Gram-positive bacteria of phenolic nor-triterpenes [2] and lipophilicity (clog *P*).

Compound	S. a.	S. e.	S. s.	E. f.	B. s.	B. a.	B. c.	B. m.	B. p.	clog P	Reference
2	>40	>40	>40	>40	40–20	>40	40–20	40	>40	5.43	
3	>40	>40	>40	>40	40–20	>40	40	>40	>40	6. 41	
5	>40	>40	>40	>40	40–20	>40	>40	>40	>40	5.26	
6	>40	5–2.5	>40	20–10	1	20–10	2.5	10–5	10–5	5.95	
10	>40	>40	>40	>40	8–4	-	>40	-	-	5.42	[21]
12	40	20	>40	>40	10	7	12	20	25	5.67	[22]
13	25	20	40	>40	13	>40	20	>40	10	5.04	[22]
14	6	>40	>40	>40	3	4	5	4	6	5.21	[23]
15	>40	>40	>40	>40	13	>40	30	>40	>40	4.67	[23]
16	>40	>40	>40	>40	14–12	-	>40	>40	>40	5.58	[20]
17	>40	>40	>40	>40	39–35	-	>40	>40	>40	6.20	[20]
20	>40	>40	>40	>40	25	-	40	>40	>40	6.86	[20]
22	30	>40	>40	-	8–4	-	10	-	-	5.79	[21]
24	>40	0.625	10–5	>40	1–0.5	5–2.5	2.5	20–10	5–2.5	7.09	[21]
25	>40	20–10	>40	>40	10–5	40–20	20–10	>40	20–10	6.41	[21]
26	>40	2	20–10	>40	5–2.5	10–5	10–5	40–20	>40	6.41	[21]
Control [3]	2.5	2.5	0.6	>20	8	-	10	-	>20	0.14	

S. a.: Staphylococcus aureus, S. e.: S. epidermidis, S. s.: S. saprophyticus, E. f.: Enterococcus faecalis, B. s.: Bacillus subtilis, B. a.: B. alvei, B. c.: B. cereus, B. m.: B. megaterium, B. p.: B. pumilus.—not assayed. [1] The experiments were carried out in triplicate, and values represent average of MIC values. [2] All compounds were inactive against Gram-negative bacteria (MIC > 40 μg/mL). Compounds (*clog P*) **1** (6.15), **4** (6.07), **7–9** (7.18, 6.90 and 6.90), **11** (6.38), **18** (6.47), **19** (5.44), **21** (7.17), and **23** (5.72) were inactive against all microorganisms assayed. [3] Cephotaxime was used as a positive control.

3.3. Structure–Activity Relationship Analysis

The influence of substitution patterns of the phenolic nor-triterpenes on their antimicrobial activity was analyzed, taking into account three regions of the triterpene skeleton, the phenolic moiety, the extended B-ring conjugation, and the triterpene scaffold (pristimerin, tingenone, or iguesterin). This analysis revealed the following trends in the structure–activity relationship (SAR) (Figure 4). Regarding the results obtained against *Bacillus* spp., the SAR studies suggest that: (a) A substituent at C-4 seems essential for the activity. Thus, comparison of the activities (MICs against *B. subtilis*) of **14**, **10**, **12**, and **1**, whose only structural difference is the substituent at C4, showed that the most effective group at this position was a carboxylic acid (**14**), followed by hydroxyl (**10**) and aldehyde (**12**) groups, while a methyl group, as in **1**, led to a loss of activity. *B. cereus*, *B. alvei*, *B. megaterium*, and *B. pumilus* showed similar behavior to *B. subtilis* (**14** vs. **12**); (b) Acetylation and methylation reduce activity, as revealed by comparison of potency of the natural compounds with their corresponding analogues (**7** vs. **24**, **8** vs. **25**, **9** vs. **26**, **10** vs. **11**, **16** vs. **17**, **18**, **19**, and **20** vs. **21**), except for compound **6** with an hydroxyl group at C6 (**6** vs. **7**). This indicates that the hydroxyl group, which could interact with the receptor as hydrogen-bond-donor (HBD), is the best functional group, suggesting that the hydrophilicity is relevant for the activity; (c) Compounds with an unconjugated double bond at C-7 showed considerable activity (e.g., **6** and **24**), and reduction at C-7/C-8 results in a partial loss of antimicrobial activity (**24** vs. **25** and **26**); (d) The β-stereochemistry at C-8 increases two-fold the activity with respect to the H-8α, as shown by comparing the MIC values of **26** and **25**; (e) A substituent at C-7, in addition to the carbonyl at C-6, is also relevant for the activity, being a ketone preferred to a hydroxyl group (**22** > **5** vs. **4**).

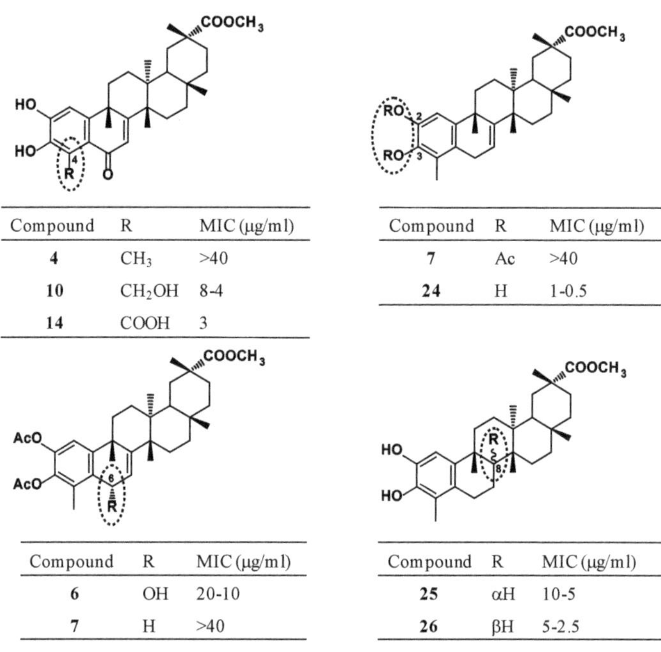

Compound	R	MIC (μg/ml)
4	CH$_3$	>40
10	CH$_2$OH	8-4
14	COOH	3

Compound	R	MIC (μg/ml)
7	Ac	>40
24	H	1-0.5

Compound	R	MIC (μg/ml)
6	OH	20-10
7	H	>40

Compound	R	MIC (μg/ml)
25	αH	10-5
26	βH	5-2.5

Figure 4. Most relevant structural features for phenolic nor-triterpenes against the most sensitive microorganism, *Bacillus subtilis*.

Furthermore, the results obtained against *Staphylococcus* spp. suggest that: (a) the presence of a carboxylic acid moiety at C-4 could strongly affect the antimicrobial activity against *S. aureus* since its substitution by an aldehyde or a methyl group reduced or fully eliminated the activity (**14** vs. **12** or **1**); (b) moreover, replacement of the carboxyl group at C-29 by its methyl ester led to a partial loss of activity (**13** vs. **12**); (c) compounds containing two carboxylic acid groups lacked activity, whereas those with one carboxylic acid showed high activity (**14** vs. **15**); (d) the reduction of the double bond at C-7/C-8 led to less activity (**24** vs. **25** and **26** against *S. epidermidis*); (e) the β-stereochemistry at the C-8 position favored the activity with respect to the H-8α (**26** vs. **25**); (f) a hydroxyl group at C-6 affected the activity against *S. epidermidis* (**6** vs. **7**). Taking into account these structural features, we propose a hypothetical compound, 4-carboxy-6α-hydroxy-pristimerol, that would have a zeylasterone A-ring (as in **14**), an unconjugated B-ring (as in **6** and **24**), and a pristimerin E-ring as being the most active against *Bacillus* and *Staphylococcus* spp. (Figure 5).

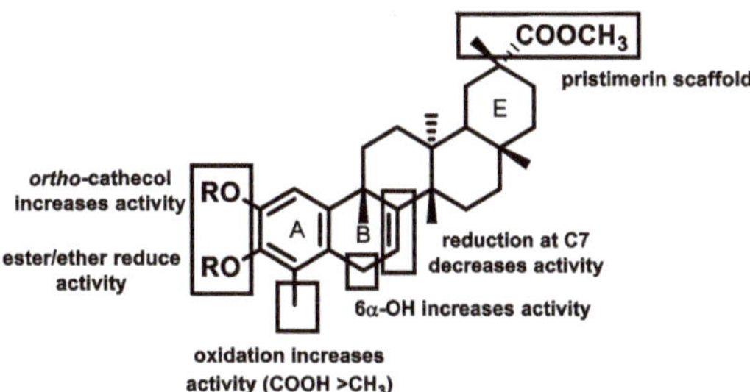

Figure 5. Structural requirements of phenolic nor-triterpenes based on SAR studies.

Some phenolic compounds primarily target the cytoplasmic membrane due to their hydrophobic nature, and preferentially partition into the lipid bilayer [33], as was observed in our previous works [22–25]. Therefore, for the purpose of correlating the antimicrobial activity with lipophilicity, clog *P* values [34] of this series of compounds were calculated, including that of the hypothetical compound, 4-carboxy-6α-hydroxy-pristimerol (clog *P* 4.88), and are shown in Table 1. Slight increases in lipophilicity of **24** (clog *P* 7.09) and **25** and **26** (clog *P* 6.41) by acetylation (**7**, **8** and **9**, respectively) (clog *P* 7.18 and 6.90) led to a suppression of the antibacterial activity. Moreover, other factors must be taken into account for the expression of the activity, such as the presence of an hydrogen-bond donor (HBD) group, strategically positioned at C-4 (e.g., **14**), C-6 (e.g., **6**), or at C-2 and C-3 (e.g., **12** and **14**) with clog *P* of 5.04, 5.95, and 5.67, respectively, which seems relevant for the activity. These observations indicated that both lipophilicity and HBD factors are involved in the antimicrobial activity of this type of compound.

4. Conclusions

In summary, the antimicrobial activity of five phenolic nor-triterpenes isolated from two *Maytenus* species and those of four derivatives revealed that compound 6 showed significant activity against Gram-positive bacteria, higher than cephotaxime, used as positive control. In order to understand and further optimize the structural requirements for effective inhibition of bacterial growth in vitro, an extensive SAR analysis of a series of nor-triterpene phenols was performed. This study suggests that the phenolic moiety and carboxyl group at C-4 on the A-ring, a nonconjugated double bond system on the B-ring, and the ring E, characteristic of pristimerin series, all contribute to the antimicrobial effectiveness. Based on these findings, we propose a hypothetical lead compound, 4-carboxy-6α-hydroxy-pristimerol. This comprehensive SAR study supports the future rational design of antimicrobial agents based on the phenolic nor-triterpene scaffold.

Supplementary Materials: The following are available online at http://www.mdpi.com/2076-3417/9/15/2957/s1, Figures S1–S4: ^{1}H and ^{13}C NMR spectra of derivatives **6–9**.

Author Contributions: Conceptualization, L.M. and I.L.B.; methodology, L.M. and I.A.J.; investigation, M.R.L. and C.P.R.; data curation, L.M. and I.A.J.; writing-original draft preparation, I.L.B.; writing-review and editing, I.L.B. and L.M.; supervision, L.M. and I.L.B.; and funding acquisition, I.L.B.

Funding: This study was supported by SAF2015-65113-C2-1-R and RTI2018-094356-B-C21 Spanish MINECO co-funded by the European Regional Development Fund (FEDER) projects.

Acknowledgments: C.P.R. thanks the Cabildo de Tenerife (Agustín de Betancourt Program).

Conflicts of Interest: The authors declare no conflict of interest

References

1. Durand, G.A.; Raoult, D.; Dubourg, G. Antibiotic discovery: History, methods and perspectives. *Int. J. Antimicrob. Agents* **2019**, *53*, 371–382. [CrossRef]

2. Rossiter, S.E.; Fletcher, M.H.; Wuest, W.M. Natural products as platforms to overcome antibiotic resistance. *Chem. Rev.* **2017**, *117*, 12415–12474. [CrossRef]

3. González, A.G.; Bazzocchi, I.L.; Moujir, L.M.; Jiménez, I.A. Ethnobotanical uses of Celastraceae. Bioactive metabolites. In *Studies in Natural Products Chemistry: Bioactive Natural Products*; (Part D); Atta-ur-Rahman, Ed.; Elsevier Science Publisher: Amsterdam, The Netherlands, 2000; Volume 23, pp. 649–738.

4. Niero, R.; de Andrade, S.F.; Cechinel, V.F. A review of the ethnopharmacology, phytochemistry and pharmacology of plants of the Maytenus genus. *Curr. Pharm. Des.* **2011**, *17*, 1851. [CrossRef]

5. Gunatilaka, A.A.L. Triterpenoid quinonemethides and related compounds (celastroloids). In *Progress in the Chemistry of Organic Natural Products*; Herz, W., Kirby, G.W., Moore, R.E., Steglich, W., Tamm, C., Eds.; Springer: New York, NY, USA, 1996; Volume 67, pp. 1–123.

6. Chen, S.-R.; Dai, Y.; Zhao, J.; Lin, L.; Wang, Y.; Wang, Y. A mechanistic overview of triptolide and celastrol, natural products from *Tripterygium wilfordii* Hook F. *Front. Pharmacol.* **2018**, *9*, 1–13. [CrossRef]

7. Yousef, B.A.; Hassan, H.M.; Zhang, L.-Y.; Jiang, Z.-Z. Anticancer potential and molecular targets of pristimerin: A mini review. *Curr. Cancer Drug Targets* **2017**, *17*, 100–108. [CrossRef]

8. Alvarenga, N.; Ferro, E.A. Bioactive triterpenes and related compounds from Celastraceae. In *Studies in Natural Products Chemistry: Bioactive Natural Products*; (Part K); Bioactive triterpenes and related compounds from Celastraceae; Atta-ur-Rahman, Ed.; Elsevier Science Publisher: Amsterdam, The Netherlands, 2006; Volume 33, pp. 239–307.

9. Bazzocchi, I.L.; Núñez, M.J.; Reyes, C.P. Diels-Alder adducts from Celastraceae species. *Phytochem. Rev.* **2018**, *17*, 669–690. [CrossRef]

10. Li, P.P.; He, W.; Yuan, P.F.; Song, S.S.; Lu, J.T.; Wei, W. Celastrol induces mitochondria-mediated apoptosis in hepatocellular carcinoma Bel-7402 cells. *Am. J. Chin. Med.* **2015**, *43*, 137–148. [CrossRef]

11. Rodrigues, A.C.B.C.; Oliveira, F.P.; Dias, R.B.; Sales, C.B.S.; Rocha, C.A.G.; Soares, M.B.P.; Costa, E.V.; Silva, F.M.A.; Rocha, W.C.; Koolen, H.H.F.; et al. In vitro and in vivo anti-leukemia activity of the stem bark of *Salacia impressifolia* (Miers) A. C. Smith (Celastraceae). *J. Ethnopharmacol.* **2019**, *231*, 516–524. [CrossRef]

12. Dai, W.; Wang, X.; Teng, H.; Li, C.; Wang, B.; Wang, J. Celastrol inhibits microglial pyroptosis and attenuates inflammatory reaction in acute spinal cord injury rats. *Int. Immunopharmacol.* **2019**, *66*, 215–223. [CrossRef]

13. Santos, V.A.F.F.M.; Santos, D.P.; Castro-Gamboa, I.; Zanoni, M.V.B.; Furlan, M. Evaluation of antioxidant capacity and synergistic associations of quinonemethide triterpenes and phenolic substances from *Maytenus ilicifolia* (Celastraceae). *Molecules* **2010**, *15*, 6956–6973.

14. Liao, L.M.; Silva, G.A.; Monteiro, M.R.; Albuquerque, S. Trypanocidal activity of quinonemethide triterpenoids from *Cheiloclinium cognatum* (Hippocrateaceae). *Z. Naturforsch C* **2008**, *63*, 207–210. [CrossRef]

15. Avilla, J.; Teixidó, A.; Velázquez, C.; Alvarenga, N.; Ferro, E.; Canela, R. Insecticidal activity of *Maytenus* species (Celastraceae) nortriterpene quinone methides against Codling Moth, *Cydia pomonella* (L.) (Lepidoptera: Tortricidae). *J. Agric. Food Chem.* **2000**, *48*, 88–92. [CrossRef]

16. Moujir, L.; Gutiérrez-Navarro, A.M.; González, A.G.; Ravelo, A.G.; Luis, J.G. The relationship between structure and antimicrobial activity in quinones from the Celastraceae. *Biochem. Syst. Ecol.* **1990**, *18*, 25–28. [CrossRef]

17. González, A.G.; Ravelo, A.G.; Bazzocchi, I.L.; Jiménes, J.; González, C.M.; Luis, J.G.; Ferro, E.A.; Gutiérrez, A.; Moujir, L. Biological study of triterpenoquinones from Celastraceae. *Il Farmaco* **1988**, *43*, 501–505.

18. Moujir, L.; Gutiérrez-Navarro, A.M.; González, A.G.; Ravelo, A.G.; Luis, J.G. Mode of action of netzahualcoyone. *Antimicrob. Agents Chemother.* **1991**, *35*, 211–213. [CrossRef]

19. Moujir, L.; Gutiérrez-Navarro, A.M.; González, A.G.; Ravelo, A.G.; Jiménez, J. Biological properties of netzahualcoyone: Conditions for activity. *Biomed. Lett.* **1991**, *46*, 7–15.

20. González, A.G.; Alvarenga, N.L.; Ravelo, A.G.; Jiménez, I.A.; Bazzocchi, I.L.; Canela, N.J.; Moujir, L. Antibiotic phenol nor-triterpenes from *Maytenus canariensis*. *Phytochemistry* **1996**, *43*, 129–132. [CrossRef]

21. Rodríguez, F.M.; López, M.R.; Jiménez, I.A.; Moujir, L.; Ravelo, A.G.; Bazzocchi, I.L. New phenolic triterpenes from Maytenus blepharodes.Semisynthesis of 6-deoxoblepharodol from pristimerin. *Tetrahedron* **2005**, *61*, 2513–2519. [CrossRef]

22. De León, L.; Beltrán, B.; Moujir, L. Antimicrobial activity of 6-oxophenolic triterpenoids. Mode of action against *Bacillus subtilis*. *Planta Med.* **2005**, *71*, 313–319. [CrossRef]

23. De León, L.; Moujir, L. Activity and mechanism of the action of zeylasterone against *Bacillus subtilis*. *J. Appl. Microbiol.* **2008**, *104*, 1266–1274. [CrossRef]

24. De León, L.; López, M.R.; Moujir, L. Antibacterial properties of zeylasterone, a triterpenoid isolated from *Maytenus blepharodes*, against *Staphylococcus aureus*. *Microbiol. Res.* **2010**, *165*, 617–626. [CrossRef]

25. López, M.R.; de León, L.; Moujir, L. Antibacterial properties of phenolic triterpenoids against *Staphylococcus epidermidis*. *Planta Med.* **2011**, *77*, 726–729. [CrossRef]

26. Shirota, O.; Morita, H.; Takeya, K.; Itokawa, H. Cytotoxic aromatic triterpenes from *Maytenus ilicifolia* and *Maytenus chuchuhuasca*. *J. Nat. Prod.* **1994**, *57*, 1675–1681. [CrossRef]

27. González, A.G.; Alvarenga, N.L.; Rodríguez, F.; Ravelo, A.G.; Jiménez, I.A.; Bazzocchi, I.L.; Gupta, M.P. New phenolic and quinone methide triterpenes from *Maytenus* species (Celastraceae). *Nat. Prod. Lett.* **1995**, *7*, 209–218. [CrossRef]

28. Ankli, A.; Heilmann, J.; Heinrich, M.; Sticher, O. Cytotoxic cardenolides and antibacterial terpenoids from *Crossopetalum gaumeri*. *Phytochemistry* **2000**, *54*, 531–537. [CrossRef]

29. Gamlath, C.; Gunaherath, K.B.; Gunatilaka, A.A.L. Studies on terpenoids and steroids. Part 10. Structures of four new natural phenolic D:A-friedo-24-noroleanane triterpenoids. *J. Chem. Soc. Perkin Trans. I* **1987**, 2849–2853. [CrossRef]

30. Kamal, G.M.; Gunaherath, K.B.; Gunatilaka, A.A.L. Studies on terpenoids and steroids. Part 10. Structures of four new natural phenolic D:A-friedo-24-noroleanane triterpenoids. *J. Chem. Soc. Perkin Trans. I* **1983**, 2845–2850. [CrossRef]

31. Gunatilaka, A.A.L.; Wimalasiri, W.R. Studies on terpenoids and steroids. Part 22. Structure and some reactions of pristimerin leucotriacetate. *J. Chem. Res. S* **1992**, *1*, 30–31. [CrossRef]

32. CLSI. *Methods for Dilution Antimicrobial Susceptibility test for Bacteria that Grow Aerobically*, 9th ed.; Approved standard M07-A9; Clinical and Laboratory Standards: Wayne, PA, USA, 2012.

33. Barzic, A.I.; Ioan, S. Antibacterial drugs-From basic concepts to complex therapeutic mechanisms of polymer systems. In *Concepts, Compounds and the Alternatives of Antibacterials*; Bobbarala, V., Ed.; Science, Technology and Medicine: London, UK, 2015.

34. Software-Predicted Lipophilicity of the Compounds was Calculated with the ALOGPS 2.1 Program. Available online: www.vcclab.org/lab/alogps/accessibleviainterneton-lineLipophilicity/AqueousSolubilityCalculationSoftware. (accessed on 20 May 2019).

Communication

Synthesis and Evaluation of the Lifespan-Extension Properties of Oleracones D–F, Antioxidative Flavonoids from *Portulaca oleracea* L.

Jeong A Yoon [1],[†], Changjin Lim [2],[†], Dong Seok Cha [3],* and Young Taek Han [1],*

[1] College of Pharmacy, Dankook University, Chenan 31116, Korea; yja9477@naver.com
[2] College of Pharmacy, CHA University, Pocheon 11160, Korea; koryoi@cha.ac.kr
[3] College of Pharmacy, Woosuk University, Jeonbuk 55338, Korea
* Correspondence: cha@woosuk.ac.kr (D.S.C.); hanyt@dankook.ac.kr (Y.T.H.);
   Tel.: +82-63-390-1578 (D.S.C.); +82-41-550-1431 (Y.T.H.)
† These authors contributed equally to this paper.

Received: 10 September 2019; Accepted: 24 September 2019; Published: 25 September 2019

Abstract: Plant-derived antioxidants have been widely used as supplementary health foods, as well as having been regarded as drug candidates for aging and aging-associated diseases. Oleracones, a novel series of flavonoids isolated from *Portulaca oleracea* L., possess potent antioxidative activities and are expected to exhibit therapeutic potential on the aging process. The current paper describes the concise sequential synthesis of oleracones D–F. Oleracones D and F were efficiently synthesized via selective intramolecular oxa-Michael addition from oleracone E. In addition, we investigated their possible lifespan-extension properties using *Caenorhabditis elegans*, which is excellently suited as an experimental model to study aging. A significant longevity effect was observed when nematodes were grown with 20 μM of oleracone E.

Keywords: oleracone; flavonoid; anti-aging; longevity; *Portulaca oleracea* L.; *Caenorhabditis elegans*; total synthesis

---

1. Introduction

Aging is a complicated biological process that is controlled by a large number of genetic and environmental factors [1–4]. Many theories have been proposed to account for the aging process, but none of them offers a fully acceptable explanation. Among all the theories, the free radical theory of aging, proposed by professor Harman in 1956 [5], has been extensively investigated. This theory continues to be revised and remains a strong theory for the aging process until now. Free radical theory explains that aging is caused by cumulative oxidative stress from free radicals, resulting in cell death and, eventually, death of the organism [6]. Hence, this theory suggests that antioxidants that can sacrificially scavenge ROS (reactive oxygen species) and/or RNS (reactive nitrogen species) are effective for delaying the aging process. Indeed, early studies have indicated that plant-derived polyphenol antioxidants, including curcumin, quercetin, and resveratrol, show therapeutic potential for aging and aging-associated diseases [7,8]. In addition, other synthetic flavonoids have been continuously reported as novel antioxidants exhibiting antiaging properties [9–12].

Portulaca oleracea L. is a well-known annual weed, one that is widely distributed in temperate and tropical regions. This plant has been used for a long time, not only as an edible potherb, but also as a traditional medicine in many countries, being used to alleviate a wide range of illnesses [13]. In addition to phytochemical research, many pharmacological studies have been performed using the extracts or single compounds from *Portulaca oleracea* L. to reveal their therapeutic properties. Interestingly, the extracts of this plant are known to possess not only a strong in vivo antioxidant capacity, but also anti-aging activity in a D-galactose-induced mice model [14,15].

Recently, oleracones, a novel series of flavonoids, were isolated from *Portulaca oleracea* L. by Chinese researchers [16,17]. They reported the potent antioxidative potential of oleracones, which allowed us to speculate that oleracones may also possess anti-aging properties [16,17]. The current paper includes the concise sequential synthesis of oleracones and an evaluation of their possible lifespan-extension properties using *Caenorhabditis elegans*, an excellent experimental model to study aging. Among the reported oleracones, we planned to synthesize two homoisoflavones, oleracones D (**1**) and F (**2**), and a dihydrochalcone, olereacone E (**3**), which exhibit potent antioxidative effects (Figure 1).

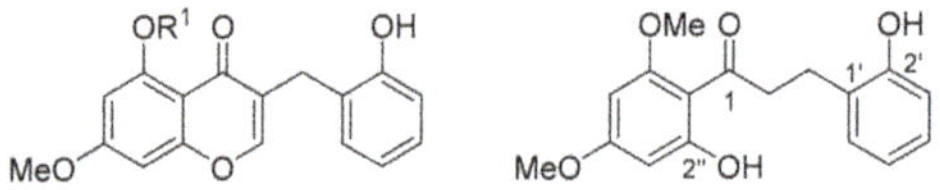

R^1= H: Oleracone D (**1**)
R^1= Me: Oleracone F (**2**)

Oleracone E (**3**)

Figure 1. The structures of oleracones D (**1**), F (**2**), and E (**3**).

2. Materials and Methods

2.1. Chemistry

2.1.1. General Information

Unless noted otherwise, all reactions were performed in dry solvents under anhydrous conditions and argon atmosphere. The reaction flasks were flame-dried before use, and all solvents for extraction and chromatography were reagent grade. All reagents were purchased from commercial suppliers and were used without further purification. TLC (thin-layer chromatography) was conducted for monitoring reaction progress with 0.25 mm silica gel plates (Merck, Kenilworth, NJ, USA). Silica gel 60 (230–400 mesh, Merck, Kenilworth, NJ, USA) was used for flash column chromatography with the indicated solvents. ^{1}H and ^{13}C spectra were recorded on a Brucker Analytik ADVANCE digital 500 (500 MHz) (Billerica, MA, USA) or BRUKER AVANCE-800 (800 MHz) (Billerica, MA, USA). ^{1}H-NMR data are reported as follows: chemical shift, multiplicity (singlet, s; doublet, d; triplet, t; quartet, q; broad b; and/or multiple resonances), coupling constant in hertz (Hz), and number of protons. Chemical shifts are stated in ppm (parts per million, δ) downfield from tetramethylsilane (TMS) and are referenced to the deuterated solvent (CDCl$_3$ and DMSO-d$_6$). IR (infrared) spectra were recorded on a FT-IR-4200 (JASCO, Tokyo, Japan) spectrometer. Low- and high-resolution mass spectra were acquired with JMS-700 (JEOL, Tokyo, Japan) equipment.

2.1.2. Experimental Section

(E)-3-(2-(Benzyloxy)phenyl)-1-(2-hydroxy-4,6-dimethoxyphenyl)prop-2-en-1-one (**6**). To a solution of 4,6-dimethoxy-2-hydroxybenzophenone 4 (700 mg, 3.57 mmol) and 2-(benzyloxy)benzaldehyde 5 (1100 mg, 5.35 mmol) in ethyl alcohol (11 mL), KOH (400 mg, 7.14 mmol) was added at room temperature. After stirring at the same temperature for 48 h, the reaction mixture was concentrated in vacuo. The reaction mixture was quenched with 1*N* aqueous solution of hydrochloric acid and extracted with EtOAc (ethyl acetate). The combined organic layer was washed with saturated sodium bicarbonate solution, dried over anhydrous magnesium sulfate, and concentrated in vacuo. Purification of the concentrated residue by flash column chromatography (EtOAc/*n*-Hexane = 1 : 40 to 1 : 10) afforded 1500 mg (75%) of chalcone 6 as a yellow powder. m.p.: 114–115 °C; R$_f$ = 0.25 (EtOAc : *n*-Hexane = 1 : 10); ^{1}H NMR (800 MHz, CDCl$_3$) δ 14.37 (s, 1H), 8.17 (d, 1H, *J* = 15.8 Hz), 8.02 (d, 1H, *J* = 15.7 Hz), 7.62 (dd, 1H, *J* = 1.5, 7.7 Hz), 7.46 (d, 2H, *J* =7.6 Hz), 7.39-7.37 (m, 2H), 7.33-7.30 (m, 2H), 7.00 (t, 1H, *J* = 7.5 Hz), 6.87 (d, 1H, *J* = 8.2 Hz), 6.10 (d, 1H, *J* = 2.3 Hz), 5.91 (d, 1H, *J* = 2.3 Hz), 5.20 (s, 2H), 3.83 (s, 3H), 3.72 (s, 3H); ^{13}C NMR (200 MHz, CDCl$_3$) δ 193.3, 168.4, 166.2, 162.6, 157.9, 138.1, 136.9,

131.4, 129.5, 128.8, 128.5, 128.1, 127.3, 125.1, 121.2, 112.9, 106.6, 93.8, 93.8, 91.3, 70.5, 55.7, 55.7; IR (thin film, neat) v_{max} 1620, 1595, 1553, 1452, 1344, 1213, 1107, 984, 816, 760 cm^{-1}; LR-MS (Low-Resolution Mass Spectroscopy) (FAB+) m/z 391 (M + H$^+$); HR-MS (High-Resolution Mass Spectroscopy) (FAB+) calculated for $C_{24}H_{23}O_5$ (M + H$^+$) 391.1545; observed 391.1541.

Oleracone E. (**3**). To a solution of chalcone **6** (500 mg, 1.28 mmol) in ethyl alcohol (10 mL), catalytic amount of 5% activated palladium on carbon was added at room temperature. After stirring for 48 h under H$_2$ atmosphere, the reaction mixture was filtered through a pad of Celite and concentrated in vacuo. Purification of the concentrated residue by flash column chromatography (EtOAc : *n*-Hexane = 1 : 10 to 1 : 3) afforded 340 mg (88%) of oleracone E (**3**) as a yellowish powder. The spectral data of **3** were consistent with previous data (Table S1). m.p.: 137–138 °C; R_f = 0.20 (EtOAc/*n*-Hexane = 1:5); ^{1}H NMR (500 MHz, DMSO-d$_6$) δ 13.63 (brs, 1H), 9.30 (brs, 1H), 7.06 (dd, 1H, J = 1.5, 7.4 Hz), 6.99 (brtd, 1H, J = 1.6, 7.7 Hz), 6.77(dd, 1H, J = 1.0, 8.0 Hz), 6.70 (td, 1H, J = 1.0, 7.4 Hz), 6.11 (d, 1H, J = 2.4 Hz), 6.09 (d, 1H, J = 2.4 Hz), 3.82 (s, 3H), 3.80 (s, 3H), 3.20 (brt, 2H, J = 7.7 Hz), 2.81 (brt, 2H, J = 7.7 Hz); ^{13}C NMR (200 MHz, DMSO-d$_6$) δ 205.4, 166.1, 166.0, 162.9, 155.5, 130.3, 127.7, 127.5, 119.5, 115.3, 106.0, 94.2, 91.2, 56.4, 56.1, 43.9, 25.5; IR (thin film, neat) v_{max} 1618, 1578, 1489, 1462, 1418, 1362, 1204, 1109, 999, 829, 758 cm^{-1}; LR-MS (FAB+) m/z 303 (M + H$^+$); HR-MS (FAB+) calculated for $C_{17}H_{19}O_5$ (M + H$^+$) 303.1232; observed 303.1226.

Oleracone F. (**2**). *N,N*-Dmethylformamide dimethyl acetal (0.053 mL, 0.40 mmol) was added to a solution of oleracone E (**3**) (100 mg, 0.33 mmol) in toluene (5.5 mL) at room temperature. The reaction mixture was refluxed for 2 h, and then cooled to room temperature. The reaction mixture was concentrated in vacuo. Purification of the concentrated residue by flash column chromatography (EtOAc : *n*-Hexane = 1 : 2) afforded 93 mg (90%) of oleracone F (**2**) as a yellow powder. The spectral data of **2** were consistent with previous data (Table S2). m.p.: 224–225 °C; R_f = 0.23 (EtOAc/*n*-Hexane = 1:2); ^{1}H NMR (500 MHz, DMSO-d$_6$) δ 9.46 (s, 1H), 7.88 (s, 1H), 7.07 (dd, 1H, J = 1.5, 7.5 Hz), 7.01 (brtd, 1H, J = 1.6, 7.7 Hz), 6.78 (dd, 1H, J = 1.0, 8.0 Hz), 6.70 (brtd, 1H, J = 1.1, 7.4 Hz), 6.61 (d, 1H, J = 2.3 Hz), 6.48 (d, 1H, J = 2.3 Hz), 3.86 (s, 3H), 3.81 (s, 3H), 3.52 (s, 2H); ^{13}C NMR (200 MHz, DMSO-d$_6$) δ 175.8, 164.1, 160.9, 160.0, 155.4, 151.9, 130.5, 127.9, 125.7, 123.8, 119.6, 115.6, 108.7, 96.4, 93.3, 56.5, 56.3, 25.6; IR (thin film, neat) v_{max} 1643, 1595, 1566, 1454, 1373, 1207, 1152, 1078, 824, 752 cm^{-1}; LR-MS (FAB+) m/z 313 (M + H$^+$); HR-MS (FAB+) calculated for $C_{18}H_{17}O_5$ (M + H$^+$) 313.1076; observed 313.1078.

Oleracone D (**1**). To a solution of oleracone F (**2**) (150 mg, 0.48 mmol) in methylene chloride (5 mL), 1.92 mL of BCl$_3$ (1 M solution in CH$_2$Cl$_2$, 1.92 mmol) was slowly added at 0 °C. After stirring for 1 h at the same temperature, the reaction mixture was quenched with H$_2$O, diluted with methylene chloride, stirred for an additional 1 h, and extracted with methylene chloride. The combined organic layer was washed with brine, dried over anhydrous magnesium sulfate, and concentrated in vacuo. Purification of the concentrated residue by flash column chromatography (EtOAc : *n*-Hexane = 1:5) afforded 114 mg (80%) of oleracone D (**1**) as a yellowish powder. The spectral data of **1** were consistent with previous data (Table S3). m.p.: 154–155 °C; R_f = 0.31 (EtOAc : *n*-Hexane = 1 : 5); ^{1}H NMR (500 MHz, DMSO-d$_6$) δ 12.76 (s, 1H), 9.47 (brs, 1H), 8.06 (s, 1H), 7.08 (dd, 1H, J = 1.4, 7.5 Hz), 7.03 (brtd, 1H, J = 1.6, 7.7 Hz), 6.81 (dd, 1H, J = 0.8, 8.0 Hz), 6.71 (brtd, 1H, J = 1.4, 7.4 Hz), 6.61 (d, 1H, J = 2.3 Hz), 6.38 (d, 1H, J = 2.3 Hz), 3.84 (s, 3H), 3.61 (s, 2H); ^{13}C NMR (200 MHz, DMSO-d$_6$) δ 181.6, 165.6, 161.6, 158.2, 155.4, 155.3, 130.4, 128.0, 125.0, 121.4, 119.5, 115.4, 105.5, 98.4, 92.8, 56.4, 25.0; IR (thin film, neat) v_{max} 1647, 1585, 1510, 1435, 1368, 1240, 1163, 1051, 822, 754 cm^{-1}; LR-MS (FAB+) m/z 299 (M + H$^+$); HR-MS (FAB+) calculated for $C_{17}H_{15}O_5$ (M + H$^+$) 299.0919; observed 299.0915.

2.2. Biology

2.2.1. *C. elegans* Maintenance

In this paper, we used wild-type N2 nematodes which were acquired from the Caenorhabditis Genetic Center (CGC; University of Minnesota, Minneapolis, MN, USA). All nematodes were nurtured at 20 °C on an NGM (Nematode Growth Medium) agar plate with *Escherichia coli* OP50 as previously reported [18]

2.2.2. Lifespan Assay

Lifespan assays were carried out under normal culture conditions. Age-synchronized nematodes were collected by embryo isolation and L1 arrest. The as-obtained L1 stage nematodes were grown on an NGM plate with or without of (**1**), (**2**), and (**3**). The survival rate of the test nematodes was determined using a dissecting microscope (SMZ1500, Nikon, Tokyo, Japan). Nematodes that failed to respond to prodding with the tip of a platinum wire were considered dead. Living nematodes were transferred to a fresh NGM plate every 2 days.

2.2.3. Statistical Analysis

The results obtained from the lifespan assay were plotted using Kaplan–Meier analysis, and the statistical significance between each group was analyzed using the log-rank test. The mean lifespan data were presented as the mean ± standard deviation.

3. Results and Discussion

The retrosynthetic analysis for oleracones is outlined in Scheme 1. Homoisoflavone skeleton was anticipated to be efficiently prepared via a deoxybenzoin route, one of the most popular synthetic methods used for the preparation of isoflavonoids [19]. Therefore, homoisoflavone **2** and its *O*-demethylated analog **1** were planned to be obtained from dihydrochalcone **3**. Oleracone E (**3**) was anticipated to be prepared via an aldol condensation between commercially available 4,6-dimethoxy-2-hydroxybenzophenone **4** and 2-(benzyloxy)benzaldehyde **5**.

R^1= H: Oleracone D (**1**)
R^1= Me: Oleracone F (**2**)

Oleracone E (**3**)

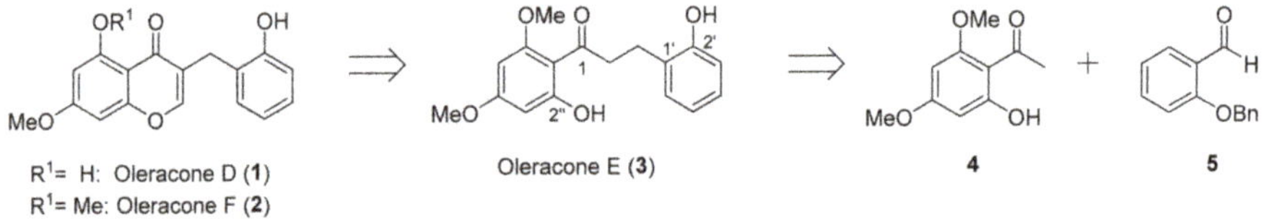

Scheme 1. Retrosynthetic analysis of oleracones D (**1**), F (**2**), and E (**3**).

Our synthesis was commenced with an aldol condensation between commercially available 4,6-dimethoxy-2-hydroxybenzophenone **4** and 2-(benzyloxy)benzaldehyde **5** to give chalcone **6** (Scheme 2). Concurrent hydrogenation and hydrogenolysis of the *O*-benzyl group in chalcone **6** using 5% activated Pd on carbon under H$_2$ atmosphere led to olereacone E (**3**) in 88% yield.

We envisaged the construction of the homoisoflavone skeleton of **1** and **2** via a selective intramolecular oxa-Michael addition reaction of an enone moiety by the phenol group at the 2″-position rather than the 2′-position of intermediate **6**, resulting from condensation reaction between **3** and a formyl reagent. According to our expectation, oleracone F (**2**) was obtained in high yield (90%) upon stirring with *N,N*-dimethylformamide dimethyl acetal [20] in toluene under reflux condition, and no regioisomeric by-products were observed. The high regioselectivity observed in the reaction was attributed to the enhanced nucleophilicity of the phenol group at the 2″-position compared to that at the 2′-position in **3**, which can be supported by the electron-withdrawing effect of the carbonyl group. Finally, selective *O*-demethylation of **2** using BCl$_3$ afforded oleracone D (**1**) in high yield (80%). The spectral data obtained for oleracones D–F (**1**–**3**) were consistent with those previously reported data [16,17].

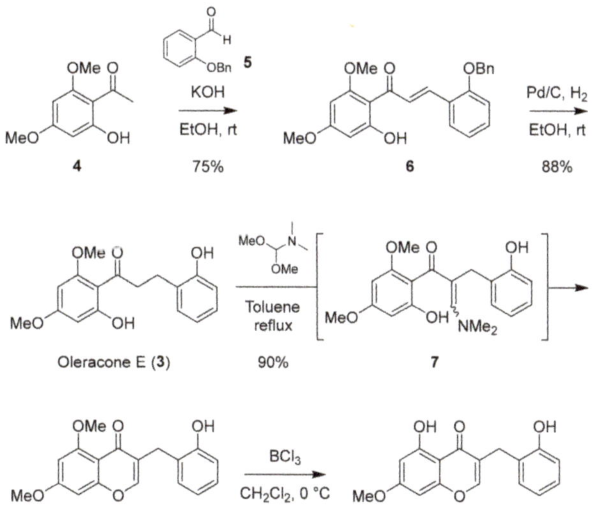

Scheme 2. Synthesis of oleracones D (**1**), F (**2**), and E (**3**).

With the oleracones D–F (**1–3**) in hand, we examined their effects on the longevity of nematodes. To test whether oleracones exhibit lifespan-extension activity, we performed a lifespan assay using wild-type N2 nematodes at 20 °C, as described previously [21]. As shown in Figure 2A,B, all of the oleracones tested could prolong the lifespan of the nematodes under standard culture conditions, although the effectiveness of each compound was somewhat variable. At the maximum concentration studied (20 μM), treatment with **3** and **2** extended the lifespan of nematodes by 13.8 and 11.8%, respectively, compared with the control, while the efficacy of **1** (4.3% extension) was inferior to that of **2** and **3**. Interestingly, the longevity effect of **1–3** on nematodes did not show any correlation with their radical scavenging activities (IC$_{50}$ values: 11.73, 13.17, and 17.78 μM, respectively) [16,17]. These results suggest the possibility that modulating aging-related factors may also cooperate with the direct radical scavenging activity that produces the anti-aging effect. However, detailed information on the underlying mechanism of the lifespan-extension properties of oleracone **1–3** remains to be defined.

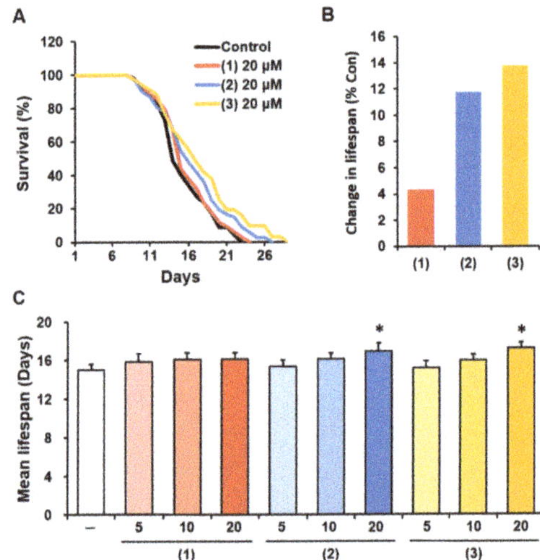

Figure 2. The longevity effect of oleracone D–F (**1–3**). (**A**) The lifespan curves obtained for compound-treated (20 μM) and untreated wild-type nematodes under normal culture conditions. (**B**) The change in lifespan observed for the compound-treated (20 μM) nematodes when compared to the control. (**C**) The mean lifespan of compound-treated (5, 10, and 20 μM) and untreated wild-type nematodes was calculated from the survival curves. The statistical difference between the curves was analyzed using a log-rank test. * $p < 0.05$ compared with the vehicle alone. All experiments were repeated in triplicate.

4. Conclusions

In summary, we concisely synthesized potent antioxidant homoisoflavones, oleracones D (**1**) and F (**2**), from dihydrochalcone olereacone E (**3**), and investigated their potential lifespan extension properties. Our results may shed light on the development of isoflavone-based anti-aging medications.

Supplementary Materials: The following are available online at http://www.mdpi.com/2076-3417/9/19/4014/s1, Table S1: Comparative analysis of ^{1}H NMR and ^{13}C NMR of oleracone E. Table S2: Comparative analysis of ^{1}H NMR and ^{13}C NMR of oleracone F. Table S3: Comparative analysis of ^{1}H NMR and ^{13}C NMR of oleracone D.

Author Contributions: Conceptualization, Y.T.H. and D.S.C.; formal analysis, J.A.Y.; investigation, D.S.C. and J.A.Y.; data curation, C.L.; draft preparation, C.L. and D.S.C.; review and editing, Y.T.H.; supervision, Y.T.H.; project administration, Y.T.H.

Funding: This work was supported by the National Research Foundation of Korea (NRF) grant (NRF-2017R1C1B1001826) and by the GRRC (Gyeonggi-do Regional Research Center) program of Gyeonggi province (GRRC-CHA2017-B02).

Acknowledgments: We thank Emeritus Young-Ger Suh (Seoul National University) for his helpful discussions in manuscript preparation.

Conflicts of Interest: The authors declare no conflict of interest.

References

1. Harman, D. The aging process. *Proc. Natl. Acad. Sci. USA* **1981**, *78*, 7124–7128. [CrossRef] [PubMed]
2. Harman, D. Free radical involvement in aging. *Drugs Aging* **1993**, *3*, 60–80. [CrossRef] [PubMed]
3. Harman, D. Aging: Prospects for further increases in the functional life span. *AGE* **1994**, *17*, 119–146. [CrossRef]
4. Harman, D. Aging and Disease: Extending Functional Life Span. *Ann. N. Y. Acad. Sci.* **1996**, *786*, 321–336. [CrossRef] [PubMed]
5. Harman, D. Free radical theory of againg. *J. Gerontol.* **1956**, *12*, 257–263. [CrossRef]

6. Harman, D. The free radical theory of aging. *Antioxi. Redox Signal.* **2003**, *5*, 557–561. [CrossRef] [PubMed]

7. Conti, V.; Izzo, V.; Corbi, G.; Russomanno, G.; Manzo, V.; De Lise, F.; Di Donato, A.; Filippelli, A. Antioxidant Supplementation in the Treatment of Aging-Associated Diseases. *Front. Pharmacol.* **2016**, *7*, 24. [CrossRef]

8. Cherniack, E. The Potential Influence of Plant Polyphenols on the Aging Process. *Complementary Med. Res.* **2010**, *17*, 181–187. [CrossRef]

9. Rosa, G.P.; Seca, A.M.L.; Barreto, M.D.C.; Silva, A.M.S.; Pinto, D.C.G.A. Chalcones and Flavanones Bearing Hydroxyl and/or Methoxyl Groups: Synthesis and Biological Assessments. *Appl. Sci.* **2019**, *9*, 2846. [CrossRef]

10. Cotelle, N.; Bernier, J.; Henichart, J.; Catteau, J.; Gaydou, E.; Wallet, J. Scavenger and antioxidant properties of ten synthetic flavones. *Free Radic. Biol. Med.* **1992**, *13*, 211–219. [CrossRef]

11. Oliveira, A.M.; Cardoso, S.M.; Ribeiro, M.; Seixas, R.S.; Silva, A.M.; Rego, A.C. Protective effects of 3-alkyl luteolin derivatives are mediated by Nrf2 transcriptional activity and decreased oxidative stress in Huntington's disease mouse striatal cells. *Neurochem. Int.* **2015**, *91*, 1–12. [CrossRef] [PubMed]

12. Catarino, M.D.; Alves-Silva, J.M.; Pereira, O.R.; Cardoso, S.M. Antioxidant capacities of flavones and benefits in oxidative-stress related diseases. *Curr. Top. Med. Chem.* **2015**, *15*, 105–119. [CrossRef] [PubMed]

13. Zhou, Y.X.; Xin, H.L.; Rahman, K.; Wang, S.J.; Peng, C.; Zhang, H. Portulaca oleracea L.: A Review of Phytochemistry and Pharmacological Effects. *BioMed Res. Int.* **2015**, *2015*, 925631. [CrossRef] [PubMed]

14. Silva, R.; Carvalho, I.S. In vitro antioxidant activity, phenolic compounds and protective effect against DNA damage provided by leaves, stems and flowers of Portulaca oleracea (*Purslane*). *Nat. Prod. Commun.* **2014**, *9*, 45–50. [CrossRef] [PubMed]

15. Ahangarpour, A.; Lamoochi, Z.; Moghaddam, H.F.; Mansouri, S.M.T. Effects of Portulaca oleracea ethanolic extract on reproductive system of aging female mice. *Int. J. Reprod. Biomed.* **2016**, *14*, 205–212. [CrossRef]

16. Yang, X.; Zhang, W.; Ying, X.; Stien, D. New flavonoids from *Portulaca oleracea* L. And their activities. *Fitoterapia* **2018**, *127*, 257–262. [CrossRef] [PubMed]

17. Yang, X.; Ying, Z.; Liu, H.; Ying, X.; Yang, G. A new homoisoflavone from Portulaca oleracea L. and its antioxidant activity. *Nat. Prod. Res.* **2018**, *32*, 1–7. [CrossRef]

18. Brenner, S. The genetics of *Caenorhabditis elegans*. *Genetics* **1974**, *77*, 71–94.

19. Balasubramanian, S.; Nair, M.G. An Efficient "One Pot" Synthesis of Isoflavones. *Synth. Commun.* **2000**, *30*, 469–484. [CrossRef]

20. Kirkiacharian, B.S.; Gomis, M. New Convenient Synthesis of Homoisoflavanones and (±)-Di-O-methyldihydroeucomin. *Synth. Commun.* **2005**, *35*, 563–569. [CrossRef]

21. Kim, Y.S.; Han, Y.T.; Jeon, H.; Cha, D.S. Antiageing properties of Damaurone D in Caenorhabditis elegans. *J. Pharm. Pharmacol.* **2018**, *70*, 1423–1429. [CrossRef] [PubMed]

 *applied sciences*

Article

A Quick, Green and Simple Ultrasound-Assisted Extraction for the Valorization of Antioxidant Phenolic Acids from Moroccan Almond Cold-Pressed Oil Residues

Duangjai Tungmunnithum [1,2,3,†], Ahmed Elamrani [4,†], Malika Abid [4], Samantha Drouet [1,2], Reza Kiani [5], Laurine Garros [1,6], Atul Kabra [7], Mohamed Addi [4,*,‡] and Christophe Hano [1,2,*,‡]

[1] Laboratoire de Biologie des Ligneux et des Grandes Cultures, INRA USC1328, Orleans University, 45067 Orléans CEDEX 2, France; duangjai.tun@mahidol.ac.th (D.T.); samantha.drouet@univ-orleans.fr (S.D.); laurine.garros@univ-orleans.fr (L.G.)

[2] Bioactifs et Cosmetiques, CNRS GDR 3711 Orleans, 45067 Orléans CEDEX 2, France

[3] Department of Pharmaceutical Botany, Faculty of Pharmacy, Mahidol University, Bangkok 10400, Thailand

[4] Laboratoire de Biologie des Plantes et des Micro-Organismes, Faculté des Sciences, Université Mohamed Ier, Oujda 60000, Maroc; a.elamrani1@ump.ac.ma (A.E.); m.abid@ump.ac.ma (M.A.)

[5] Department of Horticultural Sciences, College of Agriculture & Natural Resources, University of Tehran, P.O. Box 4111, Karaj 3158777871, Iran; kianireza37@ut.ac.ir

[6] Institut de Chimie Organique et Analytique (ICOA) UMR7311, Université d'Orléans-CNRS, 45067 Orléans CEDEX 2, France

[7] School of Pharmacy, Raffles University, Neemrana, Alwar 301705, Rajasthan, India; atul.kbr@gmail.com

[*] Correspondence: m.addi@ump.ac.ma (M.A.); hano@univ-orleans.fr (C.H); Tel.: +212-536-500-601 (M.A.); +33-237-309-753 (C.H.)

[†] These authors have equal contribution of the first authors.

[‡] These authors have equal contribution of the senior authors.

Received: 16 April 2020; Accepted: 6 May 2020; Published: 10 May 2020

Featured Application: A quick, green, and simple ultrasound-assisted microextraction was here developed and validated for a quick and simple evaluation of total phenolic content from almond oil residues for their valorization as a source of antioxidant compounds.

Abstract: Almond (*Prunus dulcis* (Mill.) D.A. Webb) is one of the most important nut crops both in terms of area and production. Over the last few decades, an important part of the beneficial actions for health associated with their consumption was attributed to the phenolic compounds, mainly accumulated in almond skin. Interestingly, after cold-pressed oil extraction, most of these antioxidant phenolic compounds are retained in a skin-enriched by-product, a so-called almond cold-pressed oil residue. In Morocco, the fifth highest ranking producer in the world, this production generates an important part of this valuable byproduct. In the present study, using a multivariate Box–Behnken design, an ultrasound-assisted extraction method of phenolic compounds from Moroccan almond cold-pressed oil residue was developed and validated. Response surface methodology resulted in the optimal extraction conditions: the use of aqueous ethanol 53.0% (*v/v*) as a green solvent, applying an ultrasound frequency of 27.0 kHz for an extraction duration of 29.4 min. The present ultrasound-assisted extraction allowed substantial gains in terms of extraction efficiency compared to conventional heat reflux extraction. Applied to three different local *Beldi* genotypes growing at three different experimental sites, the optimal conditions for ultrasound-assisted extraction led to a total phenolic content of 13.86 mg/g dry weight. HPLC analysis revealed that the main phenolic compounds from this valuable byproduct were: chlorogenic acid followed by protocatechuic acid, *p*-hydroxybenzoic acid, and *p*-coumaric acid. The accumulation of these phenolic compounds appeared to be more dependent on the genetic background than on the environmental impact here represented by the three experimental culture sites. Both in vitro cell free and cellular antioxidant assays were performed, and revealed the great potential of these extracts. In particular, correlation analysis provided evidence of the prominent roles of chlorogenic acid, protocatechuic acid, and *p*-hydroxybenzoic acid. To summarize, the validated ultrasound-assisted extraction method presented here is a quick, green, simple and efficient for the possible valorization of antioxidant phenolic compounds from Moroccan almond cold-pressed oil residues, making it possible to generate extracts with attractive antioxidant activities for future nutraceutical and/or cosmetic applications.

Keywords: almond; antioxidant; byproducts; chlorogenic acid; design of experiment; phenolic acids; ultrasound-assisted extraction

---

1. Introduction

Consumption of fruits, vegetables, nuts, and seeds has been associated with lower risks of chronic and degenerative diseases [1–4]. Particularly, given their many beneficial effects on human health, in recent decades, there has been growing interest in the consumption of nuts as a nutrient-rich food [3]. Produced and consumed worldwide, almond (*Prunus dulcis* (Mill.) D.A.Webb) is one of the most popular nuts. It can be consumed in the form of whole nuts, flour, and beverages proposed in the food industry. A large part of almond health benefits has been ascribed to their lipid profile [3,4]. Almond oil is also a sought-after and attractive component for many cosmetic formulations. Over the last few decades, the part of the beneficial actions for health, but also of the growing interest for industrial applications, ascribed to almond phenolics have become increasing [3–5].

The high antioxidant capacity of almond phenolics make it an attractive alternative to synthetic antioxidants. Synthetic antioxidants were largely used to maintain the oxidative stability of emulsions and commonly used in food products and pharmaceutical and cosmetic preparations. However, synthetic antioxidants such as butylated hydroxyanisole (BHA) or butylated hydroxytoluene (BHT) have adverse health effects, including carcinogenesis [6–8]. Therefore, the use of some of these synthetic antioxidants is now prohibited for food applications in Japan, Canada, and Europe, and they have been removed from what is generally recognized as a safe (GRAS) list. The replacement of these widely criticized synthetic molecules with natural molecules would meet the expectations of manufacturers and consumers. Therefore, it is now important to identify the natural antioxidants with a pronounced and safer radical scavenging capacity for consumers. Despite their distinct lipophilicity profile compared to BHA or BHT, some natural antioxidant phenolics have been shown to be as effective as these synthetic antioxidants in stabilizing nonpolar systems such as bulk oil or different emulsion types [9–11], in good agreement with the prediction of polar paradox theory [12]. Interestingly, after cold-pressed oil extraction, most of the antioxidant phenolic compounds accumulated in the almond skin are retained in a skin-enriched by-product [3,5,13], making this almond cold-pressed oil residue (AOR) an attractive raw material for extraction and the valorization of these natural antioxidant phenolics. In Morocco, the fifth highest ranking producer in the world, almond is the most important nut crop both in terms of area and production value. Several local genotypes, called *Beldi*, which means "from here" as opposed to acclimatized genotypes called *Romi* (i.e., from elsewhere) [14], are of special interest [15]. The almond plantations cover a total area of 150,000 ha for an average annual production estimated at 100,000 tons of unshelled products, of which 9% of this area which provides up to 14% of the Moroccan almond production is located in Eastern Morocco (Figure 1) [15]. This production generates an important part of byproducts, in particular of cold pressed almond oil residues. To date, these by-products of almond oil have been used primarily for animal feed, as litter or for energy production. However, upgrading to higher value-added sectors would significantly increase the revenue from this byproduct valuation. With an average growth rate of 5% per year since 2010 and a large profit margin, the cosmetics market is a dynamic industry. However, this sector is very competitive, with companies facing ever more restrictive environmental regulations (such as REACH in

European Union), in addition to consumer pressures that push them to innovate and gradually shift to more natural products and green production methods. The functional properties of several oleaginous species of agro-waste, including their antioxidant activity, in relation to their high concentration in phenolic compounds, have been documented [9,13,16–27]. The application of the biorefinery principle to the recovery of natural antioxidants from almond by-products as cosmetic active ingredients would therefore represent a good opportunity for the almond sector compared to their current use.

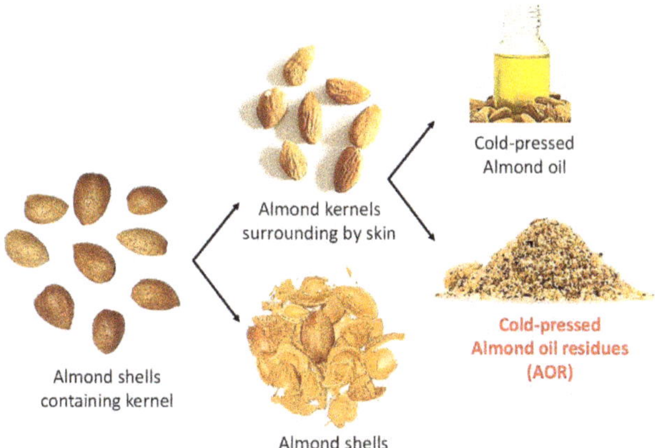

Figure 1. Parts of almond fruits leading to cold-pressed almond oil and its residue (AOR) used as byproduct in the present study to extract phenolic compounds.

For optimal valorization of these natural co-products, the development of effective extraction methods is necessary. In the past, there were many methods developed for the extraction of natural antioxidants from various natural matrices based on conventional methods such as maceration or Soxhlet extraction. More recently, green extraction methods including microwave-assisted extraction or ultrasound-assisted extraction (USAE) have been found to be particularly effective [11,21,28–30]. These green extraction technologies have also aroused great interest for industrial applications, and USAE is now considered as one of the most efficient energy saving processes in terms of duration, selectivity, and reproducibility, operating under mid-extraction conditions [28]. It is accepted that the improvement in extraction efficiency obtained using the USAE is based on both acoustic cavitation and mechanical effects [28]. Indeed, ultrasound (US) produces an acoustic cavitation effect facilitating the penetration of the extraction solvent. Therefore, easier release of the intracellular contents of the plant material is observed through greater agitation of the solvent resulting in increased surface contact between the solvent and the target compound as well as increased solubility of the target compound in the solvent of extraction [28].

Here, we report on the development and validation of a USAE method for the extraction of antioxidant phenolic acids from an enriched skin fraction made up of cold pressed AOR from *Bedli* Moroccan genotypes produced in Eastern Morocco (Figure 1).

Recently, Prgomet et al. [13] have also developed a method for comparing the polyphenol fractions from different almond byproducts including the skin using almond varieties from Portugal, but using a conventional heat reflux method. An USAE method was developed by Kahlaoui et al. [18] for the extraction of polyphenols from another almond byproduct: the hulls (the part surrounding the shell itself surrounded by the thin skin; Figure 1) from Italian and Tunisian varieties. It is thus of special interest to compare our method optimized using a different genotype, but more importantly either a green extraction method or a different (by)product. The optimal extraction conditions of this USAE using ethanol as a solvent were obtained through a multivariate technique (Behnken–Box design) coupled with response surface methodology (RSM) and then validated according to international standards of the association of analytical communities (AOAC). This USAE was applied to investigate the influence of the genetic and environment on the phenolic contents by considering three different local *Beldi* genotypes growing at three different experimental sites. Both in vitro cell free and cellular antioxidant assays were performed to evaluate the evolution of antioxidant activity of the corresponding extracts. Finally, correlations linking phytochemical profile and antioxidant activities of the extracts are presented.

2. Materials and Methods

2.1. Chemicals and Reagents

Extraction solvents (ethanol and water) used in the present study were of analytical grade (Thermo Scientific, Illkirch, France). Reagents for antioxidant assays as well as standards (chlorogenic acid, *p*-coumaric acid, protocatechuic acid and *p*-hydroxybenzoic acid) were purchased from Merck (Saint-Quentin Fallavier, France).

2.2. Plant Materials and Culture Conditions

AOR were obtained from Moroccan almonds (local ecotypes *Beldi*) grown in 3 different pilot locations in the Eastern Morocco (Sidi Bouhria (SID; 34°44′13.6″ N, 002°20′15.0″ W); Ain Sfa (AIN; 34°46′42.4″ N, 002°09′28.9″ W); Rislane (RIS; 34°44′59.8″ N, 002°26′44.7″ W)) using growing conditions as previously described by Melhaoui et al. [15]. Almonds were then triturated using an oil screw press (KOMET DD85G, IBG Monforts Oekotec GmbH & Co. KG, Monchengladbach, Germany) and the residues were ground to ca. 100–150 μm particles using a blender equipped with rotating blades (Grindomix GM 200 blender, Retsch France, Eragny, France) used as raw materials for USAE optimization.

2.3. Ultrasound-Assisted Extraction Method Development

USAE was completed with an ultrasonic bath (USC1200TH, Prolabo, Sion, Switzerland) composed of a $300 \times 240 \times 200$ mm (inner dimension) tank, with electric power of 400 W corresponding to an acoustic power of 1 W/cm^2 and maximal heating power of 400 W. The variable frequencies of this device can be selected thanks to a frequency controller, and it also has a temperature regulator as well as an automatic digital timer. Each sample was placed in 50 mL quartz tubes equipped with a vapor condenser, and was suspended in 10 mL extraction solvent. A liquid to solid ratio of 10:1 mL/g DW (dry weight) was used and extraction was performed at 45 °C.

For Extraction optimization a Box–Behnken design was used and the resulting response surface plots drawn with the help of XLSTAT2019 software (Addinsoft, Paris, France). For this purpose, three variables (aqueous Ethanol (aqEtOH) concentration (X_1), US frequency (X_2), and extraction duration (X_3)) were studied and coded at three levels (−1, 0 and +1) as described in Table 1:

Table 1. Identity, code unit, coded levels, and actual experimental values of each variable used for USAE of TPC from almond oil residues.

Variable	Code Unit	Coded Variable Levels		
		−1	0	+1
Ethanol concentration (% *v/v*) [1]	X_1	0	50	100
US frequency (kHz)	X_2	0	22.5	45
Extraction duration (min)	X_3	20	30	40

[1] % volume/volume of ethanol (analytical grade) concentration in mixture with ultrapure water (HPLC grade).

The different batches were obtained by using the DOE (design of experiment) function of XLSTAT 2019 (Addinsoft, Paris, France), which take values of selective variables at different levels

(Table 2). The experiments were carried out in triplicate. Equation of the model for the extraction of total phenolics from almond oil residues was calculated using the XLSTAT 2019 DOE analysis tool (Addinsoft, Paris, France). The corresponding response surface plots were obtained with 3D option of XLSTAT 2019 (Addinsoft, Paris, France).

Table 2. Results of Box–Behnken experimental design of USAE of TPC from AOR.

Run ID	Run Order	X_1	X_2	X_3	Experimental TPC (mg/g DW)	Predicted TPC (mg/g DW)
Obs1	10	0	+1	−1	7.93 ± 0.14	7.93
Obs2	6	+1	0	−1	6.16 ± 0.22	6.31
Obs3	1	−1	−1	0	5.03 ± 0.11	5.18
Obs4	17	0	0	0	11.33 ± 0.10	11.37
Obs5	15	0	0	0	11.37 ± 0.08	11.37
Obs6	7	−1	0	+1	7.51 ± 0.07	7.35
Obs7	12	0	+1	+1	9.06 ± 0.13	9.23
Obs8	4	+1	+1	0	6.01 ± 0.11	5.86
Obs9	18	0	0	0	11.41 ± 0.12	11.37
Obs10	5	−1	0	−1	5.52 ± 0.05	5.36
Obs11	13	0	0	0	11.32 ± 0.17	11.37
Obs12	2	+1	−1	0	5.54 ± 0.18	5.56
Obs13	8	+1	0	+1	5.18 ± 0.14	5.16
Obs14	9	0	−1	−1	7.69 ± 0.19	7.52
Obs15	11	0	−1	+1	6.88 ± 0.13	6.88
Obs16	16	0	0	0	11.44 ± 0.17	11.37
Obs17	3	−1	+1	0	7.66 ± 0.16	7.64
Obs18	14	0	0	0	11.35 ± 0.15	11.37

Experimental values are means ± RSD of 3 independent replicates.

2.4. Determination of Total Phenolic Content

After extraction, each extract was centrifuged for 15 min at 5000× g (Heraeus Biofuge Stratos, Thermo Scientific, Illkirch, France) and the resulting supernatant filtered using a syringe filter (0.45 μm, Merck Millipore, Molsheim, France) prior to analysis.

The total phenolic content (TPC) was determined spectrophotometrically using the Folin–Ciocalteu reagent (Merck, Saint-Quentin Fallavier, France) and according to the protocol adapted for a microplate reader described by Abbasi et al. [31]. Briefly, 10 μL of extract were homogenized with 180 μL of a mixture composed of 4% (*w/v*) Na_2CO_3 (prepared in NaOH 0.1 M), 0.02% (*w/v*) potassium sodium tartrate tetrahydrate and 0.02% $CuSO_4$. Following a 10-min of incubation at 25 °C, 10 μL of the Folin–Ciocalteu reagent were added, and the homogenized mixture was incubated for 30 min at 25 °C. Absorbance was measured at 650 nm with a spectrophotometer (BioTek ELX800 Absorbance Microplate Reader, BioTek Instruments, Colmar, France). A standard curve (0–40 μg/mL; R^2 = 0.998) of gallic acid (Merck, Saint-Quentin Fallavier, France) was used to express the TPC in mg of gallic acid equivalents per g DW (mg GAE/g DW).

2.5. Validation Parameters

Method validation was carried out using the recommendations of the association of analytical communities (AOAC) in terms of precision, repeatability, and recovery as described in detail in Corbin et al. [21].

For HPLC, 6-point calibration lines were obtained by means of diluted solutions of each authentic commercial standard (Merck, Saint-Quentin Fallavier, France). Each sample was injected three times, and arithmetic means were calculated to generate linear regression equations plotting was done by the peak areas (y) against the injected quantities (x) of each standard. Coefficients of determination (R^2) were used for linearity verification. The limits of detection (LOD) and of quantification (LOQ) was calculated using signal-to-noise ratios of 3:1 and 10:1, respectively.

2.6. HPLC Analysis

After extraction, each extract was centrifuged for 15 min at 3000 rpm and the resulting supernatant filtered using a syringe filter (0.45 μm, Millipore, Molsheim, France) prior to analysis. Separation and identification of the main extract constituents was done by HPLC with a Varian system (Varian, Les Ulis, France) composed of: Prostar 230 pump, Metachem Degasit, Prostar 410 autosampler, Prostar 335 Photodiode Array Detector (PAD) and driven by Galaxie version 1.9.3.2 software (Varian, Les Ulis, France). A Purospher RP-18 column (250×4.0 mm internal diameter; 5 μm) (Merck Chemicals, Molsheim, France) was used for the separation performed at a temperature set at 35 °C. The mobile phase was a mixture of: (i) A, which was acidified HPLC grade water with acetic acid (0.2% (*v/v*)), and (ii) B, which was HPLC grade methanol. During the separation run, the mobile phase composition varied according to a nonlinear gradient as follows: 8% B (0 min), 12% B (11 min), 30% B (17 min), 33% B (28 min), 100% B (30–35 min), and 8% B (36 min) at a flow rate of 1 mL/min. Between each injection, a 10-min re-equilibration time was applied. The detection of compounds was set at 295 and 325 nm (corresponding to the λmax of the main compounds). Quantification was done based on assessment of retention times of commercial standards (Merck, Saint-Quentin Fallavier, France).

2.7. Antioxidant Activities

2.7.1. In Vitro Cell Free DPPH Free Radical Scavenging Assay

The in vitro cell free DPPH (2,2-diphenyl-1-picrylhydrazyl) assay was used to evaluate the free radical scavenging activity of the samples as described by the microplate protocol of Shah et al. [32].

2.7.2. In Vitro Cell Free ABTS Antioxidant In Vitro Cell Free Assay

ABTS (2,2-azinobis(3-ethylbenzthiazoline-6-sulphonic acid)) in vitro cell free antioxidant activity of each extract was determined as described by Ullah et al. [33].

2.7.3. Cupric Ion Reducing Antioxidant Capacity (CUPRAC) In Vitro Cell Free Assay

CUPRAC assay was performed in a microplate as described by Drouet et al. [8].

2.7.4. Determination of Membrane Lipid Peroxidation Using Thiobarbituric Acid-Reactive Substances (TBARS) Assay

An *in cellulo* antioxidant assay, using yeast cells, based on the measurement of membrane lipid peroxide, was carried out with the thiobarbituric acid (TBA; Merck, Saint-Quentin Fallavier, France) method as described by Garros et al. [34].

2.8. Statistical Analysis

Means and standard deviations of three to five independent replicates were used to present the data. Model analysis (ANOVA) and 3D plots resulting from the combination of variables were performed using XLSTAT 2019 and R analysis following the manufacturer's instructions (Addinsoft, Paris, France). A Student's *t*-test was performed for comparative statistical analysis of the impact of the different cultivation sites (XLSTAT 2019, Addinsoft, Paris, France). Correlation analysis was performed with Past 3.0 (Øyvind Hammer, Natural History Museum, University of Oslo, Oslo, Norway) using the Pearson parametric correlation test and visualized using Heatmapper [35]. Principal Component Analysis (PCA) was performed with Past 3.0 (Øyvind Hammer, Natural History Museum, University of Oslo, Oslo, Norway). Significant thresholds at $p < 0.05$ or $p < 0.05$, <0.01 and <0.001 were used for all statistical tests and represented by different letters or by *, ** and ***, respectively.

3. Results and Discussion

3.1. Development of the Ultrasound-Assisted Extraction Using Box–Behnken Design

Multivariate techniques are used very effectively to optimize the extraction method from complex plant matrices such as food products and by-products [36]. Among the different multivariate techniques, when three factors are considered, the Behnken–Box design is one of the most effective techniques [36,37]. The Behnken–Box matrix is a spherical and rotating design, which, viewed on a cube, consists of the central point and the middle of the edges [36,37]. Many parameters can influence the extraction of phenolic compounds from plant matrices [38], but three parameters are very widely distinguished when developing an USAE method: the type of solvent used, the frequency of ultrasound applied and the extraction time [11,21,30].

The choice of solvent is a crucial parameter to define when developing an extraction method. Various solvents, including methanol, ethanol, or acetone, are regularly used for the extraction of plant polyphenols [28,39]. Here, given our objective of developing an extraction method in accordance with green chemistry principles for future nutraceutical and/or cosmeceutical applications of the resulting extract, EtOH was considered as an extraction solvent. First, EtOH is one of the less toxic solvents for humans and more respectful of the environment than other organic solvents such as methanol for example [38,40]. In addition, its extraction capacity can easily be modulated by addition of water, making it an ideal solvent for the extraction of a wide range of variable polarity polyphenols. Finally, these two universal solvents (i.e., EtOH and water) are already widely used for various food and/or cosmetic applications [11,28,30,38,40].

US frequency is a crucial parameter to consider because of its significant impact on the extraction efficiency. Indeed, this parameter modulates the cavitation effect as well as the diffusion coefficient of the target compound in the extraction solvent. Consequently, it improves the solubilization of the compound in the extraction solvent, thus increasing the extraction efficiency [28]. In addition, increasing US frequency can also lead to a drastic reduction in extraction time, thereby reducing energy consumption, which is in accordance with the green chemistry principles [41]. However, depending on the compound and the plant matrix subjected to the extraction, application of high US frequency can alter the native structure of the compound, which not only decreases the extraction yield, but also considerably reduces its biological activity, thus negating any valuation interest [30]. Therefore, during the development of an USAE method, US frequency must be optimized very carefully depending on the compound, and the plant matrix subjected to the extraction.

Finally, regarding the extraction time, it is important to consider that its increase does not necessarily lead to any improvement in extraction yield, since, contrarily, a prolonged exposure to US can lead to the increased degradation of the compound [30]. In addition, in order to reduce the impact of energy consumption in the green chemistry context, optimizing the extraction time also appears to be essential [41].

Having these considerations in mind, in order to develop a rapid, green and efficient USAE of TPC for the valorization of AOR, we therefore considered a Behnken–Box matrix with the following three parameters: aqEtOH concentration (X_1), ultrasound (US) frequency (X_2), and extraction duration (X_3) as described in Table 1.

Table 2 presents the experimental and predicted TPC obtained from almond oil residues for the 18 different observations (run ID) corresponding the different USAE conditions of the Behnken–Box matrix having been determined randomly (run order) after an *in silico*-assisted procedure generated by the XL-Stat2019.4.1 software.

Here, the TPC extracted from AOR ranged from: 5.03 mg/g DW (Obs3; obtained after 30 min extraction in water bath (no US application) using pure water as extraction solvent) to 11.44 mg/g DW (Obs16; obtained after 30 min at an ultrasonic bath running at a US frequency of 22.5 kHz using 50% (*v/v*) aqEtOH as extraction solvent) (Table 2). These results provide a first indication on the interest of using an ultrasound and on the choice of an extraction solvent. We noted a good repeatability of the central point (i.e., Obs4, 5, 9, 11, 16, and 18), with a mean TPC of 11.37 ± 0.05 mg/g DW corresponding to a relative standard deviation (RSD) of 0.47%, thus highlighting the high reliability of these results. Given the nature of the starting material used in the present study, this range of TPC is in fairly good agreement with the data in the literature obtained with almond and/or almond by-products from California, Portugal, Italia, and Tunisia [3,5,18,42].

A multiple regression analysis was applied to the model of the TPC as a function of the three different extraction variables. Under the described conditions, the TPC (Y_{TPC}, in mg/g DW) as a function of the three different extraction variables (Table 1) in the form of a polynomial equation was (Table 3):

$$Y_{TPC} = 11.370 - 0.354X_1 + 0.690X_2 + 0.166X_3 - 3.554X_1{}^2 - 1.756X_2{}^2 - 1.724X_3{}^2 - 0.540X_1X_2 - 0.743X_1X_3 + 0.485X_2X_3$$

Table 3. Statistical analysis of the regression coefficients of USAE of TPC from AOR.

Source	Value	SD	t	$P > |t|$
Constant	11.370	0.059	193.27	<0.0001 ***
X_1	−0.354	0.051	−6.943	0.00012 ***
X_2	0.690	0.051	13.543	<0.0001 ***
X_3	0.166	0.051	3.253	0.011 **
$X_1{}^2$	−3.554	0.069	−51.516	<0.0001 ***
$X_2{}^2$	−1.756	0.069	−25.459	<0.0001 ***
$X_3{}^2$	−1.724	0.069	−24.988	<0.0001 ***
X_1X_2	−0.540	0.072	−7.495	<0.0001 ***
X_1X_3	−0.743	0.072	−10.305	<0.0001 ***
X_2X_3	0.485	0.072	6.731	0.00015 ***

SD standard deviation; *** significant $p < 0.001$; ** significant $p < 0.01$.

The statistical analysis of the regression coefficients confirmed the relevance of our choice in the extraction variables and their respective levels for the development of the present USAE method if we refer to the level of significance with which these variables influenced the extraction (Table 3). The linear coefficients X_1 (aqEtOH concentration) and X_2 (extraction time) were statistically highly significant at $p < 0.001$, with an X_1 coefficient being negative (high EtOH concentration reduced TPC) and X_2 being positive (application of US treatment had a positive effect on TPC). An extraction duration (X_3) coefficient was also significant at $p < 0.01$, but with a coefficient value close to zero indicating that a prolonged extraction period can lead to poorer extraction yield as a consequence of degradation as described in the literature [30,41,43]. All the quadratic and interaction coefficients were statistically highly significant at $p < 0.001$, but their values negative or close to zero indicated a negative or a lower impact to the extraction efficient.

The results of the analysis of variance (ANOVA) and model fitting are presented in Table 4. An elevated F-value (567.558) and low p-value ($p < 0.0001$) indicated the statistically highly significance of the model that could predict TPC as a function of the variable values with a great precision. The low non-significant value obtained for the lack of fit confirmed this trend. The value for the determination coefficient ($R^2 = 0.997$ (with adjusted value of 0.998) for the model as well as the coefficient value (CV = 0.976) indicated the precision of the model as well as the adequacy between the model and experimental values, respectively. The model precision in the prediction of the TPC is further depicted by the predicted vs. experimental TPC plot presented in Figure S1.

To better understand the complexity of the model, 3D plots representing TPC as a function of the extraction parameters were drawn (Figure 1).

The calculated, but small, values of the linear coefficients of the second-order polynomial equation for X_2 (US frequency) and X_3 (extraction duration), as well as their interaction coefficient X_2X_3 (US frequency x duration) indicate that a controlled increase of these parameters will have a global favorable consequences for the TPC extracted from AOR. However, their small values, in association with the negative values calculated for their quadratic coefficients ($X_2{}^2$ and $X_3{}^2$, respectively), but also of all the coefficient involving aqEtOH concentration (i.e., linear coefficient X_1, quadratic coefficient $X_1{}^2$, and the interaction coefficients X_1X_2 and X_1X_3), indicate that the TPC extracted from AOR according to these extraction parameters will reach a maximum value before decreasing for high values of these parameters. These considerations were clearly observed on the 3D plots (Figure 2). For each 3D plot, a first tendency was observed with a higher TPC extracted from AOR with increased aqEtOH concentration, application of US as well as prolonged extraction time. However, after reaching a maximal value for TPC extracted from AOR, a further increase in the aqEtOH concentrations as well as application of higher US frequency and/or prolonged extraction duration resulted in a pronounced drop of the TPC (Figure 2).

Table 4. ANOVA of the predicted model used for USAE of TPC from almond residues.

Source	Sum of Square	df	Mean of Square	*F*-Value	*p*-Value
Model	106.071	9	11.786	567.558	<0.0001 ***
Lack of fit	0.166	8	0.021	-	-
Residual	0.166	8	0.021	-	-
Pure Error	0.000	0	-	-	-
Cor. Total	106.237	17	-	-	-
R^2	0.997				
R^2 adj	0.998				
CV%	0.976				

df: degree of freedom; Cor. Total: corrected total; R^2: determination coefficient; R^2 adj: adjusted R^2; CV variation coefficient value; *** significant $p < 0.001$.

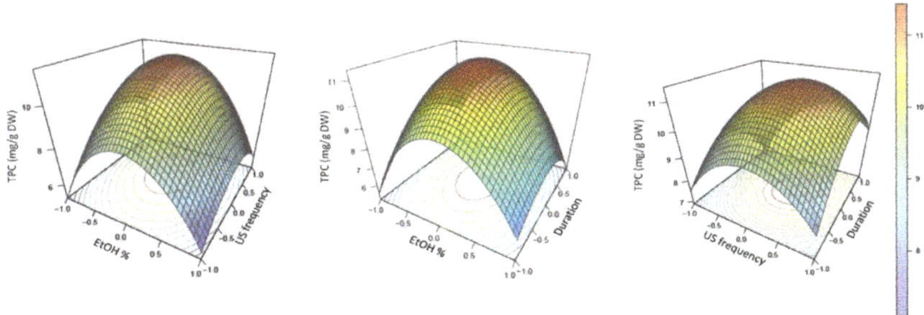

Figure 2. Predicted surface response plots of the TPC extraction yield (in mg/g DW) as a function of aqEtOH concentration and US frequency, aqEtOH concentration and extraction duration, as well as US frequency and extraction duration.

In various concentrations in mixture with water, aqEtOH solutions have been widely used as eco-friendly solvents to extract a wide range of polyphenols from plant matrices [11,21,30,38,40] including various almond products [13,18,44]. However, to obtain optimal results, the concentration of aqEtOH must be adapted because it is very dependent on the polyphenolic compound(s) as well as on the plant matrix considered [28,38,40]. Alongside, it is clearly established that, during USAE, high US frequency associated with extended extraction duration could reveal destructive through the induction of polyphenols oxidation, in particular in the presence of water [21,28,30]. Consequently, if these parameters are not finely controlled (optimized), this can lead to a sharp reduction in the extraction yield, quantitatively but also qualitatively with a drastic decay observed in the biological interest of the sample extract [11,29,30]. Using the Box–Behnken matrix for optimizing these parameters and using the resulting adjusted second order polynomial equation, optimal conditions for the extraction of phenolics from our Moroccan AOR were: 53.0% (*v*/*v*) aqEtOH as solvent, 27.0 kHz for the US frequency and an extraction duration of 29.4 min. Using these optimal conditions resulted in a TPC of 11.63 ± 0.15 mg/g DW (Figure 2). The optimal aqEtOH concentration obtained here is in line with results obtained for almond phenolics extraction very recently described [13,18]), although the starting byproduct material or the extraction method used were different from our study.

The present method was then validated in respect with the AOAC standards. According to these standards, the parameter values of this validation procedure were adequate in terms of interday and intraday precision, but also repeatability and stability (Table 5). Indeed, the RSDs of both intraday and interday precisions were of 0.05 and 0.28%, respectively. The RSDs of the repeatability corresponding to five different extraction repeats of five samples from the same batch was of 1.30%. The recovery rates at three different addition levels of chlorogenic acid in the sample before extraction were between 100.26 and 101.13% reflect the accuracy of the present method.

Table 5. Validation parameters of the developed method for quantifying TPC from almond residues.

Precision (% RSD)		Repeatability (% RSD)	Recovery [1] (%)		
Intraday	Interday		0.5	1.0	2.0
0.05	0.28	1.30	100.26 ± 0.07	100.90 ± 0.80	101.13 ± 0.50

[1] performed at 3 concentration level additions of gallic acid prior to extraction using optimal conditions with US (i.e., 0.5, 1.0, and 2.0 mg/g DW additions). Means ± SD standard deviations or % RSD of three independent extractions.

The efficiency of the present USAE method was compared with conventional heat reflux extraction (HRE) using the same conditions, in particular an aqEtOH concentration (53.0% (*v/v*) and an extraction time of 29.4 min. The difference between USAE and HRE being the application of an US frequency of 27 kHz for the present optimized UASE extraction procedure, while no US was applied for the HRE protocol operating in a classical water bath. The comparison of these extractions is depicted in Figure 2. A significant gain of 30% in TPC extracted from AOR was observed with the optimized USAE (11.63 ± 0.15 mg/g DW) as compared to conventional HRE (8.96 ± 0.21 mg/g DW) (Figure 3). Increasing the extraction time for the HRE to one hour did not achieve performance levels similar to those obtained with USAE (9.24 ± 0.37 mg/g DW). Consequently, it appears that the USAE method developed in the present study is of real interest according to the principles of green chemistry [45], not only in terms of the use of a renewable green solvent, but also in terms of reducing the energy consumption. We hypothesize that this efficiency could be partly explained by the hot spot hypothesis indicating that the cavitation bubbles, after their collapse, act as a microreactor locally generating, in the surrounding solvent, a high temperature environment and pressure leading to more efficient rupture of the plant matrix subjected to extraction and increased release as well as solubilization of phenolic compounds [28].

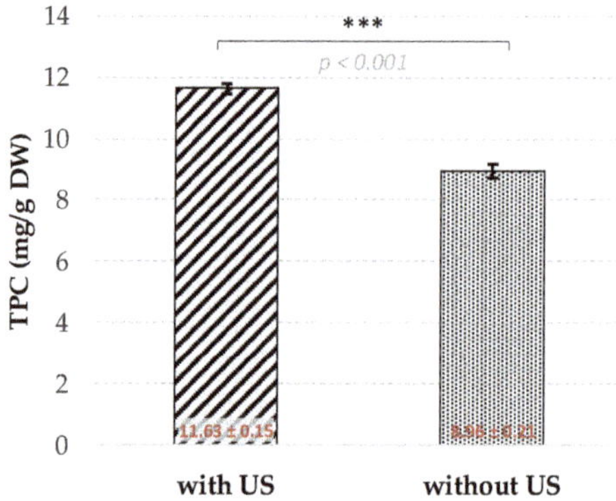

Figure 3. TPC extracted from AOR using the optimal USAE (with US) conditions and comparison with conventional heat reflux method (HRE; without US). Means ± SD standard deviations of three independent extractions; *** significant at *p* < 0.001.

3.2. Application to the Analysis of Samples from Different Cultivation Sites

The present USAE was then applied to the quantification of phenolics in samples from three different local Beldi genotypes cultivated at three different locations in Eastern Morocco. In addition to the TPC, the concentration in protocatechuic acid, *p*-hydroxybenzoic acid, chlorogenic acid, and *p*-coumaric acid, reported as the main phenolic acids possibly accumulated in almond by-products [3,5,13,18,42,44,46], were also determined by HPLC after comparison with authentic commercial standards. Figure 4a shows a typical HPLC chromatogram, recorded at 325 nm, of the AOR extract obtained after USAE and showing the sepration of these four important phenolic acids: protocatechuic acid (1), *p*-hydroxybenzoic acid (2), chlorogenic acid (3), and *p*-coumaric acid (4) (Figure 4b).

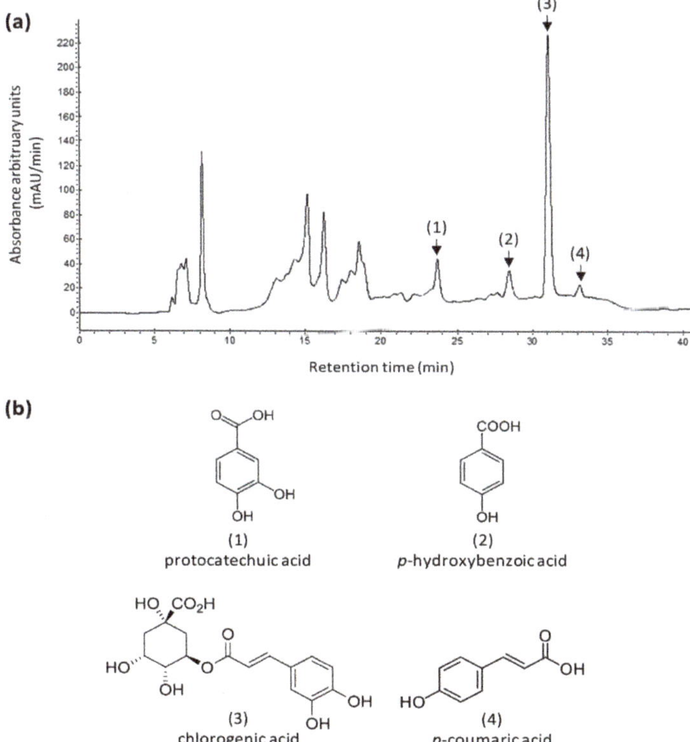

Figure 4. (a) Representative HPLC chromatogram (here with detection set at 325 nm) of an extract prepared by USAE of AOR (*Beldi* cultivar) grown in the Ain Sfa (34°46′42.4″ N, 002°09′28.9″ W) pilot location in the eastern Morocco; (b) structures and their corresponding numbers on the HPLC chromatogram of the main phenolic compounds considered in this study: protocatechuic acid (1), *p*-hydroxybenzoic acid (2), chlorogenic acid (3), and *p*-coumaric acid (4).

In order to quantify these four phenolic compounds in different samples, 6-points calibration curves of the peak areas (y) against the injected amounts (x) of protocatechuic acid and *p*-hydroxybenzoic acid at 295 nm and chlorogenic acid and *p*-coumaric acid at 325 nm were obtained with a linearity over wide ranges from 0.5 to 200 mg/L of injected solutions and R^2 greater than 0.999 (Table 6). The LODs ranged from 0.12 to 0.22 mg/mL, and LOQ from 0.38 to 0.73 mg/mL, for protocatechuic acid and chlorogenic acid, respectively (Table 6).

Table 6. Quantification parameters of the HPLC method used to quantity protocatechuic acid, *p*-hydroxybenzoic acid, chlorogenic acid, and *p*-coumaric acid after their USAE from AOR.

Compound	RT (min)	λmax (nm)	Linear Range (mg/L)	Equation	R^2	LOD (mg/L)	LOQ (mg/L)
Protocatechuic acid	23.69	295	0.5–200	y = 3.429 x + 0.814	0.9991	0.12	0.38
p-Hydroxybenzoic acid	28.46	295	0.5–200	y = 3.018 + 0.732	0.9993	0.21	0.68
Chlorogenic acid	31.02	325	0.5–200	y = 5.041x + 0.324	0.9997	0.22	0.73
p-Coumaric acid	33.07	325	0.5–200	y = 7.561x + 0.623	0.9992	0.14	0.47

Applied to the quantification of TPC, protocatechuic acid, *p*-hydroxybenzoic acid, chlorogenic acid and *p*-coumaric acid in AOR resulting from samples of three different native *Beldi* genotypes (#1 to #3) cultivated at three different pilot locations in the Eastern Morocco (Sidi Bouhria (SID); Ain Sfa (AIN); Rislane (RIS)), the results are presented in Table 7.

Table 7. Variations in TPC, and protocatechuic acid, *p*-hydroxybenzoic acid, chlorogenic acid and *p*-coumaric acid contents in AOR from samples of three different native *Beldi* genotypes produced at three different pilot locations in the Eastern Morocco.

Sample ID	TPC (mg/g DW)	Protocatechuic Acid (mg/g DW)	*p*-Hydroxybenzoic Acid (mg/g DW)	Chlorogenic Acid (mg/g DW)	*p*-Coumaric Acid (mg/g DW)
SID#1	9.35 ± 0.63 [bcd]	1.33 ± 0.10 [de]	0.78 ± 0.02 [d]	5.53 ± 0.13 [e]	0.26 ± 0.05 [ab]
SID#2	11.78 ± 1.58 [ab]	1.84 ± 0.07 [ab]	1.02 ± 0.06 [b]	7.02 ± 0.19 [bc]	0.29 ± 0.07 [ab]
SID#3	11.79 ± 1.29 [ab]	1.75 ± 0.09 [b]	0.98 ± 0.02 [b]	6.97 ± 0.04 [b]	0.29 ± 0.06 [ab]
AIN#1	8.87 ± 0.31 [d]	1.29 ± 0.06 [e]	0.75 ± 0.03 [d]	5.29 ± 0.12 [e]	0.21 ± 0.04 [b]
AIN#2	11.29 ± 1.24 [ab]	1.66 ± 0.10 [bc]	0.95 ± 0.05 [bc]	6.69 ± 0.06 [c]	0.28 ± 0.05 [ab]
AIN#3	13.86 ± 0.91 [a]	2.03 ± 0.07 [a]	1.13 ± 0.02 [a]	8.14 ± 0.10 [a]	0.26 ± 0.20 [a]
RIS#1	9.34 ± 0.27 [cd]	1.36 ± 0.09 [de]	0.75 ± 0.04 [d]	5.34 ± 0.14 [e]	0.22 ± 0.03 [b]
RIS#2	10.69 ± 0.73 [bc]	1.51 ± 0.05 [cd]	0.87 ± 0.06 [c]	6.34 ± 0.05 [d]	0.29 ± 0.02 [a]
RIS#3	11.97 ± 1.51 [ab]	1.76 ± 0.02 [b]	1.00 ± 0.10 [bc]	7.12 ± 0.08 [b]	0.30 ± 0.02 [a]

Samples were AOR from three different native *Beldi* genotypes (#1 to #3) cultivated at three different pilot locations in the Eastern Morocco: Sidi Bouhria (SID; 34°44′13.6″ N, 002°20′15.0″ W); Ain Sfa (AIN; 34°46′42.4″ N, 002°09′28.9″ W); Rislane (RIS; 34°44′59.8″ N, 002°26′44.7″ W). Values are means ± SD of three independent replicates. Different letters represent significant differences between the various extraction conditions ($p < 0.05$).

TPC ranged from 8.87 to 13.86 mg/g DW for extracts from samples AIN#1 and AIN#3, respectively; sample from genotype #3 cultivated at Ain Sfa being 56.25% richer in TPC than genotype #1 cultivated at the same location. The four quantified phenolic acids occurred for approximately 80% of the TPC. In decreasing contents: (1) chlorogenic acid was the main phenolic accumulated in the sample extracts with contents ranging from 5.29 ± 0.12 to 8.14 ± 0.10 mg/g DW for extracts from samples AIN#1 and AIN#3, respectively (sample from genotype #3 cultivated at Ain Sfa being 53.87% richer in chlorogenic acid than genotype #1 cultivated at the same location); (2) protocatechuic acid content ranged from 1.29 ± 0.06 to 2.03 ± 0.07 mg/g DW for extracts from samples AIN#1 and AIN#3, respectively (sample from genotype #3 cultivated at Ain Sfa being 57.36% richer in protocatechuic acid than genotype #1 cultivated at the same location, corresponding to the highest observed variation range); (3) *p*-hydroxybenzoic acid content ranged from 0.75 ± 0.03 to 1.13 ± 0.02 mg/g DW for extracts from samples AIN#1 and RIS#1 for the lowest content vs. sample AIN#3 for the highest content (sample from genotype #3 cultivated at Ain Sfa being 50.60% richer in *p*-hydroxybenzoic acid than genotype #1 cultivated both at Ain Sfa and Rislane); (4) *p*-coumaric acid content ranged from 0.21 ± 0.04 to 0.30 ± 0.02 mg/g DW for extracts from samples AIN#1 and AIN#3, respectively (sample from genotype #3 cultivated at Rislane being 42.85% richer in *p*-coumaric acid than genotype #1 cultivated at Ain Sfa, corresponding to the lowest observed variation range). The concentrations determined here for each phenolic compound were in the range of variations observed by Kahlaoui et al. [18] for different varieties of almond byproducts from Italia and Tunisia. Extraction of phenolic compounds from a variety of oilcakes such as hemp, canola, linseed, black cumin, sesame, fennel, sunflower, rapeseed, camelina, or milk thistle has been reported [9,11,16,17,20–22,24–27]. The TPC obtained from AOR using the present USAE method is at the top of the range compared to these other sources. Chlorogenic acid content has been reported to be high in sunflower oilcakes where its presence is problematic for the valorization of its derived protein meal by-product [26]. The other phenolic acids from AOR have been extracted from various oilcakes, such as flax, canola, and black cumin seedcakes for *p*-coumaric acid [9,20,21,24], black cumin, and camelina for *p*-hydroxybenzoic acid [9], while protocatechuic acid and *p*-hydroxybenzoic acid have been reported in camelina by-products [25]. Note that other types of phenolics such as lignans and or flavonoids in flax, sesame, and milk thistle seedcakes have been reported [16,21–24]. Interestingly, synergistic interactions between phenolic compounds could occur at concentrations found in nature for antioxidant activity [47]. Antagonism have been also described [47]. Therefore, the different compositions, but also the concentrations observed in different oilseed cakes, could result in different synergistic and/or antagonistic interactions towards their antioxidant capacity. This hypothesis is going to deserve future studies. However, it is also important to consider that these concentrations are subject to change as a result of both genetic and environmental influences as observed in milk thistle, flax, sesame, but also in some almond cultivars [34,48,49].

Indeed, it is generally accepted that the genetic background, but also the environmental conditions, such as the location (i.e., soil conditions) or the climate, could have a great influence on the accumulation of phenolic compounds [3,18,34,42,46,50]. The present preliminary results obtained from three native genotypes cultivated on the same year at three different location sites from Eastern Morocco suggested a prominent influence of genetic over environment, since the impact of the genotype was more important than the influence of the cultivation site. Indeed, for each considered cultivation sites, the genotype #1 accumulated more phenolic compounds than the genotype #3, whereas both the highest and the lowest accumulation were observed on the same location (i.e., Ain Sfa experimental site). Analyses of the variance (ANOVA) confirmed this absence of any significant influence of the cultivation site. Future works will be conducted with more genotypes as well as more experimental sites over several cultivation years to confirm or infirm this trend. However, the prominent influence of genetic background on the accumulation of phenolic compounds in almonds was reported by several authors [3,18,42,46], whereas the influence of environmental conditions on the same genotype was less studied. Bolling et al. [42] reported that the cultivation season influenced less polyphenolic accumulation than the genotype. The influence of cultivation site of the same genotype will deserve further works.

An improvement in the quantity of phenolic compounds produced in the future may also be envisaged, in the future, by combining this USAE with base or acid hydrolysis to release the wall-bound phenolics or extract further antioxidant compounds from lignin. A gain of 30% in chlorogenic acid content was reported in sunflower seed cakes after the release of wall-bound phenolics. Nonetheless, it has been stated that coupling US to base and/or acid extraction is highly destructive for some forms of phenolic compounds [22]. The use of cell wall degrading enzymes such as cellulase could be an alternative to destructive chemical hydrolysis [24]. USAE coupled with cellulase hydrolysis of phenolic compounds have been already reported [51]. Future works will be dedicated to exploring this possibility.

3.3. Determination of the Antioxidant Potential of the Extracts and Correlation Analysis

Our next goal was to ensure that the potential biological activities is retained during the USAE procedure. For this, we then determined the antioxidant potential of these nine characterized sample extracts from AOR by using both (1) in vitro cell free assays based on the chemistry of the antioxidant reaction with different mechanisms—either proton transfer or electron transfer based assays, as well as (2) *in cellulo* using eukaryotic yeast cells subjected to oxidative stress induced by UV either in the presence and absence of the extracts to have an idea of their cellular antioxidant potential. Indeed, if they were preserved, this antioxidant biological activity would be of such a nature as to be of interest for both future nutraceutical and/or cosmetic applications of these AOR extracts.

The protective antioxidant action developed by plant extracts can be influenced by many internal and external factors impacting their phytochemical compositions such as genetics (the use of different genotypes in our case) but also the environment (the use of different culture sites in our case) [3,18,34,42,46,50]. Furthermore, their antioxidant activity is generally based on complex mechanisms, which, in order to shorten, depending on the nature of the compounds present in the extract, can be based in particular on radical scavenging mechanisms. Here, to get an idea relating both to the antioxidant capacity but also to explore the possible mechanisms involved depending on the composition of the extract, we used three different in vitro cell-free assays: the DPPH, ABTS, and CUPRAC assays. These tests are based on different reaction mechanisms and could provide us a raw idea of the chemistry involved in the radical scavenging activity of the extract. Based on the chemical reaction involved, these in vitro cell free antioxidant assays can be roughly divided into different categories, with an ABTS assay based on a hydrogen atom transfer reaction (HAT), a CUPRAC assay based on an electron transfer reaction (ET), and the DPPH assay being considered as a mixed assay [52,53]. The results of these antioxidant assays expressed in μM of Trolox equivalent antioxidant capacity (TEAC) per gram DW for the nine extracts obtained after USAE of AOR are presented in Table 8.

Table 8. Variations in in vitro cell free (ABTS, DPPH and CUPRAC) and cellular (TBARS) antioxidant potential of extracts obtained from USAE of AOR from three different native *Beldi* genotypes produced at three different pilot locations in Eastern Morocco.

Sample ID	ABTS (μM TEAC/g DW [1])	DPPH (μM TEAC/g DW [1])	CUPRAC (μM TEAC/g DW [1])	TBARS (% inhibition)
SID#1	233.10 ± 12.52 [d]	323.51 ± 19.12 [a]	198.07 ± 22.97 [ab]	51.81 ± 1.13 [c]
SID#2	361.81 ± 14.48 [b]	347.40 ± 7.73 [a]	141.04 ± 2.16 [c]	66.95 ± 1.74 [a]
SID#3	366.49 ± 12.97 [b]	341.17 ± 5.49 [ab]	129.69 ± 0.32 [d]	58.73 ± 1.18 [b]
AIN#1	216.94 ± 12.32 [d]	275.84 ± 34.88 [b]	205.92 ± 17.11 [a]	50.62 ± 2.46 [c]
AIN#2	276.37 ± 13.12 [c]	326.88 ± 30.16 [ab]	164.79 ± 14.02 [b]	51.81 ± 1.45 [c]
AIN#3	401.52 ± 11.44 [a]	357.33 ± 24.24 [a]	178.73 ± 19.10 [ab]	69.12 ± 0.34 [a]
RIS#1	238.07 ± 15.86 [cd]	319.73 ± 14.74 [b]	160.93 ± 13.74 [bc]	53.13 ± 1.01 [c]
RIS#2	244.91 ± 12.02 [cd]	315.60 ± 7.43 [b]	143.76 ± 5.24 [bc]	52.63 ± 1.65 [c]
RIS#3	391.29 ± 9.64 [ab]	351.07 ± 2.89 [a]	173.47 ± 26.29 [abc]	66.85 ± 2.57 [a]

[1] TEAC: Trolox equivalent antioxidant capacity (TEAC); Samples were AOR from three different native *Beldi* genotypes (#1 to #3) cultivated at three different pilot locations in the Eastern Morocco: Sidi Bouhria (SID); Ain Sfa (AIN); Rislane (RIS). Values are means ± SD of 3 independent replicates. Different letters represent significant differences between the various extraction conditions ($p < 0.05$).

Antioxidant activity ranged from 216.94 ± 12.32 to 401.52 ± 11.44 μM TEAC/g DW for ABTS assay, and from 275.84 ± 34.88 to 357.33 ± 24.24 μM TEAC/g DW using a DPPH assay. For these two in vitro cell free antioxidant assays, the AOR extract from the genotype #3 produced at Ain Sfa showed the highest antioxidant capacity, whereas the extract obtained from the genotype #1 produced at the same location displayed the lowest antioxidant values. On the contrary, results for CUPRAC assay, ranging from 129.69 ± 0.32 to 205.92 ± 17.11 μM TEAC/g DW, showed that this genotype #1 produced at Ain Sfa possessed the highest antioxidant capacity as compared to the genotype #3 from Sidi Bouhria.

Although interesting from a strictly predictive point of view based on chemical reactions, these in vitro tests do not necessarily have a great similarity with in vivo systems. The validity of these antioxidant data must therefore be considered as limited to an interpretation within the meaning of the chemical reactivity with respect to the considered radicals generated in vitro, and have to be confirmed in vivo. In order to have an improved understanding and better reflect the in vivo situation, the antioxidant activity of these nine extracts has also been studied further for their capacity to inhibit the lipid peroxidation membrane generated by oxidative stress induced by UV-C in yeast cells. Yeast cells represent an excellent model for assessing antioxidant capacity in vivo in the context of cellular oxidative stress [54]. It is indeed an attractive and reliable eukaryotic model, whose defense and adaptation mechanisms to oxidative stress are well known and can be extrapolated to human cells presenting mechanisms certainly more complex but well conserved with this model [55,56]. Here, measured in vivo anti-lipoperoxidation activity (inhibition of malondialdehyde (MDA) formation), determined using the TBARS assay, ranged from 50.62 ± 2.46 to 69.12 ± 0.34%. Therefore, this in vivo antioxidant evaluation assay confirmed the trend observed with the HAT-based in vitro assay, and confirmed that the AOR extract from the genotype #3 produced at Ain Sfa showed the highest antioxidant capacity, particularly as compared to extracts obtained from the genotype #1 produced at the same location.

As shown in Figure 5, higher antioxidant capacity measured with HAT-based antioxidant assay appeared systematically associated with a higher accumulation of phenolics, whereas association with the ET-based antioxidant assay (i.e., CUPRAC) appeared more complex and not directly linked to the accumulation of theses phenolics (Figure 5a).

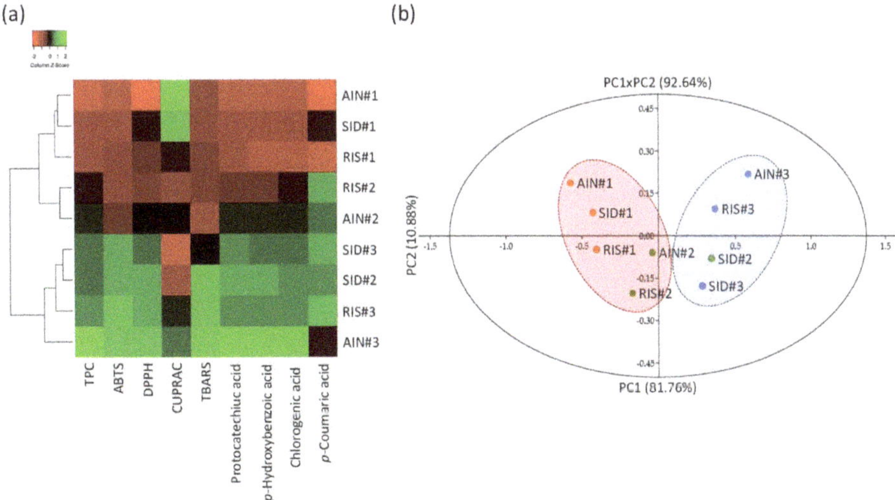

Figure 5. (**a**) hierarchical clustering analysis (HCA) showing the relation between the phytochemical composition and antioxidant activity of each extracts from AOR of Eastern Morocco obtained by USAE; (**b**) principal component analysis (PCA) showing the discrimination of the different extracts from AOR of Eastern Morocco obtained by USAE.

Principal component analysis was performed to further discriminate these nine samples (Figure 5b). The resulting biplot representation accounts for 92.64% (F1 + F2) of the initial variability of the data as shown in Figure 4b. The discrimination occurs mainly in the first dimension (PC1) which explains 85.76% of the initial variability. The loading plots (F) confirmed the strong link between phytochemical composition, in particular the presence of the phenolics, and the HAT-based as well as cellular antioxidant capacity.

In order to link the antioxidant capacity to the presence of a particular phytochemical, a Pearson correlation analysis was applied (Figure 6).

This analysis provided evidence of the strong and highly significant correlation between both HAT-based in vitro assays as well as cellular antioxidant assay and TPC of the extract, in particular with the presence of chlorogenic acid, protocatechuic acid, and *p*-hydroxybenzoic acid (Figure 6, Table S1). The presence of *p*-coumaric acid was significantly correlated with a DPPH assay only. On the contrary, none of these phytochemicals, here analyzed, were significantly correlated with the in vitro ET-based antioxidant CUPRAC assay.

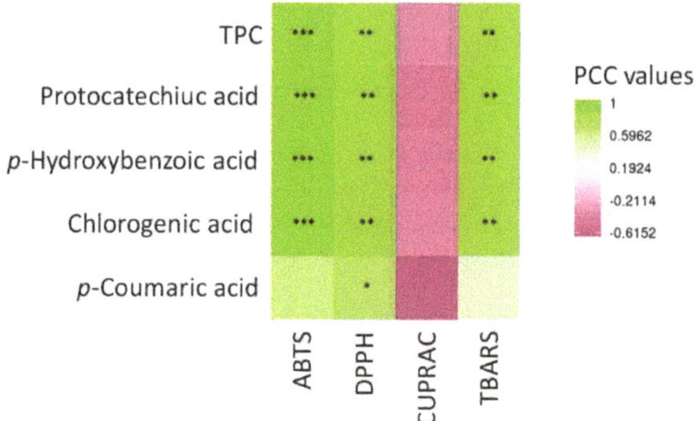

Figure 6. Pearson correlation analysis (PCC) of the relation between the main phytochemicals (protocatechuic acid, *p*-hydroxybenzoic acid, chlorogenic acid and *p*-coumaric acid) from AOR extracts obtained after USAE and the different antioxidant assays (ABTS, DPPH, CUPRAC and TBARS). *** significant $p < 0.001$; ** significant $p < 0.01$; * significant $p < 0.05$; actual PCC values are indicated in Supplementary Materials Table S1.

Altogether, these results showed a higher antioxidant activity, expressed in μM TEAC/g DW, determined with the ABTS and DPPH assays as compared to the CUPRAC assay. Therefore, these results suggested the prominence of the HAT- over the ET-based mechanism for the antioxidant action of these extracts. In good agreement, several authors have reported an antioxidant activity of extracts from various almond products based on HAT mechanism [18,42,44,46]. Similarly, a higher relation between HAT assay and phenolic acids as compared to flavonoids have been previously reported [52,57]. This observation is also in line with the results of Liang and Kitts [58] that reported a relatively stronger scavenging capacity of radicals generated by the ABTS and DPPH assays for chlorogenic acid, the main phenolic acid of our AOR extracts, and its derivatives. The authors attributed this observation to the available hydroxyl groups of these compounds. The presence of flavonoids has been also reported in almond products [13,18,42,44,46]. Here, we cannot exclude the presence of flavonoids potentially linked to the ET-based antioxidant activity evidenced by the CUPRAC assay. Prgomet et al. [13]) have reported in the presence of flavonoids in almond skin (i.e., isorhamnetin derivatives). Future works will be conducted to study in detail the flavonoid fraction of our AOR extracts. The cellular antioxidant assay using yeast further confirmed the interest of this system to study natural antioxidant from plant extracts [34,52] as also previously reported for other natural antioxidants such as thiamine and/or melatonin [55,56]. Natural antioxidants have aroused increasing interest over the past decade due to their possible use as an alternative to potentially dangerous synthetic antioxidants such as butylated hydroxyanisole (BHA) or butylated hydroxytoluene (BHT) in various food or cosmetic formulations [6–8]. Some natural antioxidant phenolics have already been shown to be as efficient in stabilizing nonpolar systems such as bulk oil or various types of emulsions as these synthetic antioxidants [7,9–11]. These preliminary results indicate a potential use as natural antioxidants of our AOR extracts generated by the present validated USAE.

4. Conclusions

P. dulcis or the so-called almond is a rich source of antioxidant phenolic compounds that are retained, after almond cold-pressed oil extraction, in a skin-enriched by-product which, thus, represent an attractive starting material for their extraction. As natural antioxidants, these phenolic compounds' almond attracted much attention as alternatives to synthetic antioxidants in foods, pharmaceutical, and cosmetic preparations. Here, using a multivariate Box–Behnken design coupled with surface response methodology, we proposed an optimized and validated USAE of these phenolic compounds from cold-pressed AOR. Optimal conditions for USAE were: aqEtOH 53.0% (*v/v*) as green solvent, US frequency 27.0 kHz and extraction duration 29.4 min. Following its optimization, the present USAE method was validated according to international standards to ensure its precision and accuracy in the quantitation of total phenolic content. The efficiency of the present USAE has allowed substantial gains in terms of extraction efficiency compared to conventional heat reflux extraction—in particular by a strong reduction in extraction time, which is of particular interest in the context of green chemistry in terms of reduction of energy consumption, together with the use of a green extraction solvent. The application of this method already makes it possible to suggest a higher impact of the genetic background than of the environment on three genotypes cultivated on three experimental sites. This method therefore opens the door to more complete studies on this subject. Finally, both in vitro cell free and cellular antioxidant assays revealed the great potential of valorization of these extracts as a source of natural antioxidants. To summarize, the present extraction method allows a quick, green, simple, and efficient validated USAE for the possible valorization of antioxidant phenolic compounds from Moroccan almond cold-pressed oil residues, making it possible to generate extracts with attractive antioxidant activities for future nutraceutical and/or cosmetic applications.

Supplementary Materials: The following are available online at http://www.mdpi.com/2076-3417/10/9/3313/s1, Figure S1: Biplot representation of the linear relation between predicted vs. measured TPC in the 18 Box–Behnken design sample extracts; Figure S2: Loading scores of the first and second axis of the principal component analysis; Table S1: Actual values for PCC (Pearson Correlation Coefficient) showing the relation between the different phytochemicals and antioxidant assays.

Author Contributions: Conceptualization, D.T., A.E., M.A. (Mohamed Addi), and C.H.; Methodology, D.T., A.E., M.A. (Malika Abid), S.D., R.K.; L.G., A.K., M.A. (Mohamed Addi), and C.H.; Software, S.D.; Validation, M.A. (Mohamed Addi), C.H., D.T.; Formal analysis, M.A. (Mohamed Addi), C.H., S.D., and D.T.; Investigation, S.D., D.T., and R.K.; Resources, M.A. (Mohamed Addi) and C.H.; Data curation, M.A. (Mohamed Addi) and C.H.; Writing—original draft preparation, C.H.; Writing—review and editing, M.A. (Mohamed Addi), D.T., A.K., and C.H.; Visualization, C.H. and S.D.; Supervision, M.A. (Mohamed Addi) and C.H.; Project administration, M.A. (Mohamed Addi) and C.H.; Funding acquisition, M.A. (Mohamed Addi) and C.H. All authors have read and agreed to the published version of the manuscript.

Funding: This research was supported by Cosmetosciences, a global training and research program dedicated to the cosmetic industry. Located in the heart of the Cosmetic Valley, this program led by University of Orléans is funded by the Région Centre-Val de Loire.

Acknowledgments: D.T. gratefully acknowledges the support of French government via the French Embassy in Thailand in the form of Junior Research Fellowship Program 2018. S.D. and L.G. acknowledge the research fellowship of the Loire Valley Region. K.R. acknowledges the research fellowship from the Ministry of Science, Research and Technology of Iran.

Conflicts of Interest: The authors declare no conflict of interest.

References

1. Oomah, D.B. Flaxseed as a functional food. *J. Sci. Food Agric.* **2001**, *81*, 889–894. [CrossRef]
2. Nayak, B.; Liu, R.H.; Tang, J. Effect of Processing on Phenolic Antioxidants of Fruits, Vegetables, and Grains—A Review. *Crit. Rev. Food Sci. Nutr.* **2015**, *55*, 887–919. [CrossRef] [PubMed]
3. Bolling, B.W. Almond polyphenols: Methods of analysis, contribution to food quality, and health promotion. *Compr. Rev. Food Sci. Food Saf.* **2017**, *16*, 346–368. [CrossRef]
4. Barreca, D.; Nabavi, S.M.; Sureda, A.; Rasekhian, M.; Raciti, R.; Silva, A.S.; Annunziata, G.; Arnone, A.; Tenore, G.C.; Süntar, İ. Almonds (Prunus Dulcis Mill. DA Webb): A Source of Nutrients and Health-Promoting Compounds. *Nutrients* **2020**, *12*, 672. [CrossRef]
5. Prgomet, I.; Gonçalves, B.; Domínguez-Perles, R.; Pascual-Seva, N.; Barros, A.I. Valorization challenges to almond residues: Phytochemical composition and functional application. *Molecules* **2017**, *22*, 1774. [CrossRef]
6. Williams, G.M.; Iatropoulos, M.J.; Whysner, J. Safety Assessment of Butylated Hydroxyanisole and Butylated Hydroxytoluene as Antioxidant Food Additives. *Food Chem. Toxicol.* **1999**, *37*, 1027–1038. [CrossRef]
7. Hano, C.; Corbin, C.; Drouet, S.; Quéro, A.; Rombaut, N.; Savoire, R.; Molinié, R.; Thomasset, B.; Mesnard, F.; Lainé, E. The lignan (+)-secoisolariciresinol extracted from flax hulls is an effective protectant of linseed oil and its emulsion against oxidative damage. *Eur. J. Lipid Sci. Technol.* **2017**, *119*, 1600219. [CrossRef]

8. Drouet, S.; Doussot, J.; Garros, L.; Mathiron, D.; Bassard, S.; Favre-Réguillon, A.; Molinié, R.; Lainé, É.; Hano, C. Selective Synthesis of 3-O-Palmitoyl-Silybin, a New-to-Nature Flavonolignan with Increased Protective Action against Oxidative Damages in Lipophilic Media. *Molecules* **2018**, *23*, 2594. [CrossRef]

9. Mariod, A.A.; Ibrahim, R.M.; Ismail, M.; Ismail, N. Antioxidant activity and phenolic content of phenolic rich fractions obtained from black cumin (Nigella sativa) seedcake. *Food Chem.* **2009**, *116*, 306–312. [CrossRef]

10. Aditya, N.P.; Hamilton, I.E.; Norton, I.T. Amorphous nano-curcumin stabilized oil in water emulsion: Physico chemical characterization. *Food Chem.* **2017**, *224*, 191–200. [CrossRef]

11. Drouet, S.; Leclerc, E.A.; Garros, L.; Tungmunnithum, D.; Kabra, A.; Abbasi, B.H.; Lainé, É.; Hano, C. A Green Ultrasound-Assisted Extraction Optimization of the Natural Antioxidant and Anti-Aging Flavonolignans from Milk Thistle Silybum marianum (L.) Gaertn. Fruits for Cosmetic Applications. *Antioxidants* **2019**, *8*, 304. [CrossRef] [PubMed]

12. Porter, W.L.; Black, E.D.; Drolet, A.M. Use of polyamide oxidative fluorescence test on lipid emulsions: Contrast in relative effectiveness of antioxidants in bulk versus dispersed systems. *J. Agric. Food Chem.* **1989**, *37*, 615–624. [CrossRef]

13. Prgomet, I.; Gonçalves, B.; Domínguez-Perles, R.; Pascual-Seva, N.; Barros, A.I. A Box-Behnken Design for Optimal Extraction of Phenolics from Almond By-products. *Food Anal. Methods* **2019**, *12*, 2009–2024. [CrossRef]

14. Delplancke, M.; Aumeeruddy-Thomas, Y. Des semis et des clones. *Rev. D'ethnoécologie* **2017**, *27*. [CrossRef]

15. Melhaoui, R.; Fauconnier, M.-L.; Sindic, M.; Addi, M.; Abid, M.; Mihamou, A.; Serghini-Caid, H.; Elamrani, A. Tocopherol Content of Almond Oils. *Commun. Appl. Biol. Sci.* **2018**, *83*, 75–77.

16. Sarkis, J.R.; Michel, I.; Tessaro, I.C.; Marczak, L.D.F. Optimization of phenolics extraction from sesame seed cake. *Sep. Purif. Technol.* **2014**, *122*, 506–514. [CrossRef]

17. Wang, F.; Hu, J.-H.; Guo, C.; Liu, C.-Z. Enhanced laccase production by Trametes versicolor using corn steep liquor as both nitrogen source and inducer. *Bioresour. Technol.* **2014**, *166*, 602–605. [CrossRef]

18. Kahlaoui, M.; Borotto Dalla Vecchia, S.; Giovine, F.; Ben Haj Kbaier, H.; Bouzouita, N.; Barbosa Pereira, L.; Zeppa, G. Characterization of Polyphenolic Compounds Extracted from Different Varieties of Almond Hulls (Prunus dulcis L.). *Antioxidants* **2019**, *8*, 647. [CrossRef]

19. Melhaoui, R.; Fauconnier, M.-L.; Sindic, M.; Addi, M.; Abid, M.; Mihamou, A.; Serghini-Caid, H.; Elamrani, A. Tocopherol content of almond oils produced in eastern Morocco. In Proceedings of the 23rd National Symposium for Applied Biological Sciences (NSABS), Brussels, Belgium, 8 February 2018; pp. 75–77.

20. Teh, S.-S.; Bekhit, A.E.-D.; Birch, J. Antioxidative polyphenols from defatted oilseed cakes: Effect of solvents. *Antioxidants* **2014**, *3*, 67–80. [CrossRef]

21. Corbin, C.; Fidel, T.; Leclerc, E.A.; Barakzoy, E.; Sagot, N.; Falguiéres, A.; Renouard, S.; Blondeau, J.; Ferroud, C.; Doussot, J.; et al. Development and validation of an efficient ultrasound assisted extraction of phenolic compounds from flax (Linum usitatissimum L.) seeds. *Ultrason. Sonochem.* **2015**, *26*, 176–185. [CrossRef]

22. Fliniaux, O.; Corbin, C.; Ramsay, A.; Renouard, S.; Beejmohun, V.; Doussot, J.; Falguières, A.; Ferroud, C.; Lamblin, F.; Lainé, E.; et al. Microwave-Assisted Extraction of Herbacetin Diglucoside from Flax (Linum usitatissimum L.) Seed Cakes and Its Quantification using an RP-HPLC-UV System. *Molecules* **2014**, 3025–3037. [CrossRef] [PubMed]

23. Bourgeois, C.; Leclerc, É.A.; Corbin, C.; Doussot, J.; Serrano, V.; Vanier, J.R.; Seigneuret, J.M.; Auguin, D.; Pichon, C.; Lainé, É.; et al. Nettle (Urtica dioica L.) as a source of antioxidant and anti-aging phytochemicals for cosmetic applications. *C. R. Chim.* **2016**, *19*, 1090–1100. [CrossRef]

24. Renouard, S.; Hano, C.; Corbin, C.; Fliniaux, O.; Lopez, T.; Montguillon, J.; Barakzoy, E.; Mesnard, F.; Lamblin, F.; Lainé, E. Cellulase-assisted release of secoisolariciresinol from extracts of flax (Linum usitatissimum) hulls and whole seeds. *Food Chem.* **2010**, *122*, 679–687. [CrossRef]

25. Terpinc, P.; Polak, T.; Makuc, D.; Ulrih, N.P.; Abramovič, H. The occurrence and characterisation of phenolic compounds in Camelina sativa seed, cake and oil. *Food Chem.* **2012**, *131*, 580–589. [CrossRef]

26. Wildermuth, S.R.; Young, E.E.; Were, L.M. Chlorogenic acid oxidation and its reaction with sunflower proteins to form green-colored complexes. *Compr. Rev. Food Sci. Food Saf.* **2016**, *15*, 829–843. [CrossRef]

27. Ahmad, B.S.; Talou, T.; Saad, Z.; Hijazi, A.; Cerny, M.; Kanaan, H.; Chokr, A.; Merah, O. Fennel oil and by-products seed characterization and their potential applications. *Ind. Crops Prod.* **2018**, *111*, 92–98. [CrossRef]

28. Lavilla, I.; Bendicho, C. Fundamentals of Ultrasound-Assisted Extraction. *Water Extr. Bioact. Compd. From Plants Drug Dev.* **2017**, 291–316. [CrossRef]

29. Medina-Torres, N.; Ayora-Talavera, T.; Espinosa-Andrews, H.; Sánchez-Contreras, A.; Pacheco, N. Ultrasound Assisted Extraction for the Recovery of Phenolic Compounds from Vegetable Sources. *Agronomy* **2017**, *7*, 47. [CrossRef]

30. Tungmunnithum, D.; Garros, L.; Drouet, S.; Renouard, S.; Lainé, E.; Hano, C. Green Ultrasound Assisted Extraction of trans Rosmarinic Acid from Plectranthus scutellarioides (L.) R.Br. Leaves. *Plants* **2019**, *8*, 50. [CrossRef]

31. Abbasi, B.H.; Siddiquah, A.; Tungmunnithum, D.; Bose, S.; Younas, M.; Garros, L.; Drouet, S.; Giglioli-Guivarc'h, N.; Hano, C. Isodon rugosus (Wall. ex Benth.) codd in vitro cultures: Establishment, phytochemical characterization and in vitro antioxidant and anti-aging activities. *Int. J. Mol. Sci.* **2019**, *20*, 452. [CrossRef]

32. Shah, M.; Ullah, M.A.; Drouet, S.; Younas, M.; Tungmunnithum, D.; Giglioli-Guivarc'h, N.; Hano, C.; Abbasi, B.H. Interactive effects of light and melatonin on biosynthesis of silymarin and anti-inflammatory potential in callus cultures of silybum marianum (L.) gaertn. *Molecules* **2019**, *24*, 1207. [CrossRef] [PubMed]

33. Ullah, M.A.; Tungmunnithum, D.; Garros, L.; Drouet, S.; Hano, C.; Abbasi, B.H. Effect of Ultraviolet-C Radiation and Melatonin Stress on Biosynthesis of Antioxidant and Antidiabetic Metabolites Produced in In Vitro Callus Cultures of Lepidium sativum L. *Int. J. Mol. Sci.* **2019**, *20*, 1787. [CrossRef] [PubMed]

34. Garros, L.; Drouet, S.; Corbin, C.; Decourtil, C.; Fidel, T.; De Lacour, J.L.; Leclerc, E.A.; Renouard, S.; Tungmunnithum, D.; Doussot, J.; et al. Insight into the influence of cultivar type, cultivation year, and site on the lignans and related phenolic profiles, and the health-promoting antioxidant potential of flax (linum usitatissimum L.) seeds. *Molecules* **2018**, *23*, 2636. [CrossRef] [PubMed]

35. Babicki, S.; Arndt, D.; Marcu, A.; Liang, Y.; Grant, J.R.; Maciejewski, A.; Wishart, D.S. Heatmapper: Web-enabled heat mapping for all. *Nucleic Acids Res.* **2016**, *44*, W147–W153. [CrossRef]

36. Ferreira, S.L.C.; Silva Junior, M.M.; Felix, C.S.A.; da Silva, D.L.F.; Santos, A.S.; Santos Neto, J.H.; de Souza, C.T.; Cruz Junior, R.A.; Souza, A.S. Multivariate optimization techniques in food analysis—A review. *Food Chem.* **2019**, *273*, 3–8. [CrossRef]

37. Ferreira, S.L.C.; Bruns, R.E.; Ferreira, H.S.; Matos, G.D.; David, J.M.; Brandão, G.C.; da Silva, E.G.P.; Portugal, L.A.; dos Reis, P.S.; Souza, A.S.; et al. Box-Behnken design: An alternative for the optimization of analytical methods. *Anal. Chim. Acta* **2007**, *597*, 179–186. [CrossRef]

38. Chemat, F.; Abert-Vian, M.; Fabiano-Tixier, A.S.; Strube, J.; Uhlenbrock, L.; Gunjevic, V.; Cravotto, G. Green extraction of natural products. Origins, current status, and future challenges. *Trends Anal. Chem.* **2019**, *118*, 248–263. [CrossRef]

39. Renouard, S.; Lopez, T.; Hendrawati, O.; Dupre, P.; Doussot, J.; Falguieres, A.; Ferroud, C.; Hagege, D.; Lamblin, F.; Laine, E.; et al. Podophyllotoxin and deoxypodophyllotoxin in juniperus bermudiana and 12 other juniperus species: Optimization of extraction, method validation, and quantification. *J. Agric. Food Chem.* **2011**, *59*. [CrossRef]

40. Chemat, F.; Vian, M.A.; Cravotto, G. Green Extraction of Natural Products: Concept and Principles. *Int. J. Mol. Sci.* **2012**, *13*, 8615–8627. [CrossRef]

41. Ameer, K.; Shahbaz, H.M.; Kwon, J.H. Green Extraction Methods for Polyphenols from Plant Matrices and Their Byproducts: A Review. *Compr. Rev. Food Sci. Food Saf.* **2017**, *16*, 295–315. [CrossRef]

42. Bolling, B.W.; Blumberg, J.B.; Oliver Chen, C.-Y. The influence of roasting, pasteurisation, and storage on the polyphenol content and antioxidant capacity of California almond skins. *Food Chem.* **2010**, *123*, 1040–1047. [CrossRef] [PubMed]

43. Beejmohun, V.; Fliniaux, O.; Grand, É.; Lamblin, F.; Bensaddek, L.; Christen, P.; Kovensky, J.; Fliniaux, M.-A.; Mesnard, F. Microwave-assisted extraction of the main phenolic compounds in flaxseed. *Phytochem. Anal.* **2007**, *18*, 275–282. [CrossRef] [PubMed]

44. Rubilar, M.; Pinelo, M.; Shene, C.; Sineiro, J.; Nuñez, M.J. Separation and HPLC-MS Identification of Phenolic Antioxidants from Agricultural Residues: Almond Hulls and Grape Pomace. *J. Agric. Food Chem.* **2007**, *55*, 10101–10109. [CrossRef] [PubMed]

45. Anastas, P.T.; Warner, J.C. *Green Chemistry: Theory and Practice*; Oxford University Press: Oxford, UK, 1998; ISBN 0198502346.

46. Monagas, M.; Garrido, I.; Lebrón-Aguilar, R.; Bartolome, B.; Gómez-Cordovés, C. Almond (Prunus dulcis (Mill.) D.A. Webb) Skins as a Potential Source of Bioactive Polyphenols. *J. Agric. Food Chem.* **2007**, *55*, 8498–8507. [CrossRef] [PubMed]
47. Freeman, B.L.; Eggett, D.L.; Parker, T.L. Synergistic and antagonistic interactions of phenolic compounds found in navel oranges. *J. Food Sci.* **2010**, *75*, C570–C576. [CrossRef]
48. Pathak, N.; Rai, A.K.; Kumari, R.; Bhat, K. V Value addition in sesame: A perspective on bioactive components for enhancing utility and profitability. *Pharmacogn. Rev.* **2014**, *8*, 147.
49. Drouet, S.; Abbasi, B.H.; Falguières, A.; Ahmad, W.; Sumaira; Ferroud, C.; Doussot, J.; Vanier, J.R.; Lainé, E.; Hano, C. Single laboratory validation of a quantitative core shell-based LC separation for the evaluation of silymarin variability and associated antioxidant activity of pakistani ecotypes of milk thistle (silybum marianum L.). *Molecules* **2018**, *23*, 904. [CrossRef]
50. Dave Oomah, B.; Mazza, G.; Kenaschuk, E.O. Flavonoid content of flaxseed. Influence of cultivar and environment. *Euphytica* **1996**, *90*, 163–167. [CrossRef]
51. Huang, D.; Zhou, X.; Si, J.; Gong, X.; Wang, S. Studies on cellulase-ultrasonic assisted extraction technology for flavonoids from Illicium verum residues. *Chem. Cent. J.* **2016**, *10*, 56. [CrossRef]
52. Nazir, M.; Tungmunnithum, D.; Bose, S.; Drouet, S.; Garros, L.; Giglioli-Guivarc'h, N.; Abbasi, B.H.; Hano, C. Differential Production of Phenylpropanoid Metabolites in Callus Cultures of Ocimum basilicum L. With Distinct in Vitro Antioxidant Activities and in Vivo Protective Effects against UV stress. *J. Agric. Food Chem.* **2019**, *67*, 1847–1859. [CrossRef]
53. Prior, R.L.; Wu, X.; Schaich, K. Standardized Methods for the Determination of Antioxidant Capacity and Phenolics in Foods and Dietary Supplements. *J. Agric. Food Chem.* **2005**, *53*, 4290–4302. [CrossRef] [PubMed]
54. Steels, E.L.; Learmonth, R.P.; Watson, K. Stress tolerance and membrane lipid unsaturation in *Saccharomyces cerevisiae* grown aerobically or anaerobically. *Microbiology* **1994**, *140 Pt 3*, 569–576. [CrossRef]
55. Wolak, N.; Kowalska, E.; Kozik, A.; Rapala-Kozik, M. Thiamine increases the resistance of baker's yeast Saccharomyces cerevisiae against oxidative, osmotic and thermal stress, through mechanisms partly independent of thiamine diphosphate-bound enzymes. *FEMS Yeast Res.* **2014**, *14*, 1249–1262. [CrossRef] [PubMed]
56. Bisquert, R.; Muñiz-Calvo, S.; Guillamón, J.M. Protective role of intracellular Melatonin against oxidative stress and UV radiation in Saccharomyces cerevisiae. *Front. Microbiol.* **2018**, *9*, 318. [CrossRef] [PubMed]
57. Prior, R.L.; Cao, G.; Martin, A.; Sofic, E.; McEwen, J.; O'Brien, C.; Lischner, N.; Ehlenfeldt, M.; Kalt, W.; Krewer, G.; et al. Antioxidant Capacity As Influenced by Total Phenolic and Anthocyanin Content, Maturity, and Variety of Vaccinium Species. *J. Agric. Food Chem.* **1998**, *46*, 2686–2693. [CrossRef]
58. Liang, N.; Kitts, D. Antioxidant Property of Coffee Components: Assessment of Methods that Define Mechanisms of Action. *Molecules* **2014**, *19*, 19180–19208. [CrossRef]

 *applied sciences*

Article

Analysis and Identification of Active Compounds from *Salviae miltiorrhizae* Radix Toxic to HCT-116 Human Colon Cancer Cells

Bohyung Kang [1], Sullim Lee [2], Chang-Seob Seo [3], Ki Sung Kang [4,*] and You-Kyung Choi [1,*]

[1] Department of Korean International Medicine, College of Korean Medicine, Gachon University, Seongnam 13120, Korea; hihi619@naver.com

[2] Department of Life Science, College of Bio-Nano Technology, Gachon University, Seongnam 13120, Korea; sullimlee@gachon.ac.kr

[3] Herbal Medicine Research Division, Korea Institute of Oriental Medicine, Daejeon 34054, Korea; csseo0914@kiom.re.kr

[4] College of Korean Medicine, Gachon University, Seongnam 13120, Korea

* Correspondence: kkang@gachon.ac.kr (K.S.K.); kosmos@gachon.ac.kr (Y.-K.C.); Tel.: +82-31-750-5402 (K.S.K.); +82-32-770-1300 (Y.-K.C.); Fax: +82-31-750-5416 (K.S.K.); +82-32-468-4021 (Y.-K.C.)

Received: 23 January 2020; Accepted: 11 February 2020; Published: 14 February 2020

Abstract: Colorectal cancer is one of the most frequently diagnosed cancers worldwide. The aim of the present study was to simultaneously analyze compounds of *Salviae miltiorrhizae* Radix (SMR) and determine their cytotoxic effects on HCT-116 human colorectal cancer cells. We established a simultaneous analysis method of five compounds (salvianic acid A, salvianolic acid B, caffeic acid, tanshinone IIA, and rosmarinic acid) contained in SMR, and found that among the various compounds in SMR, tanshinone IIA significantly decreased cell viability in a concentration-dependent manner. Hoechst staining also showed that both SMR and tanshinone IIA increased nuclear condensation, suggesting induction of apoptosis. By Western blotting, we found that tanshinone IIA induced apoptotic cell death, significantly increased Bax, but decreased Bcl-2 in the course of apoptosis. Tanshinone IIA increased the expression of cleaved caspases-7 and -8. Tanshinone IIA was shown to be an active ingredient of SMR that may be a useful chemotherapeutic strategy for patients with colorectal cancer.

Keywords: colorectal cancer; *Salviae miltiorrhizae* radix; apoptosis

---

1. Introduction

Colorectal cancer is the third-most commonly diagnosed cancer worldwide [1]. Although surgery plays a key role in the diagnosis and treatment of colorectal cancer, there are still increasing attempts to stop the progression of this cancer via the application of new synthetic and naturally-occuring compounds [2,3]. Bioactive compounds from plants have been screened for anticancer activities [4,5]. Approximately 50–60% of cancer patients in the United States utilize complementary and alternative medicines with traditional therapeutic regimens, such as radiation therapy and chemotherapy [6].

Apoptosis pathways are important targets in cancer-related therapies, and insufficient apoptosis results in uncontrolled cancer cell proliferation [7]. The use of natural phytochemicals for inhibiting cancer cell proliferation and inducing apoptosis contributes to promoting cancer cell death [8,9]. Natural phytochemicals are multiple-target molecules found in plants and microorganisms, and they exert strong anticancer activity [10,11]. Phytochemicals isolated from natural sources also exhibit various beneficial effects against inflammation, cancer, and neurodegenerative disorders [10]. This broad spectrum of biological and pharmacological activities has made natural compounds suitable candidates for treating multifactorial diseases, such as colorectal cancer.

Salviae miltiorrhizae Radix (SMR) is one of the well-known traditional herbal medicines and has been used in Asian countries [8]. Recently, there has been increasing scientific attention towards SMR for its remarkable bioactivity against cardiovascular disease, renal damage, tumor angiogenesis, and tumor cell invasion [12–14]. In the last decade, accumulating evidence has shown that SMR exerts a significant anticancer effect against promyelocytic leukemia, breast cancer, ovarian carcinomas, and hepatocellular carcinoma (HCC) [15–17]. In a recent network pharmacology-based study on the anti-HCC effect of SMR, 62 chemical compounds form SMR yielded 101 putative targets that played a critical role in HCC via multiple targets and pathways, especially the EGFR and phosphatidyl-inositol 3-kinase (PI3K)/Akt signaling pathways [18]. However, the effect of SMR and its compounds on human colon cancer cells has not been fully elucidated. The aim of the present study was to simultaneously analyze the compounds of SMR and determine their cytotoxic effects on HCT-116 human colorectal cancer cells.

2. Materials and Methods

2.1. Plant Materials

Salviae miltiorrhizae Radix (SMR) was obtained from Kwangmyungdag Medicinal Herbs (Ulsan, Korea) and identified by Dr. Goya Choi, Herbal Medicine Resources Research Center, Korea Institute of Oriental Medicine (HMRRC, KIOM; Naju, Korea). A voucher specimen (SMR-2-14-0073) was stored at the herbarium of the HMRRC, KIOM.

2.2. Chemicals and Reagents

Five reference standard compounds, salvianic acid A (98.0%), caffeic acid (99.0%), rosmarinic acid (97.0%), salvianolic acid B (98.0%), and tanshinone IIA (98.8%) were purchased from standard manufacturers: Acros Organics (Pittsburgh, PA, USA), Merck KGaA (Darmstadt, Germany), and ChemFaces Biochemical Co., Ltd. (Wuhan, China).

The solvents including methanol, acetonitrile, and water (HPLC-grade) and formic acid (≥98.0%, ACS reagent-grade) for quantitative analysis were obtained from Merck KGaA (Darmstadt, Germany) and J. T. Baker (Phillipsburg, NJ, USA), respectively.

2.3. Preparation of 70% Ethanol SMR Extract

Dried SMR (0.3 kg) was extracted with 70% ethanol (3.0 L, 3 times) for 1 h at room temperature by a Branson 8510 ultrasonicator (Denbury, CT, USA). The extract solution was filtered with 150 mm Ø filter paper (Whatman, Maidstone, Kent, UK) under vacuum, concentrated to remove the organic extract solvent (ethanol) using a Büchi rotary evaporator R-210 (Flawil, Switzerland), and then lyophilized using a Ilshin BioBase FD-5525L freeze-drier (Dongducheon, Korea) to obtain powdered extract. The yield of lyophilized 70% ethanol extract of SMR was 69.8 g (23.3%).

2.4. HPLC Analysis of Five Components in SMR

HPLC analysis was conducted using the Prominence LC-20A Series instruments (Shimadzu, Kyoto, Japan) consisting of a DGU-20A$_3$ degasser, LC-20AT solvent delivery unit, SIL-20A auto sample injector, CTO-20A column oven, and SPD-M20A photodiode array detector. All chromatographic data were obtained and analyzed with the LabSolution software (Version 5.53; SP3, Kyoto, Japan). Five components were separated using a reverse-phase SunFire™ C_{18} analytical column (4.6 × 250 mm, 5 µm; Waters, Torrance, CA, USA) at 40 °C with gradient solvent condition. The mobile phases consisted of 0.1% (v/v) aqueous formic acid (A) and 0.1% (v/v) formic acid in acetonitrile (B) and were adjusted following the gradient condition: 0–30 min, 10–60% B; 30–40 min, 60–100% B; 40–45 min, 100% B; 45–50 min, and 100–10% B. The re-equilibrium time was adjusted for 10 min. The flow rate of the mobile phase was 1.0 mL/min, and the injection volume of the standard and test solution was 10 μL each. For quantitative determination of five marker components (salvianic acid A, caffeic acid, rosmarinic acid, salvianolic acid B, and tanshinone IIA) in SMR, 200.0 mg of lyophilized SMR extract was liquefied with 20 mL of 70% methanol and sonicated for 30 min. It was also diluted 20-fold for quantification of salvianolic acid B. All samples were filtered using a membrane filter (0.2-μm, Pall Life Sciences, Ann Arbor, MI, USA) before analysis.

2.5. Cell Culture

The human colon cancer cell (HCT-116) was purchased from the ATCC (American Type Culture Collection, Manassas, VA, USA). The cell was maintained and grown in RPMI 1640 medium (Roswell Park Memorial Institute 1640; Corning, Manassas, VA, USA) contained with 10% FBS (fetal bovine serum; Gibco BRL, Carlsbad, MD, USA) and penicillin/streptomycin (Life Technologies, Waltham, MA, USA). The condition of the incubator was 37 °C and humidified atmosphere containing 5% CO_2.

2.6. Cell Viability Assay

The cell viability assay was assessed using an Ez-Cytox Kit (Dail Lab Service Co., Seoul, Korea) based on the manufacturer's instructions [19]. Briefly, the cells were seeded in a 96-well plate at 1×10^4 cells/well and then incubated. After 24 h, the cells were treated with the indicated concentrations of each sample, and the cells were then incubated for 24 h. Following incubation, Ez-Cytox solution was mixed with medium in each well and incubated for 1 h. The absorbance at 450/600 nm was determined using a SPARK 10M (Tecan Group Ltd., Männedorf, Switzerland). The cell viability of 100% was calculated from control cells.

2.7. Hoechst 33342 Cell Staining

Sample-induced nuclear condensation of HCT-116 cells was observed using Hoechst 33342 staining (Sigma Aldrich, St. Louis, MO, USA) [20]. Briefly, the cells were seeded in a 6-well plate at 4×10^5 cells per well. Following incubation for 24 h, the cells were treated with various concentrations of each sample, and the cells were then incubated for 24 h. Following incubation, Hoechst 33342 solution was added to the cells and incubated for 10 min. The stained cells were observed using a CCD camera conjugated IX50 fluorescent microscope (Olympus, Tokyo, Japan).

2.8. Western Blotting

The apoptosis signaling pathways of HCT-116 cells induced by samples were performed using Western blot analysis [21,22]. Briefly, the cells were seeded in a 6-well plate at 4×10^5 cells/well and then incubated. After 24 h, the cells were treated with the indicated concentrations of each sample, and the cells were then incubated for 24 h. Following incubation, the cells were harvested with a scraper and lysed with radio-immunoprecipitation assay buffer (Elpis Biotech, Daejeon, Korea). The protein concentrations were calculated with the Pierce BCA Protein Assay Kit (Thermo Scientific, Carlsbad, CA, USA). The protein samples were separated by electrophoresis in a SDS-PAGE. Then, the proteins were transferred to PVDF membranes (Merck Millipore, Darmstadt, Germany). The membranes were conducted blockading by 5% skim milk. Then, the membranes were probed with primary antibodies for Bax, B-cell lymphoma 2 (Bcl-2), cleaved caspase-7, cleaved caspase-8, cleaved caspase-9, glyceraldehyde 3-phosphate dehydrogenase (GAPDH) and poly(ADP-ribose) polymerase (PARP) followed by incubating with secondary antibodies for anti-rabbit IgG (Cell Signaling Technology, Inc., Danvers, MA, USA).

2.9. Statistical Analysis

All experiments were performed in triplicate, and the quantitative data were shown as mean ± SD. Statistical analysis using Student's t-test was conducted and considered statistically significant based on *p*-values less than 0.05.

3. Results and Discussion

In the present study, we analyzed five bioactive marker components found in SMR, consisting of four phenolic acids (salvianic acid A, caffeic acid, rosmarinic acid, and salvianolic acid B) and one terpenoid (tanshinone IIA). These compounds were separated with resolution >5.0 within 45 min and retention times of 5.87, 11.11, 17.72, 20.42, and 42.51 min, respectively (Figure 1). The content of rosmarinic acid, caffeic acid, salvianic acid A, salvianolic acid B, and tanshinone IIA in the samples was 3.72, 0.123, 1.27, 64.36, and 4.96 mg/g, respectively.

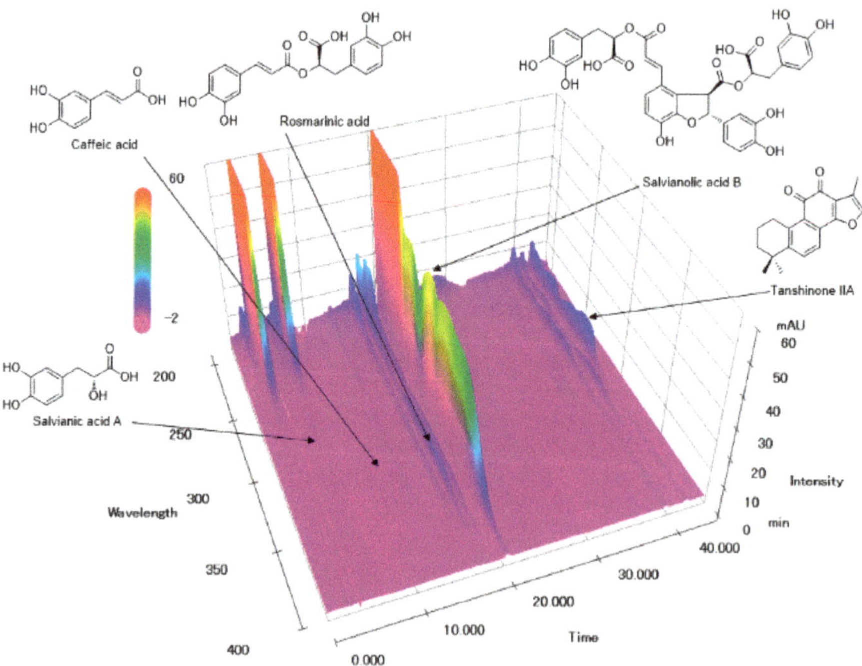

Figure 1. Three-dimensional high-performance liquid chromatogram of 70% ethanol extract of *Salviae miltiorrhizae* Radix.

We initially performed a cytotoxic evaluation using HCT-116 human colorectal carcinoma cells. As shown in Figure 2, only tanshinone IIA significantly decreased cell viability in a concentration-dependent manner, whereas 61.6 µM SMR showed approximately 50% suppression.

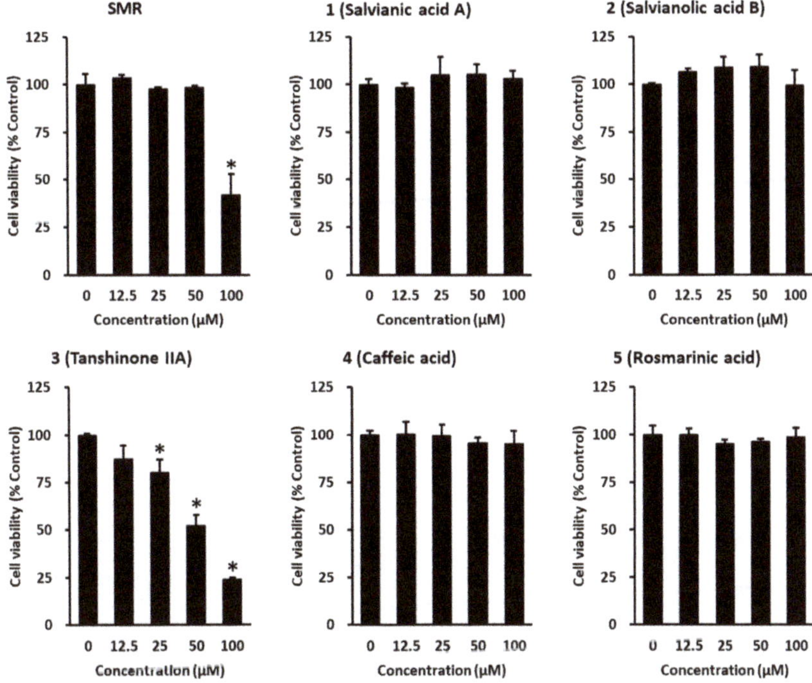

Figure 2. Cytotoxic effect of 70% ethanol extract of *Salviae miltiorrhizae* Radix (**SMR**), salvianic acid A (**1**), caffeic acid (**2**), rosmarinic acid (**3**), salvianolic acid B (**4**), and tanshinone IIA (**5**) on HCT-116 cells.

Many clinical anticancer drugs are known to exert their effects by inducing apoptosis [23]. Apoptosis is a gene-regulated response and, from the morphological point of view, is distinguished by the specific structural changes in cells, such as plasma membrane bleb formation, cell and nuclear shrinkage, oligonucleosomal DNA fragmentation, and chromatin condensation [24]. Morphological analyses showed that both SMR and tanshinone IIA decreased the number of cells and induced signs of cellular apoptosis, such as cellular shrinkage (Figure 3). Moreover, as shown in Figure 4, Hoechst staining also showed that both SMR and tanshinone IIA increased nuclear condensation, suggesting that SMR and tanshinone IIA successfully induced apoptosis, not necrosis, in human colorectal cancer cells. However, tanshinone IIA was not cytotoxic to LLC-PK1 pig kidney epithelial cell, which is normal cell lines, up to 100 μM (Supplementary Figure S1).

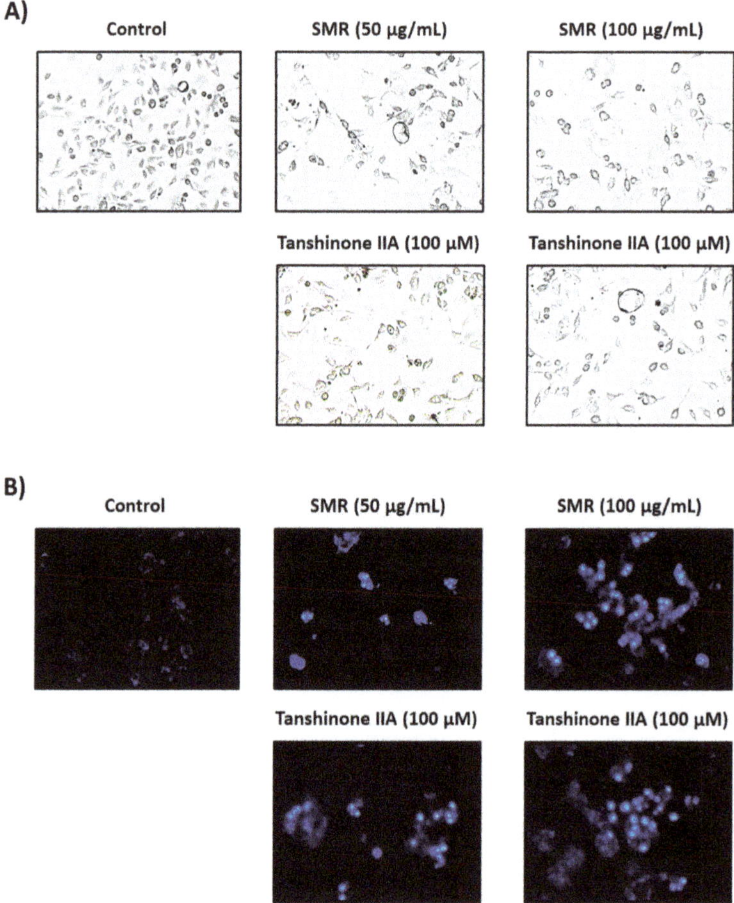

Figure 3. Effects of 70% ethanol extract of *Salviae miltiorrhizae* Radix (SMR) and tanshinone IIA on apoptosis in HCT-116 cells. (**A**) Morphology changes in HCT-116 cells. (**B**) Fluorescence microscopic images of apoptotic HCT-116 cells stained with Hoechst 33342.

Two major molecular pathways that trigger programmed cell death are the caspase-mediated intrinsic pathway, which is induced by cellular stresses, and the extrinsic pathway, which is related to the death receptor [25]. Both pathways activate the apoptotic caspases, resulting in morphological and biochemical cellular alterations related to apoptosis [26]. In addition, the extrinsic pathway controls cell turnover by decreasing mutant cells. In the extrinsic pathway, cancer cell death is triggered by the interaction with death ligands (such as tumor necrosis factor) and its death receptors. The cancer cell death-initiating complex stimulates the activation of caspase-3 and -8, which are effector and starter caspases, respectively [27,28]. The intrinsic pathway, which is typically activated in response to DNA or cellular damage, stimulates the expression of proteins in mitochondria, such as cytochrome c, which then activates caspase-3 and -9. [27,29]. It was also reported that after cleavage by caspase-9, caspase-3 inhibits reactive oxygen species production and is thus required for efficient induction of apoptosis, whereas caspase-7 is required for apoptotic cell elimination [30]. In our present study, the expressions of cleaved caspase-7 and -8 were significantly increased by tanshinone IIA, but there was no change in that of cleaved caspase-9 (Figure 4).

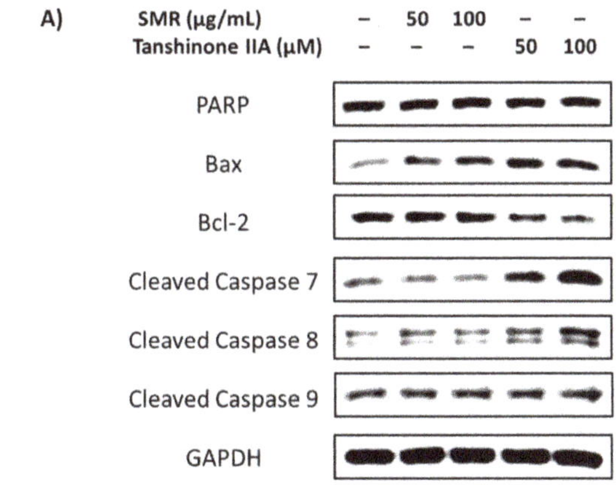

B)

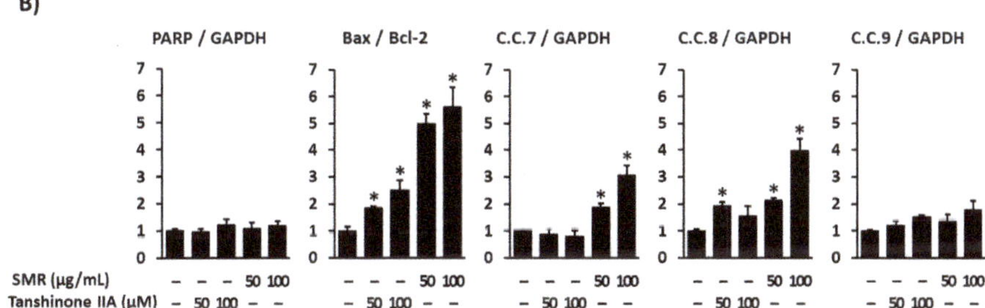

Figure 4. Effects of 70% ethanol extract of *Salviae miltiorrhizae* Radix (SMR) and tanshinone IIA on apoptosis in HCT-116 cells. (**A**) Protein expression of PARP, Bax, Bcl-2, cleaved caspase-7, cleaved caspase-8, cleaved caspase-9, and GAPDH. (**B**) Graph of relative protein expression. Data are the means of experiments performed in triplicate. Data are presented as the mean ± SD. and were analyzed using the Student's *t*-test. * $p < 0.05$ versus non-treated cells.

Furthermore, anti- and pro-apoptotic Bcl-2 members play critical roles in the mitochondria-mediated pathway. That is, the ratio of anti- and pro-apoptotic proteins (e.g., Bax/Bcl-2) is considered as a determinant of survival or apoptosis of cancer cells [31]. Earlier studies have reported that the anti-apoptotic Bcl-2 members, which consist of Bcl-xl, Bcl-2, Bcl-w, and Mcl-1, exert an important role in the resistance of cancer cells to chemotherapy. Therefore, a reduction in Bcl-2 and an increase in Bax stimulate the apoptosis process and eliminate cancer cells [32]. Our western blotting analysis results showed increased Bax expression and decreased Bcl-2 expression in cells co-treated with tanshinone IIA, which was stronger than SMR (Figure 4); however, no difference was observed in poly (ADP-ribose) polymerase (PARP) expression, which is a parameter for stress and DNA damage in cells.

In summary, we simultaneously analyzed five compounds (salvianic acid A, rosmarinic acid, salvianolic acid B, caffeic acid, and tanshinone IIA) from SMR, and determined their cytotoxic effects on HCT-116 human colon cancer cells. Among the five compounds in SMR, only tanshinone IIA significantly decreased cell viability in a concentration-dependent manner. Both SMR and tanshinone IIA increased nuclear condensation, suggesting that SMR and tanshinone IIA successfully induced apoptosis. We also found that tanshinone IIA induced apoptotic cell death and significantly increased cleaved caspases-7, -8, and Bax expression, as well as decreased Bcl-2 expression in the course of apoptosis. Taken together, our data show that tanshinone IIA is an active ingredient of SMR and may be a useful chemotherapeutic strategy for patients with colorectal cancer.

Supplementary Materials: The following are available online at http://www.mdpi.com/2076-3417/10/4/1304/s1.

Author Contributions: Conceptualization, Y.-K.C. and K.S.K.; performing experiments and analyzing data, B.K., S.L., and C.-S.S.; C.-S.S.; validation, B.K. and K.S.K.; writing—original draft preparation, Y.-K.C. and K.S.K.; writing—review and editing, K.S.K.; funding acquisition. All authors have read and agreed to the published version of the manuscript.

Funding: The present study was also supported by the Basic Science Research Program through the National Research Foundation of Korea (NRF) (2019R1F1A1059173).

Conflicts of Interest: The authors declare no conflict of interest.

References

1. Bray, F.; Ferlay, J.; Soerjomataram, I.; Siegel, R.L.; Torre, L.A.; Jemal, A. Global cancer statistics 2018: GLOBOCAN estimates of incidence and mortality worldwide for 36 cancers in 185 countries. *CA Cancer J. Clin.* **2018**, *68*, 394–424. [CrossRef] [PubMed]

2. Cheraghi, O.; Dehghan, G.; Mahdavi, M.; Rahbarghazi, R.; Rezabakhsh, A.; Charoudeh, H.N.; Iranshahi, M.; Montazersaheb, S. Potent anti-angiogenic and cytotoxic effect of conferone on human colorectal adenocarcinoma HT-29 cells. *Phytomedicine* **2016**, *23*, 398–405. [CrossRef] [PubMed]

3. Mignani, S.; Rodrigues, J.; Tomas, H.; Zablocka, M.; Shi, X.; Caminade, A.-M.; Majoral, J.-P. Dendrimers in combination with natural products and analogues as anti-cancer agents. *Chem. Soc. Rev.* **2018**, *47*, 514–532. [CrossRef]

4. Yang, E.-J.; An, J.-H.; Son, Y.K.; Yeo, J.-H.; Song, K.-S. The cytotoxic constituents of Betula platyphylla and their effects on human lung A549 cancer cells. *Nat. Prod. Sci.* **2018**, *24*, 219–224. [CrossRef]

5. Ahuja, A.; Kim, J.H.; Kim, J.H.; Yi, Y.S.; Cho, J.Y. Functional role of ginseng-derived compounds in cancer. *J. Ginseng Res.* **2018**, *42*, 248–254. [CrossRef]

6. Meeran, S.M.; Ahmed, A.; Tollefsbol, T.O. Epigenetic targets of bioactive dietary components for cancer prevention and therapy. *Clin. Epigenetics* **2010**, *1*, 101–116. [CrossRef]

7. Gezici, S.; Şekeroğlu, N. Current perspectives in the application of medicinal plants against cancer: Novel therapeutic agents. *Anticancer Agents Med. Chem.* **2019**, *19*, 101–111. [CrossRef]

8. Wang, Z.J.; Cui, L.J.; Chen, C.X.; Liu, J.; Yan, Y.P.; Wang, Z.Z. Down regulation of cinnamoyl CoA reductase affects lignin and phenolic acids biosynthesis in Salvia miltiorrhiza Bunge plant. *Plant Mol. Biol. Rep.* **2012**, *30*, 1229–1236. [CrossRef]

9. Guon, T.-E.; Chung, H.S. Induction of apoptosis with Moringa oleifera fruits in HCT116 human colon cancer cells via intrinsic pathway. *Nat. Prod. Sci.* **2017**, *23*, 227–234. [CrossRef]

10. Rahman, I.; Chung, S. Dietary polyphenols, deacetylases and chromatin remodeling in inflammation. *J. Nutrigenet Nutrigenomics* **2010**, *3*, 220–230. [CrossRef]

11. Choi, J.H.; Jang, M.; Nah, S.Y.; Oh, S.; Cho, I.H. Multitarget effects of Korean Red Ginseng in animal model of Parkinson's disease: Antiapoptosis, antioxidant, antiinflammation, and maintenance of blood-brain barrier integrity. *J. Ginseng Res.* **2018**, *42*, 379–388. [CrossRef]

12. Zhao, W.; Yuan, Y.; Zhao, H.; Han, Y.; Chen, X. Aqueous extract of Salvia miltiorrhiza Bunge-Radix Puerariae herb pair ameliorates diabetic vascular injury by inhibiting oxidative stress in streptozotocin-induced diabetic rats. *Food Chem. Toxicol.* **2019**, *129*, 97–107. [CrossRef]

13. Li, W.; Jiang, Y.H.; Wang, Y.; Zhao, M.; Hou, G.J.; Hu, H.Z.; Zhou, L. Protective Effects of Combination of Radix Astragali and Radix Salviae Miltiorrhizae on Kidney of Spontaneously Hypertensive Rats and Renal Intrinsic Cells. *Chin. J. Integr. Med.* **2019**. (Epub ahead of print). [CrossRef]

14. Zhang, L.J.; Chen, L.; Lu, Y.; Wu, J.M.; Xu, B.; Sun, Z.G.; Zheng, S.Z.; Wang, A.Y. Danshensu has anti-tumor activity in B16F10 melanoma by inhibiting angiogenesis and tumor cell invasion. *Eur. J. Pharmacol.* **2010**, *643*, 195–201. [CrossRef]

15. Gu, M.; Zhang, G.; Su, Z.; Ouyang, F. Identification of major active constituents in the fingerprint of Salvia miltiorrhiza Bunge developed by high-speed counter-current chromatography. *J. Chromatogr. A* **2004**, *1041*, 239–243. [CrossRef]

16. Li, Y.G.; Song, L.; Liu, M.; Hu, Z.B.; Wang, Z.T. Advancement in analysis of Salviae miltiorrhizae Radix et Rhizoma (Danshen). *J. Chromatogr. A* **2009**, *1216*, 1941–1953. [CrossRef]

17. Bae, W.J.; Choi, J.B.; Kim, K.S.; Syn, H.U.; Hong, S.H.; Lee, J.Y.; Hwang, T.K.; Hwang, S.Y.; Wang, Z.P.; Kim, S.W. Inhibition of proliferation of prostate cancer cell line DU-145 in vitro and in vivo using Salvia miltiorrhiza Bunge. *Chin. J. Integr. Med.* **2017**, *1*, 1. [CrossRef]

18. Luo, Y.; Feng, Y.; Song, L.; He, G.Q.; Li, S.; Bai, S.S.; Huang, Y.J.; Li, S.Y.; Almutairi, M.M.; Shi, H.L.; et al. A network pharmacology-based study on the anti-hepatoma effect of Radix Salviae Miltiorrhizae. *Chin. Med.* **2019**, *14*, 27. [CrossRef]

19. Trinh, T.A.; Park, E.-J.; Lee, D.; Song, J.H.; Lee, H.L.; Kim, K.H.; Kim, Y.; Jung, K.; Kang, K.S.; Yoo, J.-E. Estrogenic activity of Sanguiin H-6 through activation of estrogen receptor α Coactivator-binding Site. *Nat. Prod. Sci.* **2019**, *25*, 28–33. [CrossRef]

20. Kim, D.H.; Kim, D.W.; Jung, B.H.; Lee, J.H.; Lee, H.; Hwang, G.S.; Kang, K.S.; Lee, J.W. Ginsenoside Rb2 suppresses the glutamate-mediated oxidative stress and neuronal cell death in HT22 cells. *J Ginseng Res* **2019**, *43*, 326–334. [CrossRef]

21. Lee, D.; Lee, D.S.; Jung, K.; Hwang, G.S.; Lee, H.L.; Yamabe, N.; Lee, H.J.; Eom, D.W.; Kim, K.H.; Kang, K.S. Protective effect of ginsenoside Rb1 against tacrolimus-induced apoptosis in renal proximal tubular LLC-PK1 cells. *J. Ginseng Res.* **2018**, *42*, 75–80. [CrossRef] [PubMed]

22. Roy, A.; Park, H.-J.; Jung, H.A.; Choi, J.S. Estragole exhibits anti-inflammatory activity with the regulation of NF-κB and Nrf-2 signaling pathways in LPS-induced RAW 264.7 cells. *Nat. Prod. Sci.* **2018**, *24*, 13–20. [CrossRef]

23. Debatin, K.-M. Activation of apoptosis pathways by anticancer treatment. *Toxicol. Lett.* **2000**, *112*, 41–48. [CrossRef]

24. Bai, R.; Li, W.; Li, Y.; Ma, M.; Wang, Y.; Zhang, J.; Hu, F. Cytotoxicity of two water-soluble polysaccharides from Codonopsis pilosula Nannf. var. modesta (Nannf.) LT Shen against human hepatocellular carcinoma HepG2 cells and its mechanism. *Int. J. Biol. Macromol.* **2018**, *120*, 1544–1550. [CrossRef]

25. Ghasemian, M.; Mahdavi, M.; Zare, P.; Feizi, M.A.H. Spiroquinazolinone-induced cytotoxicity and apoptosis in K562 human leukemia cells: Alteration in expression levels of Bcl-2 and Bax. *J. Toxicol. Sci.* **2015**, *40*, 115–126. [CrossRef]

26. Wong, R.S.Y. Apoptosis in cancer: From pathogenesis to treatment. *JECCR* **2011**, *30*, 87. [CrossRef]

27. Adams, J.M.; Cory, S. The BCL-2 arbiters of apoptosis and their growing role as cancer targets. *Cell Death Differ* **2018**, *25*, 27. [CrossRef]

28. Waziri, P.M.; Abdullah, R.; Rosli, R.; Omar, A.R.; Abdul, A.B.; Kassim, N.K.; Malami, I.; Etti, I.C.; Sani, J.A.M.; Lila, M.A.M. Clausenidin induces caspase 8-dependent apoptosis and suppresses production of VEGF in liver cancer cells. *Asian Pac. J. Cancer Prev.* **2018**, *19*, 917.

29. Shamas-Din, A.; Kale, J.; Leber, B.; Andrews, D.W. Mechanisms of action of Bcl-2 family proteins. *Cold Spring Harb. Perspect. Biol.* **2013**, *5*, a008714. [CrossRef]

30. Brentnall, M.; Rodriguez-Menocal, L.; De Guevara, R.L.; Cepero, E.; Boise, L.H. Caspase-9, caspase-3 and caspase-7 have distinct roles during intrinsic apoptosis. *BMC Cell Biol.* **2013**, *14*, 32. [CrossRef]

31. Keskin-Aktan, A.; Akbulut, K.G.; Yazici-Mutlu, Ç.; Sonugur, G.; Ocal, M.; Akbulut, H. The effects of melatonin and curcumin on the expression of SIRT2, Bcl-2 and Bax in the hippocampus of adult rats. *Brain Res. Bull* **2018**, *137*, 306–310. [CrossRef] [PubMed]

32. Papadatos-Pastos, D.; Rabbie, R.; Ross, P.; Sarker, D. The role of the PI3K pathway in colorectal cancer. *Crit. Rev. Oncol/Hematol.* **2015**, *94*, 18–30. [CrossRef] [PubMed]

 *applied sciences*

Article

Chalcones and Flavanones Bearing Hydroxyl and/or Methoxyl Groups: Synthesis and Biological Assessments

Gonçalo P. Rosa [1], Ana M. L. Seca [1,2], Maria do Carmo Barreto [1], Artur M. S. Silva [2] and Diana C. G. A. Pinto [2,*]

[1] cE3c—Centre for Ecology, Evolution and Environmental Changes/Azorean Biodiversity Group & University of Azores, Rua Mãe de Deus, 9501-801 Ponta Delgada, Portugal
[2] QOPNA & LAQV-REQUIMTE, Department of Chemistry, University of Aveiro, 3810-193 Aveiro, Portugal
* Correspondence: diana@ua.pt; Tel.: +351-234-401-407

Received: 20 June 2019; Accepted: 14 July 2019; Published: 17 July 2019

Abstract: Chalcones and flavanones are isomeric structures and also classes of natural products, belonging to the flavonoid family. Moreover, their wide range of biological activities makes them key scaffolds for the synthesis of new and more efficient drugs. In this work, the synthesis of hydroxy and/or methoxychalcones was studied using less common bases, such as sodium hydride (NaH) and lithium bis(trimethylsilyl)amide (LiHMDS), in the aldol condensation. The results show that the use of NaH was more effective for the synthesis of 2′-hydroxychalcone derivatives, while LiHMDS led to the synthesis of polyhydroxylated chalcones in a one-pot process. During this study, it was also possible to establish the conditions that favor their isomerization into flavanones, allowing at the same time the synthesis of hydroxy and/or methoxyflavanones. The chalcones and flavanones obtained were evaluated to disclose their antioxidant, anticholinesterasic, antibacterial and antitumor activities. 2′,4′,4-Trihydroxychalcone was the most active compound in terms of antioxidant, anti-butyrylcholinesterase (IC$_{50}$ 26.55 ± 0.55 µg/mL, similar to control drug donepezil, IC$_{50}$ 28.94 ± 1.76 µg/mL) and antimicrobial activity. 4′,7-Dihydroxyflavanone presented dual inhibition, that is, the ability to inhibit both cholinesterases. 4′-Hydroxy-5,7-dimethoxyflavanone and 2′-hydroxy-4-methoxychalcone were the compounds with the best antitumor activity. The substitution pattern and the biological assay results allowed the establishment of some structure/activity relationships.

Keywords: chalcones; aldol condensation; biological activity; flavanones; cytotoxic; antioxidant; anticholinesterase

---

1. Introduction

The (*E*)-1,3-diphenylpropen-1-ones, best known as chalcones, belong to the flavonoids family and are an important class of natural products across the plant kingdom [1]. Structurally, these compounds contain two aromatic rings, bonded by a three-carbon α,β unsaturated carbonyl bridge (Figure 1), which are synthesized in plants as the C15 key intermediate in the biosynthesis of the other flavonoids [2]. Flavanones are also naturally occurring compounds and are chalcones' isomeric forms. In fact, the equilibrium between chalcones and flavanones is common in nature and is regulated by chalcone-isomerase [3].

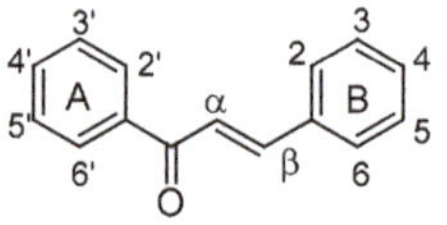

Figure 1. Basic chalcone structure.

Besides their natural occurrence, both chalcones and flavanones can be obtained synthetically and are often used as the preferred starting material for the synthesis of other polycyclic aromatic compounds [4]. Furthermore, they present great pharmacological potential with a wide variety of biological activities, including antioxidant [5], anticancer [6–8], and antimicrobial activities [9–12], and also the ability to treat cardiovascular diseases and their risk factors [13–16], among others [17].

On the other hand, cancer, neurodegenerative diseases, oxidative stress-related diseases and multi-resistant bacterial infections are, after cardiovascular diseases, the top four health problems that cause the most victims every year, leading to higher medicine consumption and putting great pressure on the national health systems of many countries [18–24]. These problems occur either due to a lack of effective medicines to treat diseases, such as neurodegenerative ones, or due to the increasing drug resistance presented by numerous pathogenic bacteria and by some cancers. Therefore, exploring well-known scaffolds as lead compounds will help in the battle against diseases that affect humanity.

Putting together the facts stated above, the synthesis of chalcone-based functionalized derivatives remains a popular research objective. The most common and efficient approach to obtain the chalcone nucleus is the aldol condensation of substituted acetophenones with proper substituted benzaldehydes in the presence of a base, namely sodium or potassium hydroxide [25–28]. Despite the efficiency of this method, when planning a synthesis some drawbacks should be considered. For instance, the protection of the reagents' hydroxyl groups should be done previously, the acetophenone hydrogen α acidity should be analyzed, and by-products can be obtained if the bases are also good nucleophilic species [29,30].

In this regard, the objective of this work is to synthesize hydroxy- and/or methoxychalcones by aldol condensation, using the less common bases sodium hydride and lithium bis(trimethylsilyl)amide. Also, this work studies their antioxidant, anticholinesterasic, antibacterial and antitumor activities, aiming to establish some potential medicinal applications. Simultaneously, a structure/activity relationship was established, and the isomeric equilibrium chalcone-flavanone was also studied.

2. Materials and Methods

2.1. General Methods

The ^{1}H, ^{13}C, HSQC and HMBC NMR spectra were measured on Bruker AMC 300 or 500 instruments, operating at 300.13 MHz and 75.47 or 500.13 and 125.75 MHz. Chemical shifts were reported relative to tetramethysilane (TMS) in δ units (ppm) and coupling constants (J) in Hz. Chromatographic purifications were carried out by prep. TLC on silica gel (Merck silica gel 60 F_{254}), the spots being visualized under a UV lamp (at 254 and/or 366 nm). Melting points were determined with a Stuart scientific SPM3 apparatus and are uncorrected. The mass spectra were acquired using ESI(+) with a Micromass Q-Tof 2^{TM} mass spectrometer.

2.2. Synthesis of Chalcones and Flavanones

Synthesis of the compounds described below follows the general scheme outlined in 3.1 (Scheme 1).

2'-Hydroxy-4,4',6'-trimethoxychalcone **1**. Compound **1** was synthesized by mixing 2'-hydroxy-4',6'-dimethoxyacetophenone (661.3 mg, 3.37 mmol) dissolved in a minimum (~15 mL) amount of tetrahydrofuran (THF) with sodium hydride (NaH) (2.5 equivalents), under nitrogen atmosphere at room temperature. After 30 min of stirring, 4-methoxybenzaldehyde (1.2 equivalents) was added to the reaction mixture and allowed to react for 3 h. The product was precipitated from the reaction mixture by pouring onto ground ice and acidifying to pH < 2 with HCl 37%. The solid was filtered and washed with water until pH > 5. The crude product was crystallized from ethanol and the desired compound **1** was obtained (928.1 mg, 88% yield).

2′-Hydroxy-4,4′,6′-trimethoxychalcone **1**: yellow crystals (ethanol); m.p. 112.4–113.6 °C (Lit. 111–115° [31]). ^{1}H NMR (300 MHz, CDCl$_3$) δ 14.40 (1H, s, 2′-O*H*), 7.79 (2H, s br, H-α, H-β), 7.56 (2H, d, *J* = 6.8 Hz, H-2, H-6), 6.92 (2H, d, *J* = 6.8 Hz, H-3, H-5), 6.11 (1H, d, *J* = 2.4 Hz, H-3′), 5.96 (1H, d, *J* = 2.4 Hz, H-5′), 3.92 (3H, s, 6′-OC*H$_3$*), 3.85 (3H, s, 4′-OC*H$_3$*), 3.83 (3H, s, 4-OC*H$_3$*); ^{13}C NMR (75 MHz, CDCl$_3$) δ 192.6 (*C*=O), 168.4 (C-2′), 166.0 (C-4′), 162.5 (C-6′), 161.4 (C-4), 142.5 (C-β), 130.1 (C-2, C-6), 128.3 (C-1), 125.1 (C-α), 114.4 (C-3, C-5), 106.3 (C-1′), 93.8 (C-3′), 91.2 (C-5′), 55.8 (6′-OCH$_3$), 55.6 (4′-OCH$_3$), 55.4 (4-OCH$_3$); TOF-ESI-MS (+) *m/z* 315 [M+H]$^+$, 337 [M+Na]$^+$, 353 [M+K]$^+$, 651 [M+Na+M]$^+$.

5,7-Dihydroxy-4′-methoxyflavanone **4**. The procedure to obtain this compound involved 3 different steps:

(a) The benzylation of the hydroxyl groups: the 4′ and 6′-hydroxyl groups in the starting material 2′,4′,6′-trihydroxyacetophenone were protected using the methodology described by Figueiredo [32]. Briefly, the 2′,4′,6′-trihydroxyacetophenone (2.7 g, 16.1 mmol), dissolved in a minimum amount of dry dimethylformamide (DMF) (~20 mL), was mixed with K$_2$CO$_3$ (6 equivalents) under constant stirring. Then, benzyl bromide (3 equivalents) was added and the reaction was performed at 150 °C under reflux for 2 h. After that, the reaction mixture was filtered to remove the K$_2$CO$_3$ and the inorganic salts washed with DMF. The filtrate was poured over crushed ice and the mixture acidified to pH < 5 with HCl 20%. The precipitated 4′,6′-dibenzyloxy-2′-hydroxyacetophenone was filtered and crystallized from ethanol (4.75 g, 85% yield).

(b) The aldol condensation: The synthesis of 4′,6′-dibenzyloxy-2′-hydroxy-4-methoxychalcone **2** was performed by dissolving 4′,6′-benzyloxy-2′-hydroxyacetophenone (1.5566 g) in dried THF and was then mixed with NaH (2.5 equivalents). After 10 min of stirring under nitrogen atmosphere at room temperature, 4-methoxybenzaldehyde (1.2 equivalents) was added. The reaction was finished after 3 h by pouring over crushed ice and addition of HCl 37% to pH < 2. The precipitate was filtered and washed with water, and the crude product was crystallized from acetone to afford 4′,6′-dibenzyloxy-2′-hydroxy-4-methoxychalcone **2** (1.6573 g, 80% yield).

(c) The benzyl group's cleavage: The final step was deprotecting the hydroxyl groups at 4′ and 6′ positions by cleavage of the benzyl groups. This procedure was adapted from the method described by Gomes et al. [33]. Briefly, 4′,6′-dibenzyloxy-2′-hydroxy-4-methoxychalcone **2** (567.0 mg) was mixed with 40 mL of a mixture of HCl/Acetic acid (1:10) under stirring at 80 °C during 13 h. The reaction was finished by pouring the mixture over crushed ice, the solid formed was washed with water until pH ~ 5 and then purified by TLC, eluting it twice in CH$_2$Cl$_2$, affording 5,7-dihydroxy-4′-methoxy-flavanone **4** (138.7 mg, 59%).

5,7-Dihydroxy-4′-methoxyflavanone **4**. pale-yellow crystals (CHCl$_3$); m.p. 191.7–193.3 °C (Lit. 193–194 °C [34]). ^{1}H NMR (300 MHz, CDCl$_3$) δ 12.05 (1H, s, 5-O*H*), 7.37 (2H, d, *J* = 8.7 Hz, H-2′, H-6′), 6.95 (2H, d, *J* = 8.7, H-3′, H-5′), 5.99 (1H, s broad, H-6), 5.98 (1H, s broad, H-8), 5.36 (1H, dd, *J* = 3.0 and 13.0 Hz, H-2), 3.83 (3H, s, 4′-OC*H$_3$*), 3.10 (1H, dd, *J* = 13.0 and 17.2 Hz, H-3a), 2.78 (1H, dd, *J* = 3.0 and 17.2 Hz, H-3b); ^{13}C NMR (75 MHz, CDCl$_3$) δ 196.2 (C-4), 164.8 (C-7), 164.3 (C-5), 163.3 (C-8a), 160.1 (C-4′), 130.3 (C-1′), 127.8 (C-2,′ C-6′), 114.1 (C-3′, C-5′), 103.1 (C-4a), 96.7 (C-6), 95.5 (C-8), 79.0 (C-2), 55.4 (4′-OCH$_3$), 43.1 (C-3); TOF-ESI-MS (+) *m/z* 287 [M+H]$^+$, 611 [M+K+M]$^+$.

4′-Hydroxy-5,7-dimethoxyflavanone **5**. The synthesis of this compound also involved the 3 steps mentioned above for compound **4**.

(a) The benzylation of the hydroxyl groups: the 4-hydroxybenzaldehyde (1.3 g, 10.6 mmol) was dissolved in a minimum of dry dimethylformamide (DMF) (~15 mL), and it was mixed with K$_2$CO$_3$ (3 equivalents) under constant stirring. Then, benzyl bromide (1.5 equivalents) was added, and the reaction was performed at 150 °C under reflux for 2 h. After that, the reaction mixture was filtered to remove the K_2CO_3 and washed with DMF. The filtrate was poured over crushed ice and HCl 20% added until pH < 5. The precipitated 4-benzyloxybenzaldehyde was filtered and crystallized from ethanol (1.8 g, 78%).

(b) The aldol condensation: The 2′-hydroxy-4′,6′-dimethoxyacetophenone (898.3 mg) was dissolved in dried THF (~15 mL) and mixed with NaH (2.5 equivalents) at room temperature and under N_2 atmosphere. After 10 min, 4-benzyloxybenzaldehyde (1.2 equivalents) was added, and the reaction was finished after 4 h by precipitation over crushed ice acidified with HCl 37% to pH < 2. 4-Benzyloxy-2′-hydroxy-4′,6′-dimethoxychalcone **3** was obtained by crystallization from ethanol (1.433 g, 80%).

(c) The benzyl group's cleavage: The chalcone **3** (330.6 mg) was mixed with 30 mL of a mixture of HCl/Acetic acid (1:10) under stirring at 55 °C during 60 h. The reaction was finished by pouring the mixture over crushed ice, the formed solid was washed with water until pH ~5 and purified by TLC using hexane/ethyl acetate (1:1) as eluent, affording 4′-hydroxy-5,7-dimethoxyflavanone **5** as pale-yellow amorphous powder (21.0 mg, 8%).

4′-Hydroxy-5,7-dimethoxyflavanone **5**: ^{1}H NMR (300 MHz, CDCl$_3$) δ 7.29 (2H, d, *J* = 8.5 Hz, H-2′, H-6′), 6.86 (2H, d, *J* = 8.5 Hz, H-3′and H-5′), 6.13 (1H, d, *J* = 2.3 Hz, H-8), 6.07 (1H, d, *J* = 2.3 Hz, H-6), 5.34 (1H, dd, *J* = 3.0 and 12.8 Hz, H-2), 3.86 (3H, s, 5-OC*H*$_3$), 3.81 (3H, s, 7-OC*H*$_3$), 3.03 (1H, dd, *J* = 12.8 and 16.6 Hz, H-3a), 2.78 (1H, dd, *J* = 3.0 and 16.6 Hz, H-3b); ^{13}C NMR (75 MHz, CDCl$_3$) δ 190.2 (C-4), 166.2 (C-7), 165.2 (C-8a), 162.3 (C-5), 156.4 (C-4′), 130.4 (C-1′), 127.9 (C-2′, C-6′), 115.7 (C-3′, C-5′), 105.8 (C-4a), 93.6 (C-8), 93.1 (C-6), 78.9 (C-2), 56.1 (5-OCH$_3$), 55.6 (7-OCH$_3$), 45.2 (C-3); TOF-ESI-MS (+) *m/z* 301 [M+H]$^+$, 323 [M+Na]$^+$, 623 [M+Na+M]$^+$.

2′,4′,4-Trihydroxychalcone **6** *and 4′,7-dihydroxyflavanone* **7**. The 2′,4′-dihydroxyacetophenone (226.7 mg) was dissolved in dried toluene and mixed with 10 mL of a 1 mol.dm^{-3} solution of LiHMDS (6.6 equivalents), under nitrogen atmosphere, at room temperature. After 30 min, 4-hydroxybenzaldehyde (1.2 equivalents) was added, and the reaction was stirred for 5 days. The reaction was finished, poured over crushed ice and acidified to pH < 2 with HCl 37%. The mixture was extracted with CH_2Cl_2 and purified by TLC, using a mixture of hexane and ethyl acetate (1:1) as eluent (twice). Compounds 2′,4′,4-trihydroxychalcone **6** and 4′,7-dihydroxyflavanone **7** were obtained as yellow amorphous powder, respectively, 18.2 mg (5%) and 8.2 mg (2%). Approximately 80% of the starting acetophenone was also recuperated.

2′,4′,4-Trihydroxychalcone **6**: ^{1}H NMR (300 MHz, acetone-d$_6$) δ 13.69 (1H, s, 2′-O*H*), 8.12 (1H, d, *J* = 8.9 Hz, H-6′), 7.85 (1H, d, *J* = 15.4 Hz, H-β), 7.77 (1H, d, *J* = 15.4 Hz, H-α), 7.75 (2H, d, *J* = 8.6 Hz, H-2, H-6), 6.94 (2H, d, *J* = 8.6 Hz, H-3, H-5), 6.48 (1H, dd, *J* = 2.4 and 8.9 Hz, H-5′), 6.38 (1H, d, *J* = 2.4 Hz, H-3′); ^{13}C-NMR (75 MHz, acetone-d$_6$) δ 192.6 (*C*=O), 167.6 (C-4′), 166.2 (C-2′), 161.1 (C-4), 145.0 (C-β), 133.2 (C-6′), 131.7 (C-2, C-6), 127.4 (C-1), 118.2 (C-α), 116.8 (C-3, C-5), 114.2 (C-1′), 108.9 (C-5′), 103.7 (C-3′); TOF-ESI-MS (+) *m/z* 257 [M+H]$^+$, 279 [M+Na]$^+$, 535 [M+Na+M]$^+$, 551 [M+K+M]$^+$.

4′,7-Dihydroxyflavanone **7**: ^{1}H NMR (300 MHz, acetone-d$_6$) δ 9.53 (1H, s, 7-O*H*), 8.60 (1H, s, 4′-O*H*), 7.74 (1H, d, *J* = 8.6 Hz, H-5), 7.42 (2H, d, *J* = 8.6 Hz, H-2′, H-6′), 6.91 (2H, d, *J* = 8.6 Hz, H-3′, H-5′), 6.59 (1H, dd, *J* = 2.3 and 8.6 Hz, H-6), 6.43 (1H, d, *J* = 2.3 Hz, H-8), 5.46 (1H, dd, *J* = 2.8 and 13.1 Hz, H-2), 3.06 (1H, dd, *J* = 13.1 and 16.7 Hz, H-3a); 2.68 (1H, dd, *J* = 2.8 and 16.7 Hz, H-3b); ^{13}C-NMR (75 MHz, acetone-d$_6$) δ 190.5 (C-4), 165.2 (C-7), 164.5 (C-8a), 158.6 (C-4′), 131.2 (C-1′), 129.4 (C-5), 128.9 (C-2′, C-6′), 116.1 (C-3′, C-5′), 115.1 (C-4a), 111.2 (C-6), 103,6 (C-8), 80.5 (C-2), 44.6 (C-3); TOF-ESI-MS (+) *m/z* 257 [M+H]$^+$, 279 [M+Na]$^+$, 513 [M+H+M]$^+$, 535 [M+Na+M]$^+$, 551 [M+K+M]$^+$.

2.3. Biological Activities

Compound 2′-hydroxy-4-methoxychalcone **8** was provided by the Laboratory of Organic Chemistry of the University of Aveiro, and the reactional conditions, yield and spectroscopic data are reported by Silva et al. [35,36].

2.3.1. DPPH Scavenging Activity

Antioxidant activity was assayed by the DPPH (1,1-diphenyl-2-picryl-hydrazyl) radical scavenging assay [37]. Serial dilutions of studied or reference compounds (Trolox and quercetin) were carried out in 96-well microplates, at different concentrations, ranging between 0.148 µg/mL and 150 µg/mL in methanol. DPPH dissolved in methanol was added to the microwells, yielding a final concentration of 45 µg/mL, and the absorbance at 515 nm was measured in a Bio Rad Model 680 Microplate Reader (Bio-Rad Laboratories, Inc., Hercules, CA, USA), after 30 min in the dark. In each assay, a control was prepared, in which the sample or standard was substituted by the same amount of solvent. Percentage of antioxidant activity (%AA) was calculated as:

$$\%AA = 100\,[1 - (A_{control} - A_{sample})/A_{control}]$$

where $A_{control}$ is the absorbance of the control, and A_{sample} is the absorbance of the chalcone/flavanone or standard. All assays were carried out in triplicate and results expressed as EC_{50}, i.e., as the concentration yielding 50% scavenging of DPPH, calculated by interpolation from the %AA vs concentration curve.

2.3.2. ABTS Scavenging Activity

To determine ABTS radical scavenging, the method of Re et al. [38] was adopted. The stock solutions included 7 mM ABTS solution and 2.4 mM potassium persulfate solution. The working solution was prepared by mixing the two stock solutions in equal quantities and allowing them to react for 12–16 h at room temperature in the dark. The solution was then diluted by mixing 1 mL ABTS solution with the amount of methanol necessary to obtain an absorbance of 0.7 at 734 nm. Serial dilutions of studied or reference compounds (trolox, quercetin) were carried out in 96-well microplates, at different concentrations, ranging between 0.146 µg/mL and 150 µg/mL in methanol. ABTS solution was then added to the microwells, and after 8 min of incubation the absorbance was taken at 405 nm in a Bio Rad Model 680 Microplate Reader (Bio-Rad Laboratories, Inc., Hercules, CA, USA). In each assay, a control was prepared, in which the sample or standard was substituted by the same amount of solvent. Percentage of antioxidant activity (%AA) was calculated as:

$$(\%) = [(Abs_{control} - Abs_{sample})]/(Abs_{control})] \times 100$$

where $Abs_{control}$ is the absorbance of ABTS radical + methanol; Abs_{sample} is the absorbance of ABTS radical + sample/standard. All assays were carried out in triplicate and results expressed as EC_{50}, i.e., as the concentration yielding 50% scavenging of ABTS, calculated by interpolation from the %AA vs concentration curve.

2.3.3. Anticholinesterasic Activity

The assay for measuring AChE and BuChE activity was modified from the assay described by Ellman et al. [39] and Arruda et al. [40]. Briefly, 3 mM 5,5′-dithiobis [2-nitrobenzoic acid] (DTNB, 5 µL), 75 mM acetylthiocholine iodide (ATCI, 5 µL) or butyrylthiocoline iodide (BuTCI, 5 µL), and sodium phosphate buffer 0.1 mol dm^{-3} (pH 8.0, 110 µL), and sample or standard (quercetin or donepezil) dissolved in buffer containing no more than 2.5% DMSO were added to the wells, and serial dilutions were carried out to obtain concentrations ranging between 0.293 µg/mL and 150 µg/mL (0.098 and 50 µg/mL for galantamine and berberine; 0.010 and 5 µg/mL for donepezil; 0.195 and 100 µg/mL for quercetin), followed by 0.25 U/mL AChE or BuChE (10 µL). The microplate was then read at 415 nm every 2.5 min for 7.5 min in a Bio Rad Model 680 Microplate Reader (Bio-Rad Laboratories, Inc., Hercules, CA, USA). For each concentration, enzyme activity was calculated as a percentage of the velocities compared to that of the assay using buffer without any inhibitor. Every experiment was done in triplicate.

2.3.4. Antimicrobial Activity

Antibacterial activity against Gram-positive *Bacillus subtilis* DSM10 and *Micrococcus luteus* DSM 20030 and Gram-negative *Escherichia coli* DSM498 was assessed by the broth microdilution method, as described by De León et al. [41]. The bacteria cultures were developed in nutrient broth (NB) at 30 °C for *B. subtilis* and *M. luteus* and at 37 °C for *E. coli*. Briefly, compounds were added to the microplates at a concentration of 200 µg/mL, and serial dilutions in NB were made until the concentration of 0.391 µg/mL. Then, the starting inoculum (1×10^5 CFU/mL) was added, and the bacterial growth was measured by the increase in optical density at 550 nm with a Bio Rad Model 680 Microplate Reader (Bio-Rad Laboratories, Inc., Hercules, CA, USA), after 24 h of growth (48 h for *M. luteus*) at the above-mentioned temperatures for each bacterial strain. Penicillin and streptomycin were used as reference compounds. The IC_{50} was calculated as the concentration of compound that inhibits 50% of bacterial growth by interpolation from the % of growth inhibition vs. concentration curve.

2.3.5. Antitumor Activity

Antitumor activity was determined by the method described in Moujir et al. [42]. A-549 (human lung carcinoma) cell line, obtained from ATCC-LGC (American Type Culture Collection), was grown as a monolayer in DMEM (Sigma) supplemented with 2% fetal bovine serum (Sigma), 1% penicillin–streptomycin mixture (10,000 UI/mL), p-hydroxybenzoic acid (2×10^4 mg/mL) and L-glutamine (200 mM). Cells were maintained at 37 °C in 5% CO_2 and 80% humidity in a SANYO CO_2 Incubator. Cytotoxicity was assessed using the colorimetric MTT [3-(4,5-dimethylthiazol-2-yl)-2,5-diphenyl tetrazolium bromide] reduction assay. Cell suspension (2×10^4 cells/well) in the lag phase of growth was incubated in a 96-well microplate with the compounds or the standards (colchicine and paclitaxel) dissolved in medium, with the concentrations ranging between 0.195 µg/mL and 200 µg/mL (0.010 µg/mL and 10 µg/mL for standards). After 48 h, MTT was added to the cells, which were then allowed to incubate for 3–4 h, and the optical density was measured using a Bio Rad Model 680 Microplate Reader (Bio-Rad Laboratories, Inc., Hercules, CA, USA) at 550 nm after dissolving the MTT formazan with DMSO (100 µL). The percentage viability (IC_{50}) was calculated from the % of inhibition vs. concentration curve. All the experiments were repeated three times.

3. Results and Discussion

3.1. Chalcones and Flavanones Synthesis

The compounds obtained in this work were synthesized by aldol condensation between 4′- and/or 6′-substituted 2′-hydroxyacetophenones and either 4-hydroxy or 4-methoxybenzaldehydes, using NaH and LiHMDS (Scheme 1).

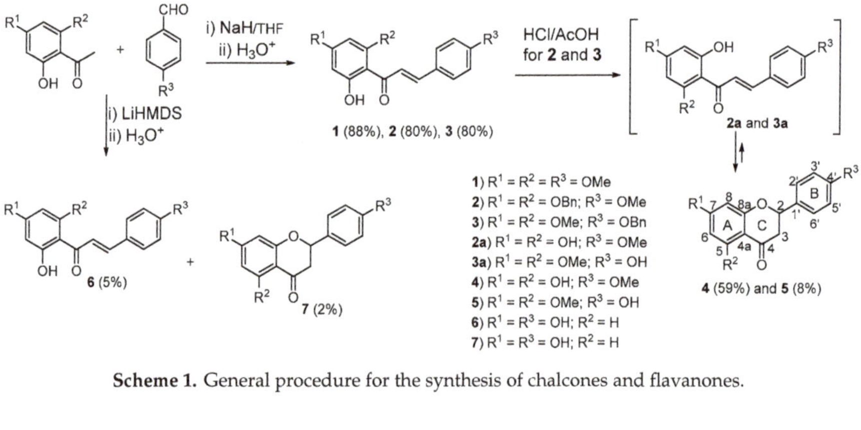

Scheme 1. General procedure for the synthesis of chalcones and flavanones.

Compounds **1** and **3** were synthesized, as far as we could confirm, using for the first time in the aldol condensation the strong non-nucleophilic base NaH, which was used with success in similar condensations [43,44]. The experimental results in the synthesis of chalcones **1–3**, with very good yields (~80%) and relatively low reaction times (3 to 4 h), proves that NaH is an effective and efficient base in the aldol condensation when it is desired to obtain non-hydroxylated chalcones, other than at the C-2′ position.

As referred above (Introduction), the hydroxyl groups present in the reagents require protection before the aldol condensation because in the strong basic conditions used, the phenoxide ions can undergo transformation into quinones and consequently prevent the formation of the desired chalcones. So, in order to obtain the desired compounds (the ones bearing hydroxyl and/or methoxyl groups), chalcones **2** and **3** need to be treated with acid to cleave the benzyl groups. The mixture used (HCl/AcOH) is one of the less harsh conditions and, in fact, removed the benzyl groups, but the obtained products were, in fact, flavanones **4** and **5** (Scheme 1). This can be explained by the acidic conditions used and the fact that those conditions favor the chalcones' isomerization into flavanones [45]. On the other hand, these extra steps, the protection and deprotection of hydroxyl groups, contribute to decrease the overall yield. For example, if we consider the synthesis of flavanones **4** and **5**, their precursors' chalcones **2** and **3** have excellent yields (80%), and the deprotection step, which produces the flavanones, is less efficient. This demonstrates that, due to this step, the flavanones' overall yield will be below 50%. At the same time, the yields (Scheme 1) allowed the assumption that the deprotection of the ring A hydroxyl groups is easier.

Since our purpose was the synthesis of chalcones bearing hydroxyl and methoxyl groups, taking into consideration that these substituents might improve the activity, and that the synthesis of flavanones **4** and **5** involves three steps, namely, (i) protection of the reagent's hydroxyl groups, (ii) aldol condensation and (iii) cleavage of the protecting groups, we envisaged the synthesis of the desired chalcones using another base. Knowing that LiHMDS was used in the synthesis of hydroxylated flavones without the protection step and with good results [46,47], we tested its use in the synthesis of chalcone **6** (Scheme 1). The experimental results showed that, using LiHMDS as the base, it is possible to synthesize polyhydroxylated chalcones in a one-pot process. However, although the desired chalcone was obtained, the yield was very low and its isomeric form, flavanone **7**, was also obtained. It should be highlighted that this reaction was accomplished at room temperature, controlled by TLC and finished after 5 days, when no more conversion was detected and 80% of the starting acetophenone was recovered. These data suggest that the conversion rate is high and that the reaction does not occur due to a lack of energy. We may suggest that using other sources of energy, such as a microwave, could originate higher yields.

The synthesized compounds' structures were confirmed by detailed analysis of their 1D and 2D NMR spectra, MS spectra (e.g., in Figures S1–S9, supplementary material) and available literature data; moreover, their purity was confirmed by UHPLC. Herein, chalcones' and flavanones' characterization is briefly discussed.

The ^{1}H NMR spectra of compounds **1–3** and **6** present the signal characteristics of chalcone structures: (i) the resonance of the AB system assigned to the olefinic protons; (ii) doublets at δ_H 7.7–7.9 ppm with $J \sim 15$ Hz, which confirms the *E* configuration of the α,β double bond; (iii) the signal at δ_C ~192 ppm assigned to the carbonyl group; (iv) the singlet at δ_H 14.4–13.7 ppm assigned to 2′-OH proton signal involvement in a hydrogen bond with the carbonyl oxygen atom; (v) two sets of doublets, with $J = 6.8$ Hz, at δ_H 6.9–7.8 ppm, assigned to the protons of the *para*-substituted aromatic ring B (Figure 1).

The ^{1}H, ^{13}C and HSQC NMR spectra of chalcone **1** also display two sets of doublets at δ_H 5.9–6.1 ppm, with a *meta*-coupling constant ($J = 2.4$ Hz), that exhibit correlation with the signals at 93.8 and 91.2 ppm, which indicates the presence of a tetra-substituted aromatic ring (ring A, Figure 1). The three sets of singlets at δ_H 3.83–3.92 ppm, showing correlation with three signals at δ_C 55.4–55.8 ppm, are characteristic of three methoxyl groups. The MS spectrum of chalcone **1** showed a signal at *m/z* 315 correspondent to [M+H]$^+$, which agrees with the molecular formula $C_{18}H_{18}O_5$.

All the spectroscopic data and the melting point are in agreement with previously published data [31], and additionally, the connectivities found in the HMBC NMR spectrum confirm compound **1** as 2′-hydroxy-4,4′,6′-trimethoxychalcone, also named flavokawain A.

The ^{1}H and ^{13}C NMR spectra of chalcone **6** show three sets of signals in the range of δ_C 6.38–8.12 ppm, characteristic of the a trisubstituted aromatic ring (ring A, Figure 1). The MS spectrum showed a signal at *m/z* 257 corresponding to [M+H]$^+$, which agrees with the molecular formula $C_{15}H_{12}O_4$. These data and the connectivities found in the HMBC spectrum confirm compound **6** as 2′,4′,4-tri-hydroxychalcone, also known as isoliquiritigenin [48].

Compounds **4**, **5** and **7** are flavanones, and this fact is well confirmed by the presence of the signals characteristic of ring C (Scheme 1): two double doublets at δ_H 2.68–3.10 and δ_H 2.68–2.78 assigned to the protons H-3, and the double doublet at δ_H 5.34–5.46 ppm assigned to H-2 in coupling with H-3; the signal at δ 196 ppm characteristic of the carbonyl group (C=O). Two sets of doublets at δ_H 7.29–7.42 (H-2′ and H-6′) and 6.86–6.95 ppm (H-3′ and H-5′), with *orto*-coupling constant of J = 8.7 Hz, indicate the presence of a *para*-substituted aromatic ring (ring B).

The presence of the methoxyl group in the ring B of compound **4** is deduced from the singlet at δ 3.83, correlating with the signal at δ_C 55.4 ppm. The MS data showed a signal at *m/z* 287 corresponding to the protonated molecule [M+H]$^+$ that is in accordance with the molecular formula $C_{16}H_{14}O_5$. The spectroscopic data and the melting point are consistent with the published data for isosakuranetin [33,49], and together with the connectivities found in HMBC spectrum, confirm that compound **4** is 5,7-dihydroxy-4′-methoxyflavanone.

The signals found in the ^{1}H and ^{13}C NMR spectra of compound **5** are very similar to the ones found for compound **4**. The differences are the non-appearance of the signal at δ_H 12.05 ppm, which confirms the absence of a 5-O*H* group; and the appearance of two singlets at δ_H 3.86 and 3.81 ppm, which means that compound **5** has two methoxyl groups. The location of the two methoxyl groups and the hydroxyl group were confirmed by the connectivities found in the HMBC spectrum of compound **5**. All the spectroscopic data are compatible with the structure of 4′-hydroxy-5,7-dimethoxyflavanone, also named naringenin 5,7-dimethyl ether. Although compound **5** is not new, [50,51], to the best of our knowledge, the full spectroscopic data are reported here for the first time.

The quasi-molecular ion at *m/z* 257, compatible with molecular formula $C_{15}H_{12}O_4$, confirms that compound **7** is an isomeric form of compound **6**. The presences of two signals at δ_H 9.53 and 8.60 ppm are assigned to 7-O*H* and 4′-O*H*, respectively, by the connectivities found in the HMBC spectrum. Moreover, the spectroscopic data are identical to that previously published for 4′,7-dihydroxyflavanone, also named liquiritigenin [52].

3.2. Biological Evaluations

The antioxidant, anticholinesterasic, antimicrobial and antitumor activities of the compounds bearing hydroxyl and methoxyl groups were studied, and the results obtained are presented and discussed in the following paragraphs. Compound **8**, 2′-hydroxy-4-methoxychalcone, was not synthesized in this work, but was included in the bioactivities study because it is, in structural terms, one of the simplest 2′-hydroxychalcones and can be used as a reference and as a starting point to analyze structure/activity relationships of the compounds that were synthesized.

3.2.1. Antioxidant Activity

In the DPPH (1,1-diphenyl-2-picryl-hydrazyl) scavenging test, the capacity of the compounds to neutralize the DPPH radical via electron transfer is measured [53,54].

Nearly all the compounds had some antioxidant activity, which was concentration-dependent (see Supplementary Materials, Figure S10), except for compounds **4** and **8**, which only had a residual activity at the highest concentration tested (150 μg/mL). Chalcone **6**, the most active compound, presented higher antioxidant activity than the isomeric flavanone **7** (Table 1), confirming that the conjugated double bond improves the molecule's ability to scavenge the DPPH radical [55].

Table 1. Antioxidant activity of the synthesized compounds **1**, **4–8**.

Compounds and References	IC$_{50}$ µg/mL (± SD)	
	DPPH	**ABTS**
1	>150	>150
4	>150	140.15 (±0.03) [a]
5	>150	85.47 (±0.15) [b]
6	26.47 (±0.70) [a]	12.72 (±0.92) [c]
7	>150	>150
8	>150	>150
Trolox	7.25 (±0.09) [b]	2.68 (±0.08) [d]
Quercetin	3.01 (±0.03) [c]	0.57 (±0.02) [e]

In each column, the letters a, b, c, d, and e indicate significant differences ($p < 0.05$).

The results also confirm that the number of free hydroxyl groups influences antioxidant activity, since compounds with a higher number of these groups present better activity [55], a fact that is also confirmed by the lower activity of the compounds with more methoxyl groups (Figure 2).

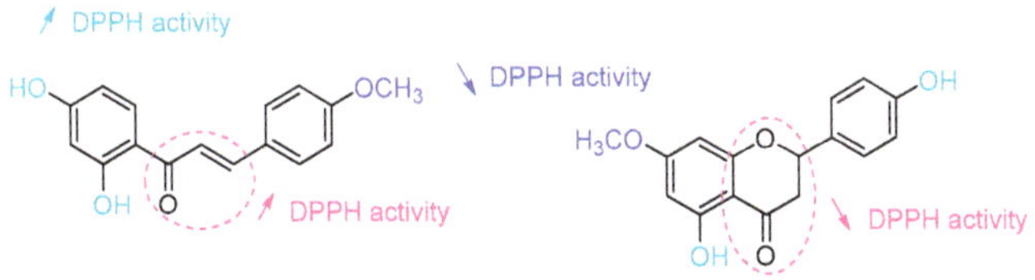

Figure 2. Structure/activity relationships established for DPPH scavenging activity.

A previous work reported that compound **6** presented a low antioxidant activity in the DPPH scavenging test [56] (86.92 ± 0.43% of DPPH inhibition at 150 µg/mL); however, since the maximum concentration used in that work was about six times lower than the maximum tested concentration in the present work, it does not conflict with the results herein reported.

Another test used was the ABTS (2,2'-azino-bis(3-ethylbenzothiazoline-6-sulphonic acid)) scavenging assay. This assay measures the ability of compounds to neutralize the ABTS radical by radical quenching via hydrogen atom transfer. However, it can also be neutralized by electron transfer on some occasions, resulting in a higher sensitivity of this method when compared with the DPPH assay [53,57]. Again, compound **6** was the one presenting the best activity, with an IC$_{50}$ of 12.72 µg/mL (Table 1), which is another result that confirms the importance of the hydroxyl groups in the chalcone scaffold for their ability to reduce and/or eliminate free radicals. This conclusion is confirmed by the lower activity of chalcones **1** and **8**, which have only one hydroxyl group. Compound **5** presents better activity than **4**, which means that the relative position between the hydroxyl and methoxyl groups is more important than the number of hydroxyl groups on the flavanone scaffold. This is confirmed by the results obtained for compound **7**, which has two hydroxyl groups, but only one substituent in ring A, and presents lower activity than compounds **4** and **5** (Figure 3). It was observed that the ABTS scavenging activity of the tested compounds is dose-dependent (Figure S11, supplementary material).

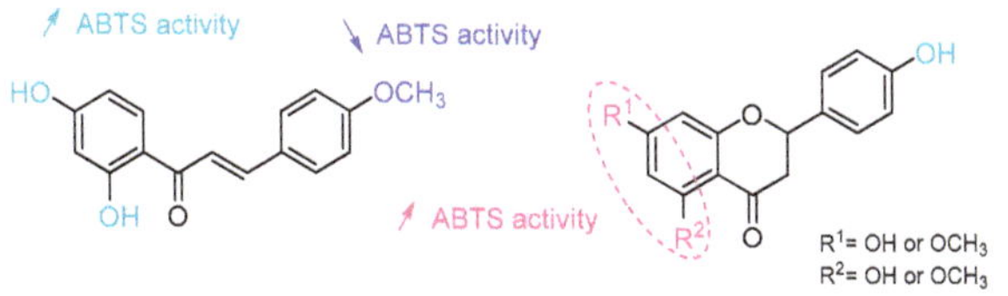

Figure 3. Structure/activity relationships established for ABTS scavenging activity.

3.2.2. Anticholinesterasic Activity

The results for acetylcholinesterase (AChE) inhibitory activity showed that compound **7** is the only one which presents some activity at the maximum concentration tested ($47.1 \pm 3.6\%$, at 150 µg/mL). This activity is less pronounced than the inhibition obtained for control compound donepezil ($95.2 \pm 0.4\%$ at 50 µg/mL), a pure competitive inhibitor of AChE used clinically in early stages of Alzheimer's disease. The results suggest that the cyclization seems to increase the inhibitory activity of the compound, since compound **7**, which is a flavanone, is more active than compound **6**, the respective chalcone. Also, the existence of free hydroxyl groups in both A and B rings seems to be important because, out of the three compounds with a flavanone structure (compounds **4**, **5** and **7**), only the one with a free hydroxyl group in both C-7 and C-4′ (compound **7**) shows activity. This is in agreement with the strong activity revealed by quercetin, whose structure also possesses free hydroxyl groups in both rings A and B, like compound **7**. So, it can be concluded that the presence of hydroxyl groups in rings A and B of the flavanone scaffold is crucial for the inhibition of AChE activity (Figure 4).

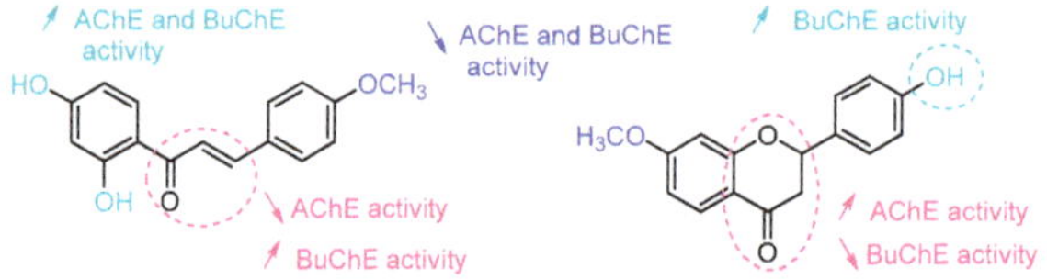

Figure 4. Structure/activity relationships established for antioxidant activity.

In terms of butyrylcholinesterase inhibition, compound **6** presented excellent activity, with a percentage of inhibition of $96.0 \pm 1.1\%$ and IC_{50} of 26.55 ± 0.55 µg/mL, which are similar values to the ones presented by donepezil (IC_{50} of 28.94 ± 1.76 µg/mL) and better than quercetin at the same concentrations. This shows that, unlike what happens with acetylcholinesterase, the presence of three hydroxyl groups in the chalcone scaffold highly favors the inhibition of butyrylcholinesterase activity. The results presented by chalcones **1** ($12.51 \pm 0.82\%$) and **8** (0%) also confirm that, like with acetylcholinesterase, the presence of methoxyl groups does not confer inhibitory activity to chalcones.

Compound **7** presents an activity very similar to the one observed for acetylcholinesterase inhibition ($46.26 \pm 1.27\%$), meaning that it is a dual inhibitor, a much-appreciated feature in the search for compounds with therapeutic potential towards Alzheimer's disease. The presence of a hydroxyl group in ring B of the flavanone scaffold increases its inhibitory effect, since flavanones **5** ($36.30 \pm 0.20\%$) and **7** present higher activity than flavanone **4** ($9.00 \pm 1.30\%$), which has a methoxyl group in said position (Figure 4). The compounds inhibited butyrylcholinesterase in a dose-dependent manner (Figure S12, supplementary material).

3.2.3. Antimicrobial Activity

As observable in Table 2, most of the compounds tested inhibited the growth of gram-positive bacteria, but none was effective against the gram-negative strain tested. This inhibition was found to be concentration-dependent (Figures S13 and S14).

The most efficient compounds against *Micrococcus luteus* were **6** and **7**, both with similar IC_{50}, which suggests that cyclization does not influence the toxicity of these molecules against this species. These results are concordant with the literature, since it is described that these compounds presented similar activity against *Mycobacterium tuberculosis* [58]. Although some compounds tested exhibit an IC_{50} value greater than 200 µg/mL, the percent of growth inhibition exhibited at the maximum concentration tested allows one to deduce some interesting structure/activity relationships, which are discussed below. The chalcone **1** had no effect against *M. luteus*, while chalcone **8** presents 41.71% of growth inhibition at 200 µg/mL. These results suggest that the presence of methoxyl groups at ring A reduces its activity against this species. The same effect is observed when comparing the activity of compounds **4** and **5** (Figure 5).

Table 2. Antimicrobial activity of synthesized compounds **1**, **4–8**.

Compounds and References	IC_{50} µg/mL (± SD)		
	M. luteus	*B. subtilis*	*E. coli*
1	>200	>200	>200
4	164.89 (±2.97) [a]	76.46 (±4.27) [a]	>200
5	>200	>200	>200
6	23.78 (±2.04) [b]	9.33 (±0.25) [b]	>200
7	23.64 (±2.24) [b]	20.48 (±3.13) [c]	>200
8	>200	74.64 (±3.92) [a]	>200
Penicillin	8.47 (±0.55) [c]	0.28 (±0.02) [d]	>40
Streptomycin	0.59 (±0.02) [d]	15.86 (±0.21) [e]	14.45 (±1.73)

In each column, the letters a, b, c, d, and e indicate significant differences ($p < 0.05$).

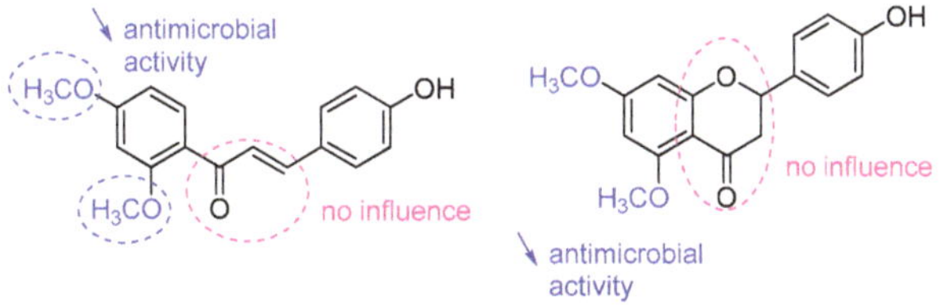

Figure 5. Structure/activity relationships established for antibacterial effect against *M. luteus*.

Compound **5** inhibits 43.08% of bacterial growth at 200 µg/mL, while this value is almost double for compound **4** (74.43 ± 1.95%), which confirms that the methoxyl groups at ring A decrease the molecule's antibacterial effect.

The strongest activity against *Bacillus subtilis* was displayed by compound **6** (IC_{50} = 9.33 µg/mL), presenting even better activity than streptomycin (Table 2). The corresponding flavanone **7** has a higher IC_{50} (20.48 µg/mL), which leads to the conclusion that the α,β unsaturated carbonyl bridge linking ring A to ring B is important in the mode of action of the compounds against this bacterial strain. The molecules with the higher numbers of hydroxyl groups are the ones with better activity, showing that these groups also play an important role in the molecule's mode of action against *B. subtilis*. Again, the compounds with two methoxyl groups in *meta* positioning (**1** and **5**) were the ones with lower or none antimicrobial activity. Comparing the results from compounds **4** and **7** (both flavanones with two hydroxyl groups), one can conclude that the presence of the methoxyl group in ring B (compound **4**) reduces the compound's activity against this bacterial strain (Figure 6).

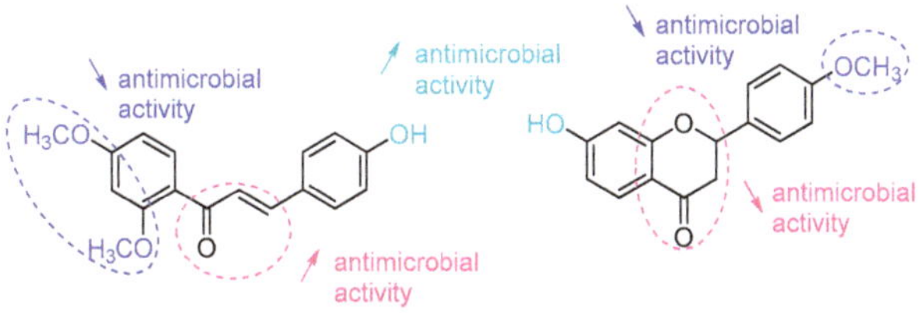

Figure 6. Structure/activity relationships established for antibacterial effect against *B. subtilis*.

3.2.4. Antitumor Activity

Considering the antitumor activity of the studied compounds, it can be noticed that compounds **5** and **8** were the most active, presenting very similar IC_{50} values. Compounds **1** and **4** showed some activity, although lower than the ones mentioned above (Table 3). However, none of the compounds tested were as active as colchicine or paclitaxel, the reference compounds used (Table 3). It was also observed that the inhibition of tumor cell growth is dose-dependent for all the compounds tested (Figure S15, supplementary material).

Table 3. Antitumor activity of the synthesized compounds 1, 4–8.

Compounds and References	IC$_{50}$ µg/mL (± SD)
	A549 Cell Line
1	101.23 (±0.42) [a]
4	135.89 (±2.61) [b]
5	93.42 (±2.42) [c]
6	>200
7	>200
8	93.31 (±2.45) [c]
Paclitaxel	5.96 (±0.48) [d]
Colchicine	2.78 (±0.71) [d]

In each column the letters a, b, c, d, and e indicate significant differences ($p < 0.05$).

Results for the activity of compounds **6** and **7** against the A549 cell line have already been reported in the literature [7,59–61]; however, their activity was tested in order to facilitate the comparison of the results, since all the compounds were tested in the same conditions. The results for compound **7** are concordant with the literature, since it is reported that the effects of this compound in A549 cells is limited to the inhibition of cell migration, without any effects on growth or cytotoxicity level [7,60]. Compound **6** did not present any activity, which was unexpected, since it is reported to possess antitumor activity against several cell lines, including A549, by inhibiting proliferation and inducing tumor cell apoptosis [59,61]. This can be explained by the fact that the authors used a much lower cell concentration than the one used in the present work, and it has been proved that lower cell concentrations are correlated with higher activities of the compounds tested [62].

From the results obtained, the only conclusions that can be drawn regarding structure/activity relationships are that neither the α,β unsaturated carbonyl bridge nor the cyclization of chalcone into flavanone have an influence in the inhibition of tumor-cell growth, since both chalcones and flavanones presented interesting activities (Figure 7). Also, the presence of methoxyl groups has some importance in the antitumor effect of the compounds, since compounds **6** and **7**, which do not have methoxyl groups, did not present any activity against the A549 cell line (Table 3). The influence of a methoxyl group in synthetic polyphenolic compounds' antitumor activity was not a surprise because it was previously detected in our group [63].

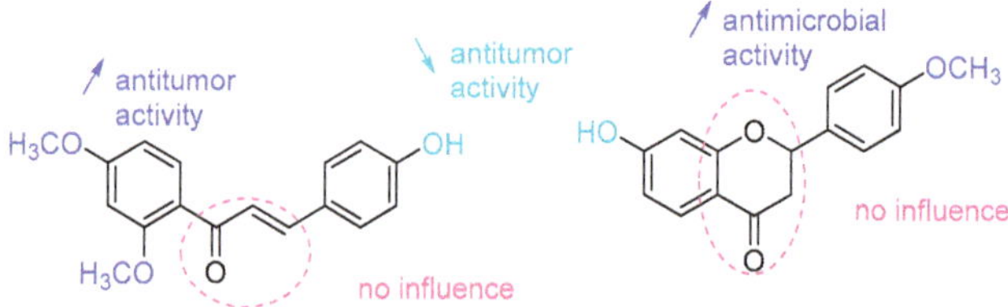

Figure 7. Structure/activity relationships established for antitumor activity.

4. Conclusions

Hydroxylated chalcone/flavanone derivatives were synthesized using the less common bases NaH or LiHMDS. Overall, it was shown that the use of NaH is efficient (~80% yield in 3–4 h) for the synthesis of chalcones not hydroxylated other than at C-2′ position. However, if the desired derivatives are polyhydroxylated chalcones, the use of LiHMDS is preferable, because the conversion rate is good and it is a one-pot procedure that avoids the protection and subsequent deprotection steps, a time-consuming procedure that also significantly decreases the total yield. Although the aim of this work was to obtain the compounds for biological evaluation, we suggest microwave or ultrasound irradiation as a better source of energy to increase the compounds' yields, an aspect that is important for their future applications.

Some of the compounds synthesized presented interesting results in their bioactivities, with chalcone **6** being the most active compound in terms of antioxidant, anti-butyrylcholinesterase and antimicrobial activity. Its isomer, flavanone **7**, showed activity against both acetylcholinesterase and butyrylcholinesterase, which is an interesting result since this dual inhibition is a much-appreciated feature in Alzheimer's disease therapy. Also, it is interesting that these compounds can be obtained in the one-pot methodology using LiHMDS.

Some important structure/activity relationships were established for all the activities tested, and the most important are highlighted and summarized in Figure 8.

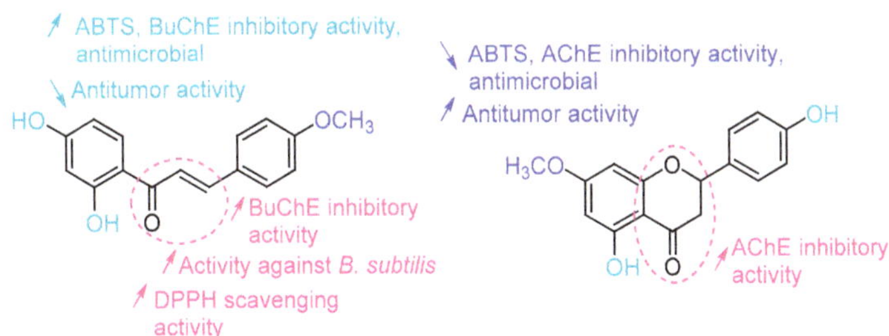

Figure 8. Summary of the most relevant structure/activity relationships established.

As shown in Figure 8, the free hydroxyl groups are essential to increase antioxidant, anti-butyrylcholinesterase and antimicrobial activities. Not only their presence but also their number increases the compounds' activity, while methoxyl groups decrease it. The flavanone scaffold increases the acetylcholinesterase inhibitory activity, while for the butyrylcholinesterase inhibition, the chalcone scaffold appears to be better. This is another interesting result because it is known that these compounds are isomers, and in biological systems they can exist in a controlled equilibrium. Regarding the antimicrobial activity, it is possible to detect that the cyclization into flavanone has no effect against the *M. luteus* strain, whereas against *B. subtilis*, the flavanone decreases the antibacterial activity. So, it seems that the α,β unsaturated carbonyl bridge linking rings A and B is important to inhibit the growth of the *B. subtilis* strain. Finally, it can be established that the presence of methoxyl groups in both chalcones and flavanones increases the antitumor activity.

Supplementary Materials: The following are available online at http://www.mdpi.com/2076-3417/9/14/2846/s1, Figure S1: 1H NMR spectrum of compound 1; Figure S2: 13C NMR spectrum of compound 1; Figure S3: HSQC NMR spectrum of compound 1; Figure S4: HMBC NMR spectrum of compound 1; Figure S5: ESI(+) Mass spectrum of compound 1; Figure S6: 1H NMR spectrum of compound 4; Figure S7: 13C NMR spectrum of compound 4; Figure S8: HSQC NMR spectrum of compound 4; Figure S9: HMBC NMR spectrum of compound 4; Figure S10: Antioxidant activity (%) presented by the compounds tested at three concentrations in the DPPH assay; Figure S11: Antioxidant activity (%) presented by the compounds tested at three concentrations in the ABTS assay; Figure S12: BuChE inhibitory activity (%) presented by the compounds tested at three concentrations; Figure S13: Inhibition of M. luteus growth (%) of the compounds tested at three concentrations; Figure S14: Inhibition of B. subtilis growth (%) of the compounds tested at three concentrations; Figure S15: Inhibition of A549 cell-line growth (%) of the compounds tested at three concentrations

Author Contributions: Conceptualization and methodology, A.M.L.S., M.C.B. and D.C.G.A.P.; formal analysis and investigation, G.P.R.; writing—original draft preparation, G.P.R.; writing—review and editing, A.M.L.S., M.C.B., A.M.S.S. and D.C.G.A.P.; supervision, A.M.L.S. and M.C.B.

Funding: This research was funded by FCT—Fundação para a Ciência e a Tecnologia, the European Union, QREN, FEDER, COMPETE, by funding the cE3c centre (FCT Unit funding (Ref. UID/BIA/00329/2013, 2015–2018) and UID/BIA/00329/2019) and the QOPNA research unit (project FCT UID/QUI/00062/2019).

Acknowledgments: Thanks are due to the University of Azores and University of Aveiro.

References

1. Miranda, C.L.; Maier, C.S.; Stevens, J.F. *Flavonoids*; John Wiley & Sons Ltd.: Chichester, UK, 2012. [CrossRef]
2. Santos, E.L.; Maia, B.H.S.; Ferriani, A.P.; Teixeira, S.D. Flavonoids: Classification, biosynthesis and chemical ecology. In *Flavonoids–from Biosynthesis to Human Health*; Justino, G.C., Ed.; Intech: London, UK, 2017; pp. 3–16, ISBN 978-953-51-3424-4.
3. Khan, M.K.; Huma, Z.E.; Dangles, O. A comprehensive review on flavanones, the major citrus polyphenols. *J. Food Compos. Anal.* **2014**, *33*, 85–104. [CrossRef]
4. Matos, M.J.; Vasquez-Rodriguez, S.; Uriarte, E.; Santana, L. Potential pharmacological uses of chalcones: A patent review (from June 2011-2014). *Expert Opin. Ther. Patents* **2014**, *25*, 3–16. [CrossRef] [PubMed]
5. Aoki, N.; Muko, M.; Ohta, E.; Ohta, S. C-Geranylated chalcones from the stems of *Angelica keiskei* with superoxide-scavenging activity. *J. Nat. Prod.* **2008**, *71*, 1308–1310. [CrossRef] [PubMed]
6. Mahapatra, D.K.; Bharti, S.K.; Asati, V. Anti-cancer chalcones: Structural and molecular target perspectives. *Eur. J. Med. Chem.* **2015**, *98*, 69–114. [CrossRef] [PubMed]
7. Wang, L.; Chen, G.; Lu, X.; Wang, S.; Hans, S.; Li, Y.; Ping, G.; Jiang, X.; Li, H.; Yang, J.; et al. Novel chalcone derivatives as hypoxia-inducible factor (HIF)-1 inhibitor: Synthesis, anti-invasive and anti-angiogenic properties. *Eur. J. Med. Chem.* **2015**, *89*, 88–97. [CrossRef] [PubMed]
8. Winter, E.; Locatelli, C.; Di Pietro, A.; Creczynski-Pasa, T.B. Recent trends of chalcones potentialities as antiproliferative and antiresistance agents. *Anticancer Agents Med. Chem.* **2015**, *15*, 592–604. [CrossRef] [PubMed]
9. Mascarello, A.; Chiaradia, L.D.; Vernal, J.; Villarino, A.; Guido, R.V.; Perizzolo, P.; Poirier, V.; Wong, D.; Martins, P.G.; Nunes, R.J.; et al. Inhibition of *Mycobacterium tuberculosis* tyrosine phosphatase PtpA by synthetic chalcones: Kinetics, molecular modeling, toxicity and effect on growth. *Bioorg. Med. Chem.* **2010**, *18*, 3783–3789. [CrossRef] [PubMed]
10. Tomar, V.; Bhattacharjee, G.; Kamaluddin, K.; Rajakumar, S.; Srivastava, K.; Puri, S.K. Synthesis of new chalcone derivatives containing acridinyl moiety with potential antimalarial activity. *Eur. J. Med. Chem.* **2010**, *45*, 745–751. [CrossRef] [PubMed]
11. Rizvi, S.U.F.; Siddiqui, H.L.; Johns, M.; Detorio, M.; Schinazi, R.F. Anti-HIV-1 and cytotoxicity studies of piperidyl-thienyl chalcones and their 2-pyrazoline derivatives. *Med. Chem. Res.* **2012**, *21*, 3741–3749. [CrossRef]
12. Abdullah, M.I.; Mahmood, A.; Madni, M.; Masood, S.; Kashif, M. Synthesis, characterization, theoretical, anti-bacterial and molecular docking studies of quinoline based chalcones as a DNA gyrase inhibitor. *Bioorg. Chem.* **2014**, *54*, 31–37. [CrossRef]
13. Birari, R.B.; Gupta, S.; Mohan, C.G.; Bhutani, K.K. Antiobesity and lipid lowering effects of glycyrrhizachalcones: Experimental and computational studies. *Phytomedicine* **2011**, *18*, 795–801. [CrossRef] [PubMed]
14. Kantevari, S.; Addla, D.; Bagul, P.K.; Sridhar, B.; Banerjee, S.K. Synthesis and evaluation of novel 2-butyl-4-chloro-1-methylimidazole embedded chalcones and pyrazoles as angiotensin converting enzyme (ACE) inhibitors. *Bioorg. Med. Chem.* **2011**, *19*, 4772–4781. [CrossRef] [PubMed]
15. Sashidhara, K.V.; Palnati, G.R.; Sonkar, R.; Avula, S.R.; Awasthi, C.; Bhatia, G. Coumarinchalcone fibrates: A new structural class of lipid lowering agents. *Eur. J. Med. Chem.* **2013**, *64*, 422–431. [CrossRef] [PubMed]
16. Mahapatra, D.K.; Asati, V.; Bharti, S.K. Chalcones and their role in management of diabetes *mellitus*: Structural and pharmacological perspectives. *Eur. J. Med. Chem.* **2015**, *92*, 839–865. [CrossRef] [PubMed]
17. Singh, P.; Anand, A.; Kumar, V. Recent developments in biological activities of chalcones: A mini review. *Eur. J. Med. Chem.* **2014**, *85*, 758–777. [CrossRef] [PubMed]
18. World Health Organization. Available online: http://www.who.int/news-room/fact-sheets/detail/cancer (accessed on 19 January 2019).
19. Siegel, R.L.; Miller, K.D.; Jemal, A. Cancer statistics. *CA Cancer J. Clin.* **2017**, *67*, 7–30. [CrossRef]
20. Cassini, A.; Högberg, L.D.; Plachouras, D.; Quattrocchi, A.; Hoxha, A.; Simonsen, G.S.; Colomb-Cotinat, M.; Kretzschmar, M.E.; Devleesschauwer, B.; Cecchini, M.; et al. Attributable deaths and disability-adjusted life-years caused by infections with antibiotic-resistant bacteria in the EU and the European Economic Area in 2015: A population-level modelling analysis. *Lancet Infect. Dis.* **2019**, *19*, 56–66. [CrossRef]

21. Tillotson, G.S.; Zinner, S.H. Burden of antimicrobial resistance in an era of decreasing susceptibility. *Expert Rev. Anti Infect. Ther.* **2017**, *15*, 663–676. [CrossRef]

22. Maresova, P.; Mohelska, H.; Dolejs, J.; Kuca, K. Socio-economic aspects of Alzheimer's disease. *Curr. Alzheimer Res.* **2015**, *12*, 903–911. [CrossRef]

23. Findley, L.J. The economic impact of Parkinson's disease. *Parkinsonism Relat. Disord.* **2007**, *13*, 8–12. [CrossRef]

24. Saeidnia, S.; Abdollahi, M. Toxicological and pharmacological concerns on oxidative stress and related diseases. *Toxicol. Appl. Pharmacol.* **2013**, *273*, 442–455. [CrossRef] [PubMed]

25. Bukhari, S.N.A.; Jasamai, M.; Jantan, I.; Ahmad, W. Review of methods and various catalysts used for chalcone synthesis. *Mini Rev. Org. Chem.* **2013**, *10*, 73–83. [CrossRef]

26. Ugwu, D.I.; Ezema, B.E.; Okoro, U.C.; Eze, F.U.; Ekoh, O.C.; Egbujor, M.C.; Ugwuja, D.I. Synthesis and pharmacological applications of chalcones: A review. *Int. J. Chem. Sci.* **2015**, *13*, 459–500.

27. Hsieh, C.T.; Hsieh, T.J.; El-Shazly, M.; Chuang, D.W.; Tsai, Y.H.; Yen, C.T.; Wu, S.F.; Wu, Y.C.; Chang, F.R. Synthesis of chalcone derivatives as potential anti-diabetic agents. *Bioorg. Med. Chem. Lett.* **2012**, *22*, 3912–3915. [CrossRef] [PubMed]

28. Perrin, C.L.; Chang, K.L. The complete mechanism of an aldol condensation. *J. Org. Chem.* **2016**, *81*, 5631–5635. [CrossRef] [PubMed]

29. Clayden, J.; Greeves, N.; Warren, S. *Organic Chemistry*, 2nd ed.; Oxford University Press: Oxford, UK; New York, NY, USA, 2012; pp. 141–162, ISBN 978-0199270293.

30. Mellado, M.; Madrid, A.; Martinéz, U.; Mella, J.; Salas, C.O.; Cuellar, M. Hansch's analysis application to chalcone synthesis by Claisen-Schmidt reaction based DFT methodology. *Chem. Pap.* **2018**, *72*, 703–709. [CrossRef]

31. Dharmaratne, H.R.W.; Nanayakkara, N.P.D.; Khan, I.A. Kavalactones from *Piper methysticum,* and their ^{13}C NMR spectroscopic analyses. *Phytochemistry* **2002**, *59*, 429–433. [CrossRef]

32. Figueiredo, A.G.P.R. Transformações de Cromona-3-Carbaldeído e de (*E*)-*o*-Dihidroxi-2-Estirilcromonas. Ph.D. Thesis, University of Aveiro, Aveiro, Portugal, July 2008.

33. Gomes, A.; Neuwirth, O.; Freitas, M.; Couto, D.; Ribeiro, D.; Figueiredo, A.G.P.R.; Silva, A.M.S.; Seixas, R.S.G.R.; Pinto, D.C.G.A.; Tomé, A.C.; et al. Synthesis and antioxidant properties of new chromone derivatives. *Bioorg. Med. Chem.* **2009**, *17*, 7218–7226. [CrossRef]

34. Cavaleiro, J.A.S. Isosakuranetin, a flavonone from *Artemisia compestris* subsp. *maritima. Fitoterapia* **1986**, *57*, 278–279.

35. Silva, A.M.S.; Tavares, H.R.; Barros, A.I.N.R.A.; Cavaleiro, J.A.S. NMR and structural and conformational features of 2'-hydroxychalcones and flavones. *Spectrosc. Lett.* **1997**, *30*, 1655–1667. [CrossRef]

36. Silva, A.M.S.; Cavaleiro, J.A.S.; Tarrago, G.; Marzin, C. Synthesis and characterization of ruthenium(II) complexes of 2'-hydroxychalcones. *New J. Chem.* **1999**, *23*, 329–335. [CrossRef]

37. Blois, M. Antioxidant determinations by the use of a stable free radical. *Nature* **1958**, *181*, 1199–1200. [CrossRef]

38. Re, R.; Pellegrini, N.; Proteggente, A.; Pannala, A.; Yang, M.; Rice-Evans, C. Antioxidant activity applying an improved ABTS radical cation decolorization assay. *Free Radic. Biol. Med.* **1999**, *26*, 1231–1237. [CrossRef]

39. Ellman, G.; Courtney, K.; Andres, V.; Featherstone, R. A new and rapid colorimetric determination of acetylcholinesterase activity. *Biochem. Pharmacol.* **1961**, *7*, 88–95. [CrossRef]

40. Arruda, M.; Viana, H.; Rainha, N.; Neng, N.; Rosa, J.; Nogueira, J.; Barreto, M.C. Anti-acetylcholinesterase and antioxidant activity of essential oils from *Hedychium gardnerianum* Sheppard ex Ker-Gawl. *Molecules* **2012**, *17*, 3082–3092. [CrossRef] [PubMed]

41. De León, L.; Beltrán, B.; Moujir, L. Antimicrobial activity of 6-oxophenolic triterpenoids. Mode of action against *Bacillus subtilis. Planta Med.* **2005**, *71*, 313–319. [CrossRef]

42. Moujir, L.L.; Gouveia, V.; Toubarro, D.; Barreto, M.C. Determination of Cytotoxicity Against Tumour Cell Lines. In *Determination of Biological Activities*; Barreto, M.C., Simões, N., Eds.; Azores University: Ponta Delgada, Portugal, 2012; pp. 41–62.

43. Rocha, D.H.A.; Pinto, D.C.G.A.; Silva, A.M.S. One-pot synthesis of 3-methylflavones and their transformation into (*E*)-3-styrilflavones via Wittig reactions. *Synlett* **2013**, *24*, 2683–2686. [CrossRef]

44. Gan, L.S.; Zeng, L.W.; Li, X.R.; Zhou, C.X.; Li, J. New homoisoflavonoid analogues protect cells by regulating autophagy. *Bioorg. Med. Chem. Lett.* **2017**, *27*, 1441–1445. [CrossRef]

45. Aköz, B.E.; Ertan, R. Chemical and structural properties of chalcones I. *FABAD J. Pharm. Sci.* **2011**, *36*, 223–242.

46. Cushman, M.; Nagarathnan, D. A method for the facile synthesis of ring-A hydroxylated flavones. *Tetrahedron Lett.* **1990**, *31*, 6497–6500. [CrossRef]

47. Vaz, P.A.A.M.; Pinto, D.C.G.A.; Rocha, D.H.A.; Silva, A.M.S.; Cavaleiro, J.A.S. New synthesis of 3-aroylflavone derivatives; Knoevenagel condensation and oxidation versus one-pot synthesis. *Synlett* **2012**, *23*, 2353–2356. [CrossRef]

48. Aida, K.; Tawata, M.; Shindo, H.; Onaya, T.; Sasaki, H.; Yamaguchi, T.; Chin, M.; Mitsuhashi, H. Isoliquiritigenin: A new aldose reductase inhibitor from glycyrrhizae radix. *Planta Med.* **1990**, *56*, 254–258. [CrossRef] [PubMed]

49. Suksamrarn, A.; Chotipong, A.; Suavansri, T.; Boongird, S.; Timsuksai, P.; Vimuttipong, S.; Chuaynugul, A. Antimycobacterial activity and cytotoxicity of flavonoids from the flowers of *Chromolaena odorata*. *Arch. Pharm. Res.* **2004**, *27*, 507–511. [CrossRef] [PubMed]

50. Shoja, M. Crystal structure of 4′-hydroxy-5,7-dimethoxyflavanone, $C_{17}H_{16}O_5$. *Z. Kristallogr. New Cryst. Struct.* **1997**, *212*, 127–128. [CrossRef]

51. Engels, N.S.; Waltenberger, B.; Michalak, B.; Huynh, L.; Tran, H.; Kiss, A.K.; Stuppner, H. Inhibition of pro-inflammatory functions of human neutrophils by constituents of *Melodorum fruticosum* leaves. *Chem. Biodivers.* **2018**, *15*, e1800269. [CrossRef] [PubMed]

52. Khamsan, S.; Liawruangrath, S.; Teerawutkulrag, A.; Pyne, S.G.; Garson, M.J.; Liawruangrath, B. The isolation of bioactive flavonoids from *Jacaranda obtusifolia* H.B.K. ssp. *rhombifolia* (G.F.W. Meijer) Gentry. *Acta Pharm.* **2012**, *62*, 181–190. [CrossRef] [PubMed]

53. Badarinath, A.V.; Rao, K.M.; Chetty, C.M.S.; Ramkanth, S.; Rajan, T.V.S.; Gnanaprakash, K. A review of *in vitro* antioxidant methods: Comparisons, correlations and considerations. *Int. J. PharmTech Res.* **2010**, *2*, 1276–1285.

54. Nimse, S.B.; Pal, D. Free radicals, natural antioxidants and their reaction mechanisms. *RSC Adv.* **2015**, *5*, 27986–28006. [CrossRef]

55. Rice-Evans, C.A.; Miller, N.J.; Paganga, G. Antioxidant properties of phenolic compounds. *Trends Plant Sci.* **1997**, *2*, 152–159. [CrossRef]

56. Rossi, M.; Caruso, F.; Crespi, E.J.; Pedersen, J.Z.; Nakano, G.; Duong, M.; Mckee, C.; Lee, S.; Jiwrajka, M.; Caldwell, C.; et al. Probing antioxidant activity of 2′-hydroxychalcones: Crystal and molecular structures, *in vitro* antiproliferative studies and *in vivo* effects on glucose regulation. *Biochimie* **2013**, *95*, 1954–1963. [CrossRef]

57. Cerretani, L.; Bendini, A. Rapid assays to evaluate the antioxidant capacity of phenols in virgin olive oil. In *Olives and Olive Oil in Health and Disease Prevention*; Preedy, V.R., Watson, R.R., Eds.; Elsevier: Albany, NY, USA, 2010; pp. 625–635. ISBN 9780123744203.

58. Chokchaisiri, R.; Suaisom, C.; Sriphota, S.; Chindaduang, A.; Chuprajob, T.; Suksamrarn, A. Bioactive flavonoids of the flowers of *Butea monosperma*. *Chem. Pharm. Bull.* **2009**, *57*, 428–432. [CrossRef] [PubMed]

59. Hsu, Y.L.; Kuo, P.O.; Chiang, L.C.; Lin, C.C. Isoliquiritigenin inhibits the proliferation and induces apoptosis of human non-small lung cancer A549 cells. *Clin. Exp. Pharmacol. Physiol.* **2004**, *31*, 414–418. [CrossRef] [PubMed]

60. Wang, Y.; Xie, S.; Liu, C.; Wu, Y.; Liu, Y.; Cai, Y. Inhibitory effect of liquiritigenin on migration via downregulation ProMMP-2 and PI3K/Akt signaling pathway in human lung adenocarcinoma A549 cells. *Nutr. Cancer* **2012**, *64*, 627–634. [CrossRef] [PubMed]

61. Tang, Z.H.; Li, T.; Tong, Y.G.; Chen, X.J.; Chen, X.P.; Wang, Y.W.; Lu, J.J. A systematic review of the anticancer properties of compound isolated from licorice (Gancao). *Planta Med.* **2015**, *81*, 1670–1687. [CrossRef] [PubMed]

62. Lieberman, M.M.; Patterson, G.M.L.; Moore, R.E. *In vitro* bioassays for anticancer drug screening: Effects of cell concentration and other assay parameters on growth inhibitory activity. *Cancer Lett.* **2001**, *173*, 21–29. [CrossRef]
63. Marinho, J.; Pedro, M.; Pinto, D.C.G.A.; Silva, A.M.S.; Cavaleiro, J.A.S.; Sunkel, C.E.; Nascimento, M.S.J. 4′-Methoxy-2-styrylchromone a novel microtubule-stabilizing antimitotic agente. *Biochem. Pharmacol.* **2008**, *75*, 826–835. [CrossRef]

Article

Volatile Secondary Metabolites with Potent Antidiabetic Activity from the Roots of *Prangos pabularia* Lindl.—Computational and Experimental Investigations

Sodik Numonov [1,2,3], Farukh S. Sharopov [1,2,4], Sunbula Atolikhshoeva [1], Abduahad Safomuddin [5], Mahinur Bakri [1], William N. Setzer [6,7], Azizullo Musoev [1], Mizhgona Sharofova [3], Maidina Habasi [1,2,*] and Haji Akber Aisa [1,2,*]

[1] Key Laboratory of Plant Resources and Chemistry in Arid Regions, Xinjiang Technical Institute of Physics and Chemistry, Chinese Academy of Sciences, Urumqi 830011, China; sodikjon82@gmail.com (S.N.); shfarukh@mail.ru (F.S.S.); sunbula87@mail.ru (S.A.); mahinur@ms.xjb.ac.cn (M.B.); azizullo.dr@gmail.com (A.M.)

[2] Research Institution "Chinese-Tajik Innovation Center for Natural Products", Ayni St. 299/2, Dushanbe 734063, Tajikistan

[3] Center for Research in Innovative Technologies, Academy of Sciences of the Republic of Tajikistan, Ayni St. 299/3, Dushanbe 734063, Tajikistan; mijgona72@mail.ru

[4] Department of Pharmaceutical Technology, Avicenna Tajik State Medical University, Rudaki 139, Dushanbe 734003, Tajikistan

[5] Department of Chemistry, National University of Tajikistan, Rudaki 17/2, Dushanbe 734003, Tajikistan; asafomuddin@rambler.ru

[6] Department of Chemistry, University of Alabama in Huntsville, Huntsville, AL 35899, USA; wsetzer@chemistry.uah.edu

[7] Aromatic Plant Research Center, 230 N 1200 E, Suite 100, Lehi, UT 84043, USA

* Correspondence: maidn@ms.xjb.ac.cn (M.H.); haji@ms.xjb.ac.cn (H.A.A.); Tel.: +86-991-3835679 (M.H. & H.A.A.); Fax: +86-991-3838957 (M.H. & H.A.A.)

Received: 14 May 2019; Accepted: 31 May 2019; Published: 10 June 2019

Abstract: (1) Background: Almost 500 million people worldwide are suffering from diabetes. Since ancient times, humans have used medicinal plants for the treatment of diabetes. Medicinal plants continue to serve as natural sources for the discovery of antidiabetic compounds. *Prangos pabularia* Lindl. is a widely distributed herb with large reserves in Tajikistan. Its roots and fruits have been used in Tajik traditional medicine. To our best knowledge, there are no previously published reports concerning the antidiabetic activity and the chemical composition of the essential oil obtained from roots of *P. pabularia*. (2) Methods: The volatile secondary metabolites were obtained by hydrodistillation from the underground parts of *P. pabularia* growing wild in Tajikistan and were analyzed by gas chromatography (GC) and gas chromatography-mass spectrometry (GC-MS). Protein tyrosine phosphatase 1B (PTP-1B) inhibition assay and molecular docking analysis were carried out to evaluate the potential antidiabetic activity of the *P. pabularia* essential oil. (3) Results: The main constituents of the volatile oil of *P. pabularia* were 5-pentylcyclohexa-1,3-diene (44.6%), menthone (12.6%), 1-tridecyne (10.9%), and osthole (6.0%). PTP-1B inhibition assay of the essential oil and osthole resulted in significant inhibitory activity with an IC_{50} value of 0.06 ± 0.01 and 0.93 ± 0.1 µg/mL. Molecular docking analysis suggests volatile compounds such as osthole inhibit PTP-1B, and the results are also in agreement with experimental investigations. (4) Conclusions: Volatile secondary metabolites and the pure isolated compound (osthole) from the roots of *P. pabularia* exhibited potent antidiabetic activity, twenty-five and nearly two times more than the positive control (3-(3,5-dibromo-4-hydroxybenzoyl)-2-ethylbenzofuran-6-sulfonic acid-(4-(thiazol-2-ylsulfamyl)-phenyl)-amide)) with an IC_{50} value of 1.46 ± 0.4 µg/mL, respectively.

Keywords: *Prangos pabularia* Lindl.; volatile oil; PTP-1B; osthole; 5-pentylcyclohexa-1,3-diene; antidiabetic activity

1. Introduction

Diabetes, or diabetes mellitus, is a chronic metabolic disease associated with high blood sugar levels over a prolonged period [1]. In 2017, according to the International Diabetes Federation report, approximately 425 million adults (20–79 years) had diabetes worldwide with 3.2 to 5.0 million deaths from the disease [2]. Unfortunately, these numbers are gradually increasing in most countries.

From ancient times, humans have used medicinal plants for the prevention and the therapy of diabetes mellitus. Medicinal plants serve as natural sources for the discovery of compounds with antidiabetic activities. In this relation, Tajikistan has a rich flora with around 4550 species of higher plants that represent great interest for the discovery of alternative medications for the treatment of diabetes [3]. Tajikistan is known for its diversity of environmental conditions, including climate, high altitudes, mountainous soil and minerals, and a relatively large number of sunny days per year, factors that can affect plant development as well as biosynthesis and accumulation of secondary metabolites [4,5]. Essential oils have applications in medicine, pharmaceutical, food, and cosmetic industries. They possess possible health benefits with antioxidant, antimicrobial, antitumor, anticarcinogenic, anti-inflammatory, anti-atherosclerosis, antimutagenic, antiplatelet aggregation, and angiogenesis-inhibitory activities. Therefore, extensive research has been directed toward the use of medicinal plants to control diabetes mellitus and its complications [6].

PTP-1B (protein-tyrosine phosphatase 1B) belongs to the protein tyrosine phosphatase (PTP) family. It is also known as tyrosine-protein phosphatase non-receptor type 1 that is encoded by the *PTPN1* gene [7,8]. PTP-1B is localized on the cytoplasmic face of the endoplasmic reticulum and contains the essential catalytic cysteine [9]. The PTP-1B enzyme catalyzes the hydrolysis of phosphotyrosine from specific proteins [10]. PTP-1B is considered to be a promising potential therapeutic target for treatment of various diseases, including diabetes, obesity, and cancer [11]. It inactivates the insulin signal transduction cascade by dephosphorylating phosphotyrosine residues in the insulin-signaling pathway [12]. Natural, synthetic, as well as semi-synthetic compounds have shown prominent antidiabetic activities by inhibiting PTP-1B activity [13].

Prangos pabularia Lindl., a member of the Apiaceae, is a widely distributed herb up to 150 cm high with a thick cylindrical root and has large reserves in Tajikistan [14]. *P. pabularia*, locally known as "Yugan", is a well-known species of the genus in Tajikistan. It typically grows in mountainous areas and limestone slopes at altitudes 780 to 3600 m above sea level [15]. Its roots and fruits are valued in Tajik traditional medicine and are widely used as general tonics as well as for treatment of vitiligo [4,15]. *P. pabularia* has been used to treat leukoplakia, digestive disorders, scars, and bleeding [16]. The root extracts of *P. pabularia* have been examined for cytotoxic activity; the dichloromethane extract of *P. pabularia* roots demonstrated notable cytotoxicity on the HeLa carcinoma cell line [17]. *P. ferulacea* root is used as an effective wound healing agent in traditional medicine of the western north of Iran [18].

Members of the *Prangos* genus are natural sources of phytochemicals, including coumarins and terpenoids. Individual isolated pure compounds such as osthole and isoimperatorin showed the highest inhibitory potency against the growth of human carcinoma cell lines. Osthole exhibited the greatest cytotoxicity and was found to induce apoptosis in PC-3, H1299, and SKNMC cells at low micromolar concentrations. Thus, osthole can be considered to be a promising lead in anticancer drug discovery and development [17]. Several new compounds have been isolated from the essential oil of *Prangos* species. A new bisabolene derivative was isolated from essential oil of the fruits of Turkish endemic *Prangos uechtritzii* [19]. The 3,7(11)-Eudesmadien-2-one, a new eudesmane type sesquiterpene ketone was isolated from *Prangos heyniae* H. Duman & M.F. Watson essential oil [20].

The (2*S*)-3,5-Nonadiyne-2-yl acetate was isolated from *Prangos platychlaena* ssp. *platychlaena* fruit essential oils [21].

Recently, we reported that the roots of *P. pabularia* are good sources of biologically active secondary metabolites (coumarins) such as heraclenol, heraclenin, imperatorin, osthole, yuganin A, and others. Yuganin A showed potent effects on the proliferation of B16 melanoma cells [22]. In a continuation of this investigation, the current report presents the promising antidiabetic activity and the chemical composition of volatile secondary metabolites of the underground parts of *P. pabularia* growing wild in Tajikistan. There are several reports on the composition of the essential oils isolated from leaves, fruits, and umbels of *P. pabularia* growing in Iran and Turkey [23–25], but, until now, there has been no published reports on antidiabetic activity and volatile secondary metabolites of the underground parts of *P. pabularia*.

2. Results and Discussions

2.1. Chemical Composition of Essential Oils

Volatile secondary metabolites were obtained by hydrodistillation of *P. pabularia* roots growing wild in Tajikistan and were analyzed by gas chromatography (GC) and gas chromatography-mass spectrometry (GC-MS). Identification of the oil components was based on their Kovats retention indices (RI) determined by reference to a homologous series of *n*-alkanes and by comparison of their mass spectral fragmentation patterns with those reported in the literature [26] and stored in the MS library. Forty-two compounds were identified in the volatile oil accounting for 97.3% of the composition; 5-Pentylcyclohexa-1,3-diene (44.6%), menthone (12.6%), 1-tridecyne (10.9%), and osthole (6.0%) were identified as major constituents of the volatile oil obtained from roots of *P. pabularia* (Table 1). The structure of the osthole was established on the basis one-dimensional (1D) NMR and electrospray ionization (ESI)-MS spectroscopic studies, respectively [22]. The chemical structures of the main components of the essential oil from the roots of *P. pabularia* are presented in Figure 1. The GC-MS chromatogram of the volatile oil of *P. pabularia* is presented in Figure 2.

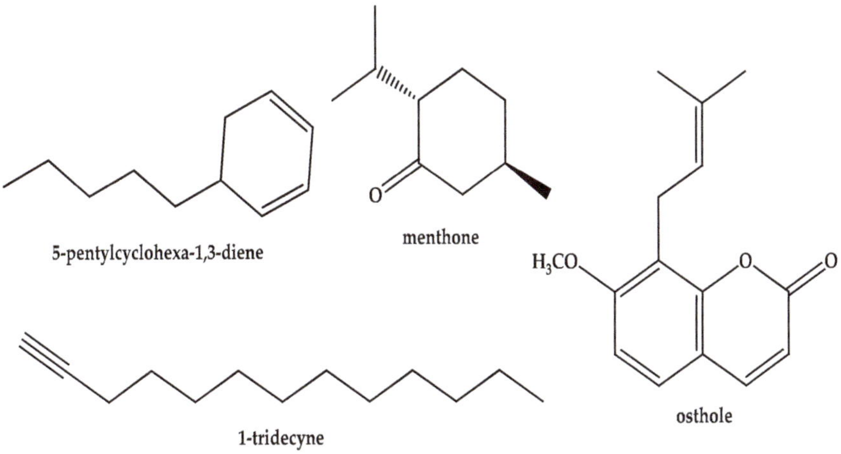

Figure 1. Structures of the main components of *Prangos pabularia* essential oil.

Table 1. Chemical composition of the essential oil of the roots of *Prangos pabularia* growing wild in Tajikistan.

RT	RI	Compound	MS	Fragmentations, *m/z* (%)	%
5.487	934	(*E*)-1,3-Nonadiene	124.13	67 (100%); 54 (82.37%); 124 (41.49%); 68 (36.07%); 81 (35.96%)	1.5
5.556	937	1-Nonen-4-yne	122.11	79 (100%); 77 (49.67%); 78 (21.32%); 80 (19.82%)	0.8
5.674	941	α-Pinene	136.13	93 (100%); 91 (74.23%); 79 (52.77%); 77 (51.98%); 92 (42.0%)	0.1
6.006	953	Camphene	136.13	93 (100%); 91 (81.41%); 77 (56.16%); 79 (54.03%); 92 (31.5%)	Tr
6.106	957	Propylbenzene	120.09	91 (100%); 65 (18.7%); 120 (17.78%); 92 (10.48%); 78 (8.87%)	0.2
6.162	959	(3*E*)-2-Methylocten-5-yne	122.11	79 (100%); 91 (86.08%); 77 (84.95%); 93 (59.7%); 122 (35.51%)	1.1
6.656	977	Sabinene	136.14	93 (100%); 77 (69.92%); 79 (68.13%); 91 (58.51%); 53 (30.37%)	0.2
6.974	989	β-Pinene	136.13	93 (100%); 91 (28.2%); 69 0(26.9%); 79 (23.51%); 77.1 (22.1%)	0.1
7.068	993	3-Octanol	130.14	59 (100%); 83 (81.4%); 55 (78.3%); 101 (29.96); 57 (23.3%)	Tr
7.324	1002	α-Phellandrene	136.13	93 (100%); 122 (84.73%); 91 (81.47%); 79 (75.95%); 107 (52.5%)	0.1
7.468	1008	δ-3-Carene	136.13	93 (100%); 91 (95.41%); 77(56.16%); 79 (54.03%); 92 (31.5%)	0.1
7.849	1022	1,9-Decadiyne	134.11	79 (100%); 77 (52.95%); 67 (51.87%); 91 (49.70%); 81 (37.7%)	0.3
7.943	1025	Limonene	136.13	93 (100%); 68 (80.91%); 67 (80.89%); 79 (80.07%); 91 (73.16%)	0.4
7.999	1027	1,8-Cineole	154.14	81 (100%); 111 (90%); 67 (87.77%); 108 (87.37%); 55 (85.72%)	0.2
8.162	1033	(*Z*)-β-Ocimene	136.13	93 (100%); 91 (86.50%); 79 (63.32%); 77 (43.30%); 92 (36.2%)	0.1
8.437	1044	(*E*)-β-Ocimene	136.13	93 (100%); 91 (86.50%); 79 (63.32%); 77 (43.30%); 80 (37.2%)	0.1
8.693	1053	5-Butylcyclohexa-1,3-diene	136.13	79 (100%); 91 (84.32%); 77 (79.63%); 93 (49.2%); 136 (36.48%)	0.1
11.399	1154	Menthone	154.14	112 (100%); 55 (77.25%); 69 (69.36%); 139 (44.49%); 97 (34.09%)	12.6
11.543	1160	5-Pentylcyclohexa-1,3-diene	150.14	79 (100%); 91 (65.34%); 77 (58.97%); 93 (58.47%); 94 (35.13%)	44.6
11.612	1162	Viridene	150.14	138 (100%); 124 (39.8%); 93 (25.5%); 137 (26.1)	0.3
11.649	1164	*iso*-Menthone	154.14	112 (100%); 55 (91.02%); 69 (65.32%); 95 (37.37%); 139 (35.89%)	1.0
11.831	1170	2-Methoxy-3-(1-methylpropyl)pyrazine	166.11	138 (100%); 124 (39.81%); 151 (37.35%); 93 (35.52%); 137 (26.1%)	1.8
11.974	1176	*neoiso*-Pulegol	150.14	67 (100%); 55 (73.02%); 53 (62.77%); 69 (56.57%); 79 (53.38%)	0.8
13.449	1231	Unidentified	-	55 (100%); 57 (73.89%); 56 (68.30%); 71 (65.04%); 69 (59.13%)	1.7
13.649	1239	Pulegone	152.12	81 (100%); 67 (93.29%); 109 (57.3%); 152 (48.89%); 82. (38.27%)	1.3
13.731	1242	Unidentified	-	138 (100%); 108 (27.65%); 95 (19.27%); 109 (17.38%); 54 (16.08%)	0.7
14.018	1252	Unidentified	-	138 (100%); 108 (35.28%); 95 (27.12%); 109 (15.38%); 54 (10.2%)	0.2
14.081	1255	*cis*-Piperitone epoxide	168.12	55 (100%); 69 (80.71%); 67 (49.99%); 125 (44.00%); 53 (43.72%)	1.3
14.118	1256	(4*Z*)-Decen-1-ol	156.15	55 (100%); 70 (64.85%); 69 (62.42%); 56 (57.42%); 83 (49.55%)	0.5
14.543	1272	1-Decanol	158.17	55 (100%); 70 (64.85%); 69 (62.42%); 56 (57.42%); 83 (49.55%)	2.0
15.299	1300	1-Tridecyne	180.19	81 (100%); 55 (79.41%); 67 (78.15%); 69 (47.9%); 68 (40.50%)	10.9
17.012	1365	Piperitenone oxide	166.10	67 (100%); 138 (63.48%); 68 (62.28%); 53 (38.29%); 79 (36.54%)	0.7
17.143	1371	1-Undecanol	172.18	55 (100%); 69 (73.24%); 56 (61.43%); 83 (55.86%); 70 (49.46%)	0.3

Table 1. *Cont.*

RT	RI	Compound	MS	Fragmentations, *m/z* (%)	%
17.868	1399	3-Dodecyn-2-ol	182.17	55 (100%); 67 (99.69%); 69 (76.48%); 95 (72.69%); 68 (56.92%)	2.1
18.237	1414	β-Longipinene	190.17	91 (100%); 77 (72.64%); 93 (62.14%); 161 (58.85%); 105 (57.88%)	0.6
18.406	1421	β-Caryophyllene	204.19	91 (100%); 79 (86.59%); 93 (67.92%); 77 (66.1%); 105 (55.12%)	1.0
18.718	1433	γ-Elemene	204.19	121 (100%); 177 (98.85%); 91 (58.5%); 93 (57.4%); 107 (43.1%)	0.1
18.781	1436	*trans*-α-Bergamotene	204.19	93 (100%); 91 (80.84%); 119 (75.05%); 77 (57.71%); 69 (48.76%)	1.6
19.331	1459	(7Z)-Dodecen-1-ol	184.18	67 (100%); 55 (62.01%); 81 (56.91%); 54 (49.91%); 82 (47.42%)	0.2
19.768	1477	γ-Gurjunene	204.19	91 (100%); 77 (87.25%); 79 (83.86%); 93 (79.86%); 161 (76.27%)	0.1
20.056	1489	β-Selinene	204.19	79 (100%); 91 (75.67%); 67 (75.38%); 93 (63.4%); 105 (61.85%)	0.5
20.193	1494	(3Z,6E)-α-Farnesene	204.19	119 (100%); 91 (69.2%); 79 (60.9%); 81 (58.5%); 77 (55.6%)	0.5
20.231	1496	Valencene	204.19	91 (100%); 79 (97.61%); 105 (83.71%); 93 (80.93%); 77 (75.81%)	0.8
20.274	1498	α-Selinene	204.19	93 (100%); 91 (83.3%); 69 (68.3%); 105 (50.3%); 77 (49.2%)	0.3
33.581	2141	Osthole	244.11	244 (100%); 201 (93.94%); 229 (93.95%); 131 (65.60%); 189 (63.98%)	6.0
		Monoterpene hydrocarbons			1.3
		Oxygenated monoterpenoids			17.8
		Sesquiterpene hydrocarbons			5.5
		Fatty acid derived			19.8
		Others			53.0
		Total Identified			97.3

*RI: retention indices.

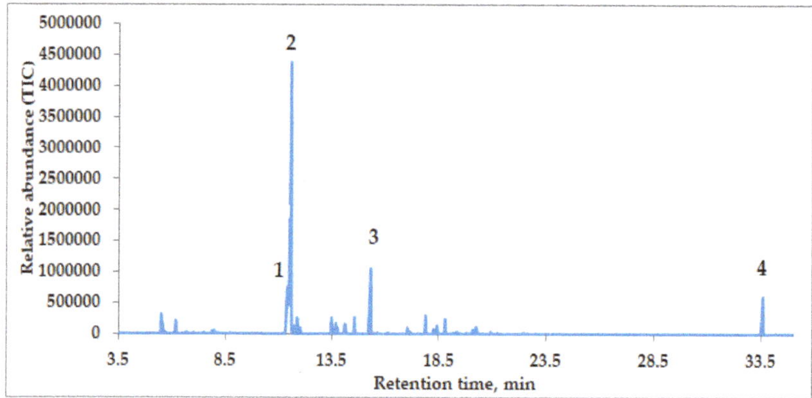

Figure 2. The GC-MS chromatogram of the volatile oil of *Prangos pabularia*. 1. menthone; 2. 5-pentylcyclohexa-1,3-diene; 3. 1-tridecyne; 4. osthole.

2.2. NMR Data of Osthole

^{1}H NMR (400 MHz, CDCl$_3$): δ 6.23 (1H, d, *J* = 9.4 Hz, H-3), 7.61 (1H, d, *J* = 9.4 Hz, H-4), 7.28 (1H, d, *J* = 8.6 Hz, H-5), 6.83 (1H, d, *J* = 8.6 Hz, H-6), 3.53 (1H, d, *J* = 7.3 Hz, H-11), 5.22 (1H, m, H-12), 1.67 (3H, s, H-14), 1.84 (3H, s, H-115), 3.92 (3H, s, OCH3-7); ^{13}C NMR (100 MHz, CDCl$_3$): δ 161.49 (C-2), 113.11 (C-3), 143.87 (C-4), 126.32 (C-5), 107.47 (C-6), 160.34 (C-7), 118.10 (C-8), 152.95 (C-9), 113.14 (C-10), 22.06 (C-11), 121.25 (C-12), 132.75 (C-13), 18.06 (C-14), 25.91 (C-15), 56.17 (OCH3-7).

Acorenone, (*E*)-anethol, β-bisabolenal, β-bisabolenol, β-bisabolene, bicyclogermacrene, δ-3-carene, chrysanthenyl acetate, β-caryophyllene, elemol, 3,7(11)-eudesmadien-2-one, geranial, germacrene D, α-humulene, kessane, limonene, *p*-menth-3-ene, nerolidol, (*Z*)-β-ocimene, (*E*)-β-ocimene, α-pinene, β-pinene, α-phellandrene, β-phellandrene, sabinene, γ-terpinene, α- terpinolene, *m*-tolualdehyde, and 2,3,6-trimethyl benzaldehyde were reported as major components (≥10%) in the essential oil of *Prangos* species (Table 2).

Table 2. The major compounds reported as chemical composition of essential oil from *Prangos* species.

Prangos Species	Plant Part	Major Components of the Essential Oil	Ref.
P. acaulis	aerial parts	δ-3-carene (25.5%), α-terpinolene (14.8%), α-pinene (13.6%), limonene (12.9%), myrcene (8.1%)	[27]
P. asperula	fruits	δ-3-carene (16.1%), β-phellandrene (14.7%), α-pinene (10.5%), α-humulene (7.8%), germacrene-D (5.4%)	[28]
P. asperula	fresh aerial parts	sabinene (43.5%), β-phellandrene (36.1%), (*E*)-nerolidol (15.2%), *p*-menth-3-ene (14.9%), (*E*)-nerolidol (14.7%), *p*-menth-3-ene (13.3%) in stem and leaves, α-phellandrene (11.9%) in fruits, β-myrcene (9.2%) in stem and leaves; α-terpinene (8.3%) in fruits, β-phellandrene (7.9%) in flowers	[29]
P. asperula	aerial parts	2,3,6-trimethyl benzaldehyde (18.4%), δ-3-carene (18.0%) and α-pinene (17.4%)	[30]
P. asperula	fruit	sabinene (43.5%)	[31]
P. ferulacea	fruits	α-pinene (57%) (vegetative stage), γ-terpinene (30.2–33.3%) and α-pinene (16.7–12.8%)	[32]
P. ferulacea	aerial parts	(*E*)-anethol (95.5%) (flowering stage)	[16]
P. ferulacea	fruits	chrysanthenyl acetate (26.5%), limonene (19.6%), α-pinene (19.5%)	[33]
P. ferulacea	aerial parts	β-pinene (43.1%), α-pinene (22.1%) and δ-3-carene (16.9%)	[34]
P. ferulacea	aerial parts	2,3,6-trimethyl benzaldehyde (66.6%)	[35]

Table 2. *Cont.*

Prangos Species	Plant Part	Major Components of the Essential Oil	Ref.
P. ferulacea	aerial parts	β-caryophyllene (48.2%), α-humulene (10.3%) and spathulenol (9.4%)	[36]
P. ferulacea	aerial parts	α-pinene (36.6%) and β-pinene (31.1%)	[37]
P. ferulacea	roots	β-phellandrene (32.1%), *m*-tolualdehyde (26.2%), and δ-3-carene (25.8%)	[18]
P. ferulaceae	fruits and umbels	α-pinene and (*Z*)-β-ocimene	[38]
P. heyniae	fruits	β-bisabolenal (18.0–53.3%), β-bisabolenol (2.3–14.6%) and β-bisabolene (10.1–12.1%)	[39]
P. heyniae	aerial parts	β-bisabolenal (1.4–70.7%), (8.2%), elemol (3.4–46.9%), kessane (26.9%), β-bisabolene (14.4%), germacrene D (10.3–12.1%), germacrene B 3,7(11)-eudesmadien-2-one (16.1%) and β-bisabolenol (8.4%)	[20]
P. latiloba	aerial parts	geranial (26.8%)	[31]
P. pabularia	flowering aerial parts	α-pinene (32.4%), δ-3-carene (12.4%), germacrene D (8.1%), limonene (6.4%) and bicyclogermacrene (6.2%)	[40]
P. pabularia	fruit	bicyclogermacrene (21%), (*Z*)-β-ocimene (19%), α-humulene (8%), α-pinene (8%), spathulenol (6%), suberosin (2%)	[24]
P. peucedanifolia	flowering aerial parts	α-pinene (38.1%), bicyclogermacrene (11.3%) and δ-3-carene (9.2%)	[40]
P. uloptera	aerial parts	δ-3-carene (26.3%), α-pinene (15.4%), β-myrcene (9.0%), *p*-cymene (8.6%)	[41]
P. uloptera	aerial parts	β-caryophyllene (18.2%), germacrene D (17.2%) and limonene (8.7%)	[42]
P. uloptera	seed	α-pinene (41.9%) and β-cedrene (4.0%)	[42]
P. uloptera	aerial parts	α-pinene (20.3%), (*E*)-β-ocimene (19.6%), β-caryophyllene in fresh aerial parts; β-caryophyllene (13.9%), α-pinene (13.6%), caryophyllene-oxide (11.6%) in dried aerial parts; (9.9%), δ-3-carene (8.0%), germacrene D (6.0%) in fresh aerial parts; spathulenol (7.8%) and germacrene D (4.7%) in dried aerial parts	[43]
P. uloptera	fruits	germacrene D (17.6%), acorenone (16.9%), α-pinene (14.9%), and α-humulene (8.2%)	[44]

Razavi reported that the composition of the essential oils isolated from leaves, fruits, and umbels of *P. pabularia* collected from Iran were dominated by spathulenol, α-bisabolol, and α-pinene [25]. Bicyclogermacrene, (*Z*)-β-ocimene, α-humulene, α-pinene, and spathulenol were reported as the main constituents of the essential oil of *P. pabularia* fruits collected from Turkey [24]. The chemical composition of the root essential oil of *P. pabularia* differed from those from leaves, fruits, and umbels with regard to predominance of sesquiterpenes and monoterpenes. In 2016, Tabanca and co-authors reported that suberosin (1.8%) was identified in the essential oil obtained from fruits of *P. pabularia*. In present work, 5-pentylcyclohexa-1,3-diene (44.6%), menthone (12.6%), 1-tridecyne (10.9%), and osthole (6%) (an isomer of suberosin) were identified as the dominant constituents of the volatile oil of the roots of *P. pabularia*. These major volatile compounds were not identified from the other *Prangos* species (Table 2). Therefore, it confirms the different chemical composition from *P. pabularia*. Recently, we reported that osthole was isolated from the chloroform extract of the roots of *P. pabularia*, and its structure was elucidated by spectroscopic means, namely, high resolution electrospray ionisation mass spectrometry (HR-ESIMS) and one-dimensional (1D) and two-dimensional (2D) nuclear magnetic resonance (NMR) spectroscopy [22]. In addition, osthole was isolated from the hexane extract of the fruits of *P. asperula* [30].

Both osthole and suberosin were found in *Arracacia tolucensis* var. *multifida* volatile oil [45]. The essential oil with the high coumarin content showed moderate in-vitro antibacterial activity against representative Gram-positive and Gram-negative bacteria [45].

2.3. Antidiabetic Activity of Essential Oil and Isolated Compound (Osthole)

The effect of the obtained essential oil and the pure compound (osthole) from *P. pabularia* roots for its *in vitro* inhibition of the enzyme PTP-1B was determined. The essential oil induced a PTP-1B enzymatic inhibition in a concentration-dependent manner with IC_{50} values 0.06 ± 0.01 µg/mL ($p < 0.02$), which is more than 25 times more potent than the positive control (3-(3,5-dibromo-4-hydroxybenzoyl)-2-ethylbenzofuran-6-sulfonic acid-(4-(thiazol-2-ylsulfamyl)-phenyl)-amide) with IC_{50} 1.46 ± 0.4 µg/mL ($p < 0.05$). The individual compound (osthole) also exhibited strong inhibitory activity against PTP-1B, with IC_{50} values 0.93 ± 0.1 µg/mL ($p < 0.01$); it was also more effective than the positive control. The dose response curves of the inhibition of the PTP-1B enzyme of *P. pabularia* essential oil and osthole are shown in Figure 3.

Wang and co-authors presented a strategy based on GC-MS coupled with molecular docking for analysis, identification, and prediction of PTP-1B inhibitors in the Himalayan cedar essential oil. β-Pinene (49.3%), α-pinene (29.4%), α-terpineol (4.1%), and β-caryophyllene (3.7%) were the main components of Himalayan cedar oil that inhibited PTP-1B with IC_{50} value 120.71 ± 0.26 µg/mL. The docking results of the PTP-1B inhibitory activity of caryophyllene oxide was also in agreement with its *in vitro* activity [46]. The IC_{50} value for PTP-1B inhibition for caryophyllene oxide was in the range 25.8–31.3 µM [46,47]. New terpenoids cedrodorols A-B from *Cedrela odorata* showed inhibitory PTP-1B activity with IC_{50} values 13.09 and 3.93 µg/mL, respectively [48]. In another study, Bharti and co-authors reported the *in vivo* antidiabetic activity of *Cymbopogon citratus* essential oil with major compounds, geranial (42.4%), neral (29.8%), myrcene (8.9%), and geraniol (8.5%), which were fully supported by molecular docking predictions [49].

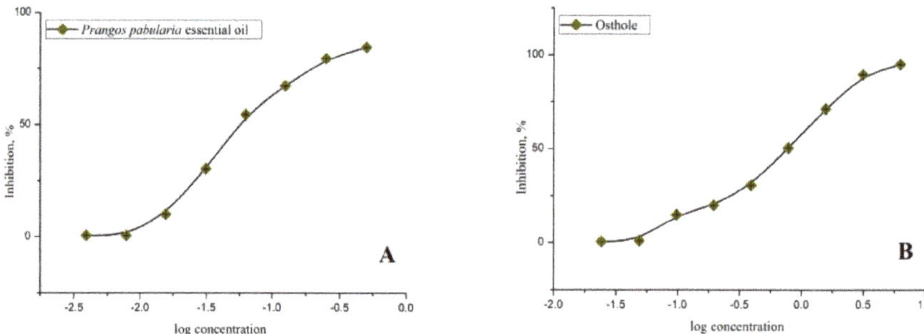

Figure 3. Dose response curve of inhibition of protein-tyrosine phosphatase 1B (PTP-1B) enzyme for *Prangos pabularia* essential oil (**A**) ($IC_{50} = 0.06 \pm 0.01$ µg/mL) and the pure compound osthole (**B**) ($IC_{50} = 0.93 \pm 0.01$ µg/mL).

Hong-Jen Liang investigated the hypoglycemic effects of osthole in diabetic db/db mice, and the main mechanisms of these effects were elucidated using an *in vitro* cell-based assay and *in vivo* assays using a diabetic db/db mouse model. Results showed that osthole significantly alleviated hyperglycemia by activating PPARα/γ in a dose-dependent manner based on the results of the transition transfection assay [50,51]. Wei-Hwa Lee reported that the western blot analysis revealed osthole to significantly induce phosphorylation of AMP-activated protein kinase (AMPK) and acetyl-CoA carboxylase (ACC) as well as increase translocation of glucose transporter 4 (GLUT4) to plasma membranes and glucose uptake in a dose-dependent manner [50]. These results suggest that the increase in the AMP:ATP ratio by osthole had triggered activation of the AMPK signaling pathway, leading to increases in plasma membrane GLUT4 concentration and glucose uptake level [52]. Other research has clearly shown that osthole lowered fasting blood glucose (FBG) and improved insulin secretion. This may indicate partial recovery from pancreatic damage, as indicated from histological characteristics [53].

2.4. Molecular Docking

A molecular docking analysis was carried out on the 12 most abundant components from the root essential oil of *P. pabularia* using the Molegro Virtual Docker program [54]. The MolDock "rerank" docking energies as well as the scaled molecular docking energies are summarized in Table 3. There are two ligand binding regions in human PTP-1B—the catalytic site and an allosteric site (Figure 4). Nearly all of the ligands examined docked preferentially to the allosteric binding site in PDB 1T48, and the best docking ligand was osthole. In the allosteric binding site, the coumarin rings are located in a hydrophobic sandwich formed by Phe280 and Leu192 (Figure 5). Additionally, the aromatic residues Trp291 and Phe196 form face-to-edge π interactions with the coumarin moiety. Residues Ala189 and Glu276 surround the isopropylidene group of osthole. There are no apparent hydrogen-bonding interactions in the docked osthole in the allosteric binding site.

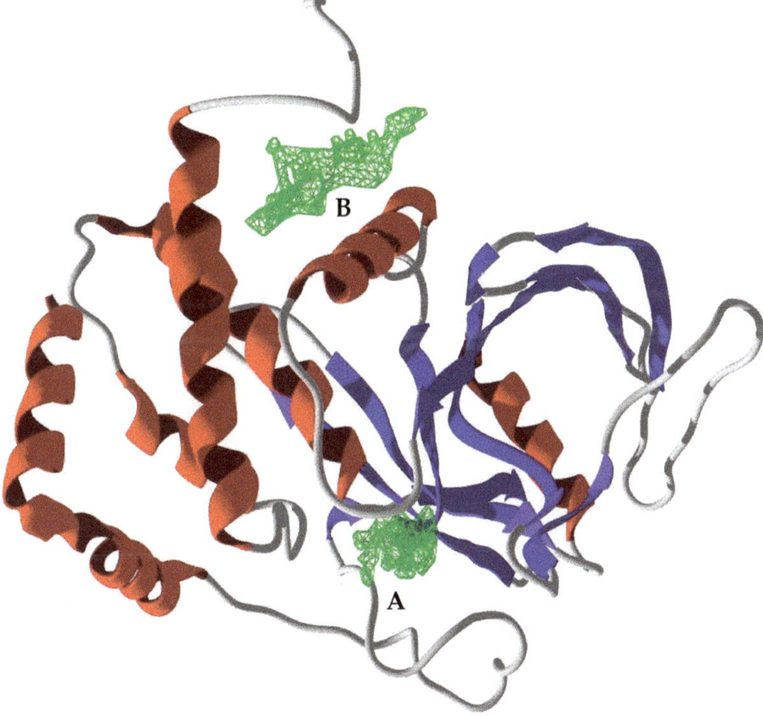

Figure 4. Ribbon structure of human protein tyrosine phosphatase 1B (PTP-1B, PDB 1T48). The catalytic site (**A**) and the allosteric binding site (**B**) are shown as green cross-hatched areas.

The active site of PTP-1B is composed of highly polar residues, including Arg24, Lys41, Arg47, and Asp48, as well as the phosphate-binding loop, Cys215-Arg22 [55]. Therefore, the active site of PTP-1B is not a likely target for the hydrophobic essential oil components of *P. pabularia*. Nevertheless, the lowest-energy docked pose of osthole with the active site of PTP-1B (PDB 2HB1) has a scaled docking energy of −103.0 kJ/mol. This docking pose of osthole shows π-stacking of the coumarin moiety with Phe182 and Tyr46 and is held close to the catalytic site residues of Cys215 and Arg221 (Figure 6). In addition, there are hydrogen-bonding interactions between the osthole carbonyl oxygen and the side chains of Lys120 and Arg221.

Table 3. MolDock "rerank" docking energies (kJ/mol) and molecular-weight-scaled docking energies (in parentheses) for *Prangos pabularia* root essential oil major components with human PTP-1B.

Ligand	1T48	1T49	2BGE	2CMB	2F71	2HB1	3CWE
	Allosteric Site	Allosteric Site	Active Site	Active Site	Active Site	Active Site	Active Site
(3E)-2-Methylocten-5-yne	−64.6 (−93.9)	−60.2 (−87.4)	−62.5 (−90.8)	−61.1 (−88.7)	−60.0 (−87.2)	−57.8 (−84.1)	−62.0 (−90.1)
(E)-1,3-Nonadiene	−67.8 (−97.9)	−58.5 (−84.5)	−63.0 (−91.1)	−63.5 (−91.8)	−65.9 (−95.2)	−62.2 (−89.9)	−64.1 (−92.7)
1-Decanol	−76.5 (−102.0)	−66.8 (−89.0)	−71.0 (−94.6)	−73.0 (−97.4)	−72.7 (−96.9)	−69.8 (−93.1)	−70.4 (−93.9)
1-Tridecyne	−86.4 (−110.4)	−72.7 (−92.8)	−73.2 (−93.4)	−79.0 (−100.9)	−76.7 (−97.9)	−71.8 (−91.6)	−73.4 (−93.7)
2-Methoxy-3-(1-methylpropyl)pyrazine	−76.5 (−100.4)	−64.3 (−84.4)	−65.6 (−86.0)	−66.2 (−86.8)	−62.9 (−82.5)	−66.0 (−86.5)	−71.7 (−94.0)
3-Dodecyn-2-ol	−84.2 (−107.1)	−72.4 (−92.1)	−79.2 (−100.8)	−74.2 (−94.4)	−80.5 (−102.4)	−77.3 (−98.3)	−77.4 (−98.4)
5-Pentyl-1,3-cyclohexadiene	−72.3 (−98.1)	−66.2 (−89.9)	−70.7 (−95.9)	−68.9 (−93.5)	−68.5 (−92.9)	−67.8 (−92.0)	−68.7 (−93.2)
cis-Piperitone epoxide	−76.0 (−99.3)	−65.5 (−85.6)	−65.8 (−86.0)	−67.4 (−88.1)	−69.0 (−90.2)	−73.1 (−95.4)	−75.0 (−98.0)
Menthone	−72.3 (−97.2)	−60.6 (−81.5)	−59.0 (−79.3)	−55.2 (−74.2)	−59.7 (−80.3)	−65.6 (−88.3)	−68.6 (−92.2)
Osthole	−103.4 (−119.3)	−87.1 (−100.5)	−79.3 (−91.5)	−81.5 (−94.0)	−85.4 (−98.6)	−89.3 (−103.0)	−85.9 (−99.1)
Pulegone	−71.8 (−97.0)	−63.6 (−85.8)	−43.4 (−58.6)	−57.3 (−77.3)	−59.0 (−79.7)	−69.1 (−93.4)	−70.6 (−95.4)
trans-α-Bergamotene	−74.9 (−91.7)	−65.4 (−80.1)	−67.2 (−82.3)	−68.6 (−84.1)	−62.9 (−77.0)	−81.5 (−99.8)	−68.2 (−83.5)
Co-crystallized ligand	−142.7 (−127.1)	−137.2 (−127.7)	−105.6 (−127.7)	−162.0 (−132.0)	−147.8 (−138.3)	−97.2 (−107.3)	−129.2 (−117.9)

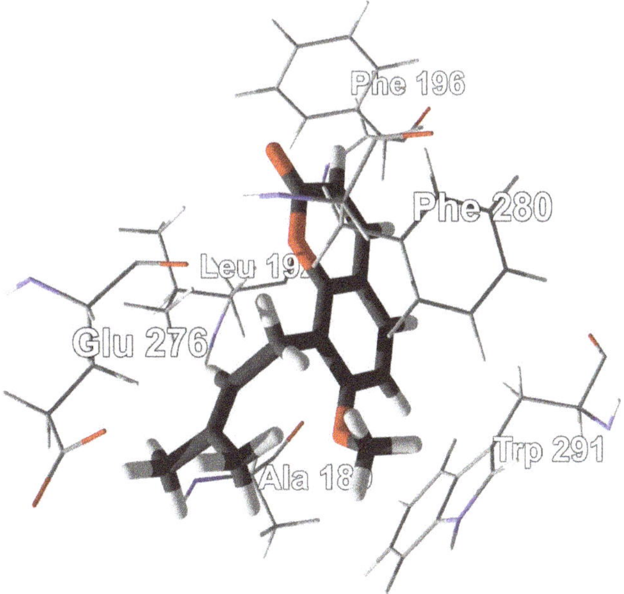

Figure 5. The allosteric binding site of human protein tyrosine phosphatase 1B (PTP-1B, PDB 1T48) with the lowest-energy docked pose of osthole.

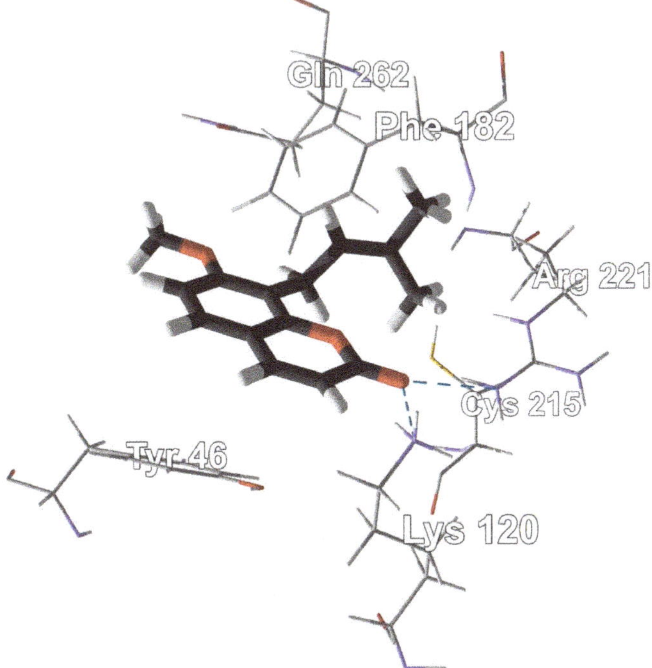

Figure 6. The active site of human protein tyrosine phosphatase 1B (PTP-1B, PDB 2HB1) with the lowest-energy docked pose of osthole. Hydrogen-bonding interactions are indicated by the blue dashed lines.

Ala and co-workers noted that " … the active site of PTP-1B possesses very few desirable drug-design features" and that the highly charged portions of the active site " … significantly increases the difficulty of designing potent inhibitors with acceptable membrane permeability" [55]. Thus, we conclude that the likely binding site for the *P. pabularia* essential oil components is the hydrophobic allosteric binding site, which is more consistent with the greater exothermic docking energies with the allosteric binding site (PDB 1T48, Figure 4B) than with the enzyme active site (Figure 4A). Furthermore, the docking energy for osthole is the most exothermic of the ligands examined, and this compound represents 6.0% of the essential oil composition.

The most abundant component, 5-pentylcyclohexa-1,3-diene (44.6%), is a hydrocarbon, and although the docking energies are somewhat lower for the allosteric site than those for other essential oil components, the abundance of this compound may be a factor in the PTP-1B inhibitory activity of *P. pabularia* root essential oil. Wiesmann and co-workers pointed out that allosteric inhibitors " … prevent formation of the active form of the enzyme by blocking mobility of the catalytic loop" [56].

3. Materials and Methods

3.1. Plant Material

The roots of *P. pabularia* Lindl. were collected from the Yovon region (38 18′47″ N, 69 02′ 35″ E and 950 m above sea level) of Tajikistan in April 2017. The plant was authenticated by Doctor Farukh Sharopov, and the voucher sample (No. TAS 23659-1) was deposited in the herbarium of the Xinjiang Technical Institute of Physics and Chemistry Urumqi, Chinese Academy of Science. The air-dried sample was chopped into small pieces and hydrodistilled for 3 h to give the yellow essential oil with 0.1% yield. Osthole was isolated from the roots of *Prangos pabularia* by silica gel column (100–200 mesh) by elution of hexane ethyl acetate (20:1) [22].

3.2. Gas Chromatography

The quantification of the essential oil of *P. pabularia* was carried out by gas chromatography using Shimadzu GC-2010 (Shimadzu, Kyoto, Japan) plus gas chromatograph, non-polar Phenomenex ZB-5 fused bonded column (30 m length × 0.25 mm inner diameter and 0.25 μm film thickness) and flame ionization detector (FID). Helium was the carrier gas, and the flow rate = 1.5 mL/min with split mode. The following temperature program was used: Initial temperature 120 °C held for 2 min, temperature increased at a rate of 8 °C/min until 320 °C and then held for 10 min at 320 °C. Injector and detector and injector temperatures were 310 °C and 320 °C, respectively. GC Solution software (version 2.53, Shimadzu, Kyoto, Japan) was used for recording and integration. The percentages of each component are reported as raw percentages based on peak area without standardization.

3.3. Gas Chromatographic-Mass Spectral Analysis

Compound identification of *P. pabularia* essential oil was carried out by gas chromatography-mass spectrometry using Agilent 6890 GC, Agilent 5973 (Agilent Technologies, Palo Alto, CA, USA) mass selective detector with electron ionization mass spectrometry (EIMS), (electron energy = 70 eV, scan range = 45–400 amu, and scan rate = 3.99 scans/s), with HP-5ms capillary column (30 m length × 0.25 mm inner diameter and 0.25 μm phase thickness). Helium was the carrier gas with a flow rate of 1 mL/min. Oven temperature program: Hold at 40 °C for 10 min, increase at 3 °C/min up to 200 °C, and then increase at 2 °C/min to 220 °C. The injector and the interface temperatures were 200 °C and 280 °C, respectively. A 1% *w/v* solution of the essential oil in CH_2Cl_2 was prepared, and 1 μL was injected with a splitless injection mode. Identification of the oil components was based on their Kovats indices determined by reference to a homologous series of *n*-alkanes and by comparison of their mass spectral fragmentation patterns with those reported in the literature (Adams 2007) and stored in the MS databases (NIST 17, WILEY 10, FFNSC versions 1.2, 2, and 3).

3.4. NMR and HR-ESIMS Analysis

NMR spectra were recorded on a Varian MR-400 (400 MHz for ^{1}H and 100 MHz for ^{13}C) spectrometer (Palo Alto, CA, USA) in CDCl$_3$. TMS (δ 0.00) signal was used as an internal standard for ^{1}H NMR shifts, and CDCl$_3$ (77.160 ppm vs. TMS) signal was used as a reference for ^{13}C NMR shifts. The HR-ESIMS data were collected with a QStar Elite mass spectrometer (AB SCIEX, Framingham, MA, USA).

3.5. Antidiabetic Activity: PTP-1B Enzymatic Assay

The *P. pabularia* essential oil was screened for PTP-1B inhibition using pNPP (*p*-nitrophenyl phosphate disodium salt) as the substrate. Both the essential oil sample and the enzyme were pre-incubated at room temperature for 5 min before use. A buffer solution (178 µL of 20 mM HEPES, 150 mM NaCl, and 1 mM EDTA) was added to each well of a 96-well plate. The PTP-1B protein solution (1 µL at a concentration 0.115 mg/mL) was added to the buffer solution, and then 1 µL of the test solution and the positive control solution were added. The pNPP substrate (20 µL of 35 mM) was added and mixed for 10 min. The plate was incubated for 30 min in the dark, and the reaction then terminated by adding 10 µL of 3 M NaOH. The absorbance was then determined at 405 nm wavelength using a Spectra Max MD5 plate reader (Molecular Devices, USA). The system without the enzyme solution was used as a blank. Inhibition (%) = [(OD$_{405}$ − OD$_{405}$ blank)/OD$_{405}$ blank] × 100. The IC$_{50}$ was calculated from the percent inhibition values.

3.6. Molecular Docking

Molecular docking of PTP-1B with the major components found in *P. pabularia* root essential oil was carried out on the X-ray crystal structures from the Protein Data Bank (PDB): 1T48, 1T49 [56], 2BGE [57], 2CMB [55], 2F71 [58], 2HB1 [59], and 3CWE [60]. The water molecules and the co-crystallized ligands were removed from the protein crystal structures. Molecular docking for the essential oil components with each of the protein structures was carried out using Molegro Virtual Docker, v. 6.0.1 (Molegro ApS, Aarhus, Denmark) as previously described [61]. A total of 12 major essential oil components were used in the docking study. The three-dimensional ligand structures were built using Spartan '18 for Windows, v. 1.2.0 (Wavefunction Inc., Irvine, CA, USA). For each docking simulation, a maximum of 1500 iterations with a maximum population size of 50 and 100 runs per ligand was carried out. MolDock re-rank scores were used to sort the poses generated for each ligand. In order to account for the bias toward high molecular weights, the following scheme was used: DS$_{norm}$ = 7.2 × E$_{dock}$/MW$^{1/3}$, where DS$_{norm}$ is the normalized docking score, E$_{dock}$ is the MolDock re-rank score, MW is the molecular weight, and 7.2 is a scaling constant to bring the average DS$_{norm}$ values comparable to E$_{dock}$ [62].

4. Conclusions

This study reports the chemical profiles of the essential oils from the roots of *P. pabularia* containing 5-pentylcyclohexa-1,3-diene, menthone, 1-tridecyne, and osthole as major compounds. The high coumarin (osthole) content of the essential oil is particularly interesting in regard to the biological activities of the plant. Our investigations clarified the use of essential oil from the roots of *P. pabularia* for the development of formulations based on enzyme inhibition of PTP-1B. Based on molecular docking, we conclude that secondary volatile metabolites (especially osthole) are likely responsible for the inhibition of PTP-1B. Furthermore, the experimental data are also in agreement with the computational investigation. The anti-diabetic activity of essential oil is related to the dominant volatile compounds, which may be acting synergistically. Further confirmation of anti-diabetic activity of the essential oil from *P. pabularia* needs more research efforts (especially *in vivo* antihyperglycemic activity), which may be applied in food, agriculture, and medicinal industries as a source of anti-diabetic agent.

Author Contributions: S.N., F.S.S., A.M., S.A. performed the phytochemical investigation, designed, and wrote the manuscript; W.N.S., F.S.S., S.N., A.S., M.B. analyzed data; W.N.S., S.N., M.S., M.H. studied the bio-pharmacological activities; H.A.A., W.N.S., made a critical revision of the manuscript.

Funding: The authors are grateful for financial support to the Chinese Academy of Sciences President's International Fellowship Initiative (Grant No. 2019PB0043), Central Asian Drug Discovery & Development Center of Chinese Academy of Sciences (Grant No. CAM 201808), Foreign young scholar (Grant No. 2018FYB0004) and CAS "Light of West China" Program 2018-YDYLTD-001.

Acknowledgments: W.N.S. participated in this project as part of the activities of the Aromatic Plant Research Center (APRC, https://aromaticplant.org/).

Conflicts of Interest: The authors declare no conflict of interest.

References

1. WHO. *About Diabetes*; World Health Organization: Geneva, Switzerland, 2014.
2. International Diabetes Federation. *IDF Diabetes Atlas*; International Diabetes Federation: Brussels, Belgium, 2017.
3. Sharopov, F.S.; Zhang, H.; Wink, M.; Setzer, W.N. Aromatic medicinal plants from Tajikistan (Central Asia). *Medicines* **2015**, *2*, 28–46. [CrossRef] [PubMed]
4. Sharopov, F.; Setzer, W.N. Medicinal plants of Tajikistan. In *Vegetation of Central Asia and Environs*; Egamberdieva, D., Öztürk, M., Eds.; Springer: Basel, Switzerland, 2018; pp. 163–210.
5. Sharopov, F. Phytochemistry and Bioactivities of Selected Plant Species with Volatile Secondary Metabolites. Ph.D. Thesis, University of Heidelberg, Heidelberg, Germany, 2015.
6. Yen, H.-F.; Hsieh, C.-T.; Hsieh, T.-J.; Chang, F.-R.; Wang, C.-K. *In vitro* anti-diabetic effect and chemical component analysis of 29 essential oils products. *J. Food Drug Anal.* **2015**, *23*, 124–129. [CrossRef] [PubMed]
7. Meshkani, R.; Taghikhani, M.; Al-Kateb, H.; Larijani, B.; Khatami, S.; Sidiropoulos, G.K.; Hegele, R.A.; Adeli, K. Polymorphisms within the protein tyrosine phosphatase 1B (PTPN1) gene promoter: Functional characterization and association with type 2 diabetes and related metabolic traits. *Clin. Chem.* **2007**, *53*, 1585–1592. [CrossRef] [PubMed]
8. Numonov, S.; Edirs, S.; Bobakulov, K.; Qureshi, M.N.; Bozorov, K.; Sharopov, F.; Setzer, W.N.; Zhao, H.; Habasi, M.; Sharofova, M.; et al. Evaluation of the antidiabetic activity and chemical composition of *Geranium collinum* root extracts—Computational and experimental investigations. *Molecules* **2017**, *22*, 983. [CrossRef] [PubMed]
9. Tonks, N.K.; Neel, B.G. Combinatorial control of the specificity of protein tyrosine phosphatases. *Curr. Opin. Cell Biol.* **2001**, *13*, 182–195. [CrossRef]
10. Bakke, J.; Haj, F.G. Protein-tyrosine phosphatase 1B substrates and metabolic regulation. *Semin. Cell Dev. Biol.* **2015**, *37*, 58–65. [CrossRef] [PubMed]
11. Venkataraghavan, R.; Brindha, P.; Ivo, R.S. A review on protein tyrosine phosphatases—an important target for various diseases. *Asian J. Pharm. Clin. Res.* **2018**, *11*, 11–16.
12. Florez, J.C.; Agapakis, C.M.; Burtt, N.P.; Sun, M.; Almgren, P.; Rastam, L.; Tuomi, T.; Gaudet, D.; Hudson, T.J.; Daly, M.J.; et al. Association testing of the protein tyrosine phosphatase 1B gene (PTPN1) with type 2 diabetes in 7,883 people. *Diabetes* **2005**, *54*, 1884–1891. [CrossRef] [PubMed]
13. Verma, M.; Gupta, S.J.; Chaudhary, A.; Garg, V.K. Protein tyrosine phosphatase 1B inhibitors as antidiabetic agents—A brief review. *Bioorg. Chem.* **2017**, *70*, 267–283. [CrossRef] [PubMed]
14. Jumaev, B.B.; Mirzoev, B.; Nigmonov, M.; Safarov, E.H.; Abdullaev, A.; Karimov, K.H. Physiological and biochemical characteristics leaf different layers of vegetative and generative shoots *Prangos pabularia* (*Prangos pabularia* Lindl.). *Rep. Acad. Sci. Repub. Tajikistan* **2014**, *57*, 695–700.
15. Sadikov, Y.J. Medicinal plant of Tajikistan. *News Tajik Acad. Sci.* **2003**, *5*, 41–48.
16. Razavi, S.M. Chemical composition and some allelopathic aspects of essential oils of (*Prangos ferulacea* L.) Lindl at different stages of growth. *J. Agric. Sci. Technol.* **2012**, *14*, 349–356.
17. Shokoohinia, Y.; Hosseinzadeh, L.; Alipour, M.; Mostafaie, A.; Mohammadi-Motlagh, H.-R. Comparative evaluation of cytotoxic and apoptogenic effects of several coumarins on human cancer cell lines: Osthole induces apoptosis in p53-deficient H1299 cells. *Adv. Pharmacol. Sci.* **2014**, *2014*, 8. [CrossRef] [PubMed]
18. Yousefi, K.; Hamedeyazdan, S.; Hodaei, D.; Fathiazad, F. An *in vitro* ethnopharmacological study on *Prangos ferulacea:* A wound healing agent. *BioImpacts* **2017**, *7*, 75–82. [CrossRef]

19. Başer, K.H.; Demirci, B.; Demirci, F.; Bedir, E.; Weyerstahl, P.; Marschall, H.; Duman, H.; Aytaç, Z.; Hamann, M.T. A new bisabolene derivative from the essential oil of *Prangos uechtritzii* fruits. *Planta Medica* **2000**, *66*, 674–677. [CrossRef]

20. Özek, G.; Bedir, E.; Tabanca, N.; Ali, A.; Khan, I.A.; Duran, A.; Başer, K.H.C.; Özek, T. Isolation of eudesmane type sesquiterpene ketone from *Prangos heyniae* H.Duman & M.F.Watson essential oil and mosquitocidal activity of the essential oils. *Open Chem.* **2018**, *16*, 453–467.

21. Tabanca, N.; Wedge, D.E.; Li, X.C.; Gao, Z.; Ozek, T.; Bernier, U.R.; Epsky, N.D.; Baser, K.H.; Ozek, G. Biological evaluation, overpressured layer chromatography separation, and isolation of a new acetylenic derivative compound from *Prangos platychlaena* ssp. *platychlaena* fruit essential oils. *J. Planar Chromatogr.—Mod. TLC* **2018**, *31*. [CrossRef]

22. Numonov, S.; Bobakulov, K.; Numonova, M.; Sharopov, F.S.; Setzer, W.N.; Khalilov, Q.; Begmatov, N.; Habasi, M.; Aisa, H.A. New coumarin from the roots of *Prangos pabularia*. *Nat. Prod. Res.* **2017**, *32*, 2325–2332. [CrossRef]

23. Ozek, G.; Ozek, T.; Iscan, G.; Baser, K.H.C.; Hamzaoglu, E.; Duran, A. Comparison of hydrodistillation and microdistillation methods for the analysis of fruit volatiles of *Prangos pabularia* Lindl., and evaluation of its antimicrobial activity. *S. Afr. J. Bot.* **2007**, *73*, 563–569. [CrossRef]

24. Tabanca, N.; Tsikolia, M.; Ozek, G.; Ozek, T.; Abbas, A.; Bernier, U.R.; Duran, A.; Baser, K.H.; Khan, I.A. The identification of suberosin from *Prangos pabularia* essential oil and its mosquito activity against *Aedes aegypti*. *Rec. Nat. Prod.* **2016**, *10*, 311–325.

25. Razavi, S.M. Chemical and allelopathic analyses of essential oils of *Prangos pabularia* Lindl. from Iran. *Nat. Prod. Res.* **2011**, *26*, 2148–2151. [PubMed]

26. Adams, R. *Identification of Essential Oil Components by Gas Chromatography/Mass Spectrometry*, 4th ed.; Allured Publishing Corp: Carol Stream, IL, USA, 2007.

27. Meshkatalsadat, M.H.; Bamoniri, A.; Batooli, H. The bioactive and volatile constituents of *Prangos acaulis* (DC) Bornm extracted using hydrodistillation and nanoscale injection techniques. *Dig. J. Nanomater. Biostruct.* **2010**, *5*, 263–266.

28. Sajjadi, S.E.; Mehregan, I. Chemical composition of the essential oil of *Prangos asperula* Boiss. subsp. *haussknechtii* (Boiss.) Herrnst. et Heyn fruits. *DARU J. Pharm. Sci.* **2003**, *11*, 23–28.

29. Mneimne, M.; Baydoun, S.; Nemer, N.; Apostolides, N.A. Chemical composition and antimicrobial activity of essential oils isolated from aerial parts of *Prangos asperula* Boiss. (Apiaceae) growing wild in Lebanon. *Med. Aromat. Plants* **2016**, *5*, 2–5.

30. Sajjadi, S.E.; Zeinvand, H.; Shokoohinia, Y. Isolation and identification of osthol from the fruits and essential oil composition of the leaves of *Prangos asperula* Boiss. *Res. Pharm. Sci.* **2009**, *4*, 19–23.

31. Lingan, K. A review on major constituents of various essential oils and its application. *J. Transl. Med. (Sunnyvale)* **2018**, *8*, 1000201. [CrossRef]

32. Baser, K.H.C.; Ermin, N.; Adigüzel, N.; Aytaç, Z. Composition of the essential oil of *Prangos ferulacea* (L.) Lindl. *J. Essent. Oil Res.* **2011**, *8*, 297–298. [CrossRef]

33. Massumi, M.A.; Fazeli, M.R.; Alavi, S.H.R.; Ajani, Y. Chemical constituents and antibacterial activity of essential oil of *Prangos ferulacea* (L.) Lindl. fruits. *Iran. J. Pharm. Sci.* **2007**, *3*, 171–176.

34. Delnavazi, M.R.; Soleimani, M.; Hadjiakhoondi, A.; Yass, N. Isolation of phenolic derivatives and essential oil analysis of *Prangos ferulacea* (L.) Lindl. aerial parts. *Iran. J. Pharm. Res.* **2017**, *16*, 207–215. [PubMed]

35. Ercan, S.F.; Bas, H.; Koc, M.; Pandir, D.; Öztemiz, S. Insecticidal activity of essential oil of *Prangos ferulacea* (Umbelliferae) against *Ephestia kuehniella* (Lepidoptera: Pyralidae) and *Trichogramma embryophagum* (Hymenoptera: Trichogrammatidae). *Turkish J. Agric. For.* **2013**, *37*, 719–725. [CrossRef]

36. Mohibi, Z.; Heshmati, G.A.; Sefidkon, F.; Chahouki, Z.M.A. The influence of plant growth satge, individuals of species, and extraction methods on the essential oil content and the chemical composition of *Prangos ferulacea* (L.) Lindl. *Appl. Ecol. Environ. Res.* **2017**, *15*, 1765–1776. [CrossRef]

37. Amiri, H. Chemical composition and antibacterial activity of essential oil of *Prangos ferulacea* (L.) Lindl. *J. Med. Plants* **2007**, *1*, 36–41.

38. Razavi, S.M.; Nazemiyeh, H.; Zarrini, G.; Asna-Asharii, S.; Dehghan, G. Chemical composition and antimicrobial activity of essential oil of *Prangos ferulaceae* (L.) Lindl from Iran. *Nat. Prod. Res.* **2009**, *24*, 530–533. [CrossRef] [PubMed]

39. Başer, K.H.C.; Özek, T.; Demirci, B.; Duman, H. Composition of the essential oil of *Prangos heyniae* H. Duman et M. F. Watson, a new endemic from Turkey. *Flavour Fragr. J.* **2000**, *15*, 47–49. [CrossRef]

40. Kiliç, Ö.; Bengü, A.Ş.; Özdemir, F.A.; Çelik, Ş. Essential oil composition of two *Prangos* Lindl. (Apiaceae) species from Turkey. *Prog. Nutr.* **2017**, *19*, 69–74.

41. Abolghasemi, M.M.; Piryaei, M. Fast determination of *Prangos uloptera* essential oil by nanoporous silica-polypyrrole SPME fiber. *Chemija* **2012**, *23*, 244–249.

42. Sefidkon, F.N. Chemical composition of the oil of *Prangos uloptera* DC. *J. Essent. Oil Res.* **2001**, *13*, 84–85. [CrossRef]

43. Alikhah-Asl, M.; Azarnivand, H.; Jafari, M.; Arzani, H.; Amin, G.; Zare-Chahouki, M.A. Variations of essential oils in fresh and dried aerial parts of *Prangos uloptera*. *J. Nat. Prod.* **2012**, *5*, 5–9.

44. Hayta, Ş. Essential oil composition of the fruit of *Prangos uloptera* (Apiaceae) DC. from Turkey. *Pharm. Chem. J.* **2018**, *5*, 1–5.

45. Figueroa, M.; Rivero-Cruz, I.; Rivero-Cruz, B.; Bye, R.; Navarrete, A.; Mata, R. Constituents, biological activities and quality control parameters of the crude extract and essential oil from *Arracacia tolucensis* var. *multifida*. *J. Ethnopharmacol.* **2007**, *113*, 125–131. [CrossRef]

46. Wang, H.; Gu, D.; Wang, M.; Guo, H.; Wu, H.; Tian, G.; Li, Q.; Yang, Y.; Tian, J. A strategy based on gas chromatography-mass spectrometry and virtual molecular docking for analysis and prediction of bioactive composition in natural product essential oil. *J. Chromatogr. A* **2017**, *1501*, 128–133. [CrossRef]

47. Saifudin, A.; Tanaka, K.; Kadota, S.; Tezuka, Y. Chemical constituents of *Blumea balsamifera* of Indonesia and their protein tyrosine phosphatase 1B inhibitory activity. *Nat. Prod. Commun.* **2012**, *7*, 815–818. [CrossRef]

48. Wu, W.B.; Huan, Z.; Dong, S.; Yue, M. New triterpenoids with protein tyrosine phosphatase 1B inhibition from *Cedrela odorata*. *Asian Nat. Prod. Res.* **2014**, *16*, 1–8. [CrossRef]

49. Bharti, S.K.; Kumar, A.; Prakash, O.; Krishnan, S.; Gupta, A.K. Essential oil of *Cymbopogon citratus* against diabetes: Validation by *in vivo* experiments and computational studies. *J. Bioanal. Biomed.* **2013**, *5*, 194–203.

50. Zhang, Z.-R.; Leung, W.N.; Cheung, H.Y.; Chan, C.W. Osthole: A review on its bioactivities, pharmacological properties, and potential as alternative medicine. *Evid.-Based Complement. Altern. Med.* **2015**, *2015*, 919616. [CrossRef]

51. Liang, H.-J.; Suk, F.-M.; Wang, C.-K.; Hung, L.-F.; Liu, D.-Z.; Chen, N.-Q.; Chen, Y.-C.; Chang, C.-C.; Liang, Y.-C. Osthole, a potential antidiabetic agent, alleviates hyperglycemia in db/db mice. *Chem.-Biol. Interact.* **2009**, *181*, 309–315. [CrossRef]

52. Lee, W.-H.; Lin, R.-J.; Lin, S.-Y.; Chen, Y.-C.; Lin, H.-M.; Liang, Y.-C. Osthole enhances glucose uptake through activation of AMP-activated protein kinase in skeletal muscle cells. *J. Agric. Food Chem.* **2011**, *59*, 12874–12881. [CrossRef]

53. Yao, Y.; Zhao, X.; Xin, J.; Wu, Y.; Li, H. Coumarins improved type 2 diabetes induced by high-fat diet and streptozotocin in mice via antioxidation. *Can. J. Physiol. Pharm.* **2018**, *96*, 765–771. [CrossRef]

54. Thomsen, R.; Christensen, M.H. MolDock: A New Technique for High-Accuracy Molecular Docking. *J. Med. Chem.* **2006**, *49*, 3315–3321. [CrossRef]

55. Ala, P.J.; Gonneville, L.; Hillman, M.C.; Becker-Pasha, M.; Wei, M.; Reid, B.G.; Klabe, R.; Yue, E.W.; Wayland, B.; Douty, B.; et al. Structural basis for inhibition of protein-tyrosine phosphatase 1B by isothiazolidinone heterocyclic phosphonate mimetics. *J. Biol. Chem.* **2006**, *281*, 32784–32795. [CrossRef]

56. Wiesmann, C.; Barr, K.J.; Kung, J.; Zhu, J.; Erlanson, D.A.; Shen, W.; Fahr, B.J.; Zhong, M.; Taylor, L.; Randal, M.; et al. Allosteric inhibition of protein tyrosine phosphatase 1B. *Nat. Struct. Mol. Biol.* **2004**, *11*, 730–737. [CrossRef]

57. Black, E.; Breed, J.; Breeze, A.L.; Embrey, K.; Garcia, R.; Gero, T.W.; Godfrey, L.; Kenny, P.W.; Morley, A.D.; Minshull, C.A.; et al. Structure-based design of protein tyrosine phosphatase-1B inhibitors. *Bioorg. Med. Chem. Lett.* **2005**, *15*, 2503–2507. [CrossRef]

58. Klopfenstein, S.R.; Evdokimov, A.G.; Colson, A.-O.; Fairweather, N.T.; Neuman, J.J.; Maier, M.B.; Gray, J.L.; Gerwe, G.S.; Stake, G.E.; Howard, B.W.; et al. 1,2,3,4-Tetrahydroisoquinolinyl sulfamic acids as phosphatase PTP1B inhibitors. *Bioorganic Med. Chem. Lett.* **2006**, *16*, 1574–1578. [CrossRef]

59. Wan, Z.-K.; Lee, J.; Xu, W.; Erbe, D.V.; Joseph-McCarthy, D.; Follows, B.C.; Zhang, Y.-L. Monocyclic thiophenes as protein tyrosine phosphatase 1B inhibitors: Capturing interactions with Asp48. *Bioorg. Med. Chem. Lett.* **2006**, *16*, 4941–4945. [CrossRef]

60. Han, Y.; Belley, M.; Bayly, C.I.; Colucci, J.; Dufresne, C.; Giroux, A.; Lau, C.K.; Leblanc, Y.; McKay, D.; Therien, M.; et al. Discovery of [(3-bromo-7-cyano-2-naphthyl)(difluoro)methyl]phosphonic acid, a potent and orally active small molecule PTP1B inhibitor. *Bioorg. Med. Chem. Lett.* **2008**, *18*, 3200–3205. [CrossRef]

61. Byler, K.G.; Ogungbe, I.V.; Setzer, W.N. In-silico screening for anti-Zika virus phytochemicals. *J. Mol. Graph. Model.* **2016**, *69*, 78–91. [CrossRef]

62. Setzer, M.S.; Byler, K.G.; Ogungbe, I.V.; Setzer, W.N. Natural products as new treatment options for trichomoniasis: A molecular docking investigation. *Sci. Pharm.* **2017**, *85*, 5. [CrossRef]

Article

Evaluation of the *In Vitro* Wound-Healing Activity and Phytochemical Characterization of Propolis and Honey

Alexandra M. Afonso [1], Joana Gonçalves [1,2,†], Ângelo Luís [1,2,*,†], Eugenia Gallardo [1,2] and Ana Paula Duarte [1,2,*]

[1] Centro de Investigação em Ciências da Saúde (CICS-UBI), Universidade da Beira Interior, Avenida Infante D. Henrique, 6200-506 Covilhã, Portugal; alexandrafonso9@gmail.com (A.M.A.); janitagoncalves@hotmail.com (J.G.); egallardo@fcsaude.ubi.pt (E.G.)

[2] Laboratório de Fármaco-Toxicologia, UBIMedical, Universidade da Beira Interior, Estrada Municipal 506, 6200-284 Covilhã, Portugal

*   Correspondence: angelo.luis@ubi.pt (Â.L.); apcd@ubi.pt (A.P.D.); Tel.: +351-275-329-002 (Â.L.); +351-275-329-099 (A.P.D.)

†   These authors contributed equally to the work.

Received: 17 February 2020; Accepted: 4 March 2020; Published: 7 March 2020

Abstract: Honey and propolis are natural substances produced by *Apis mellifera* that contain flavonoids, phenolic acids, and several other phytochemicals. The aim of this study was to phytochemically characterize three different types of honey and propolis, both separately and mixed, and to evaluate their wound-healing activity. Total phenolic compounds and flavonoids were determined using the Folin–Ciocalteu's and aluminum chloride colorimetric methods, respectively. The antioxidant activity was evaluated by both the DPPH free radical scavenging assay and β-carotene bleaching test, and the anti-inflammatory activity was determined by a protein denaturation method. To evaluate the wound-healing activity of the samples, NHDF cells were subjected to a wound scratch assay. The obtained results showed that dark-brown honey presents a higher concentration of phenolic compounds and flavonoids, as well as higher antioxidant and anti-inflammatory activities. Propolis samples had the highest concentrations in bioactive compounds. Examining the microscopic images, it was possible to verify that the samples promote cell migration, demonstrating the wound-healing potential of honey and propolis.

Keywords: honey; propolis; phenolic compounds; antioxidant; wound-healing activity; NHDF cells

---

1. Introduction

Honey is a natural substance produced by the western honeybee (*Apis mellifera*) from nectar and exudates of flowers and trees. Honey contains flavonoids, phenolic compounds, and numerous sugars in its composition, mainly glucose and fructose. It also contains small amounts of minerals, vitamins, proteins, and enzymes [1–5].

Propolis is a resinous substance produced by the same species of bees, being composed of wax, resinous secretions from plants, tree buds, pollen, and salivary enzymes. Propolis contains essential oils and organic compounds such as phenolic acids and flavonoids [4,6–8]. Similarly to honey, propolis also contains several minerals, vitamins, and enzymes [9]. This substance is used to protect the beehive, acting as a natural sealant, antiseptic, and embalming agent [1,4,6–8,10].

The color and flavor of both honey and propolis vary according to the plant species used in their production, health state of the bees, season, and the environmental conditions to which the beehive is exposed [2–5]. The color of the honey can vary from deep brown to yellow, and propolis can be found in variations of green, red, or brown [6,7,11]. Differences in their composition result in different biological activities [1,4,7,12]. Honey and propolis have been used in traditional medicine for thousands of years mainly in burns and infected lesions [4,10]. Propolis was used primarily as an oral antiseptic and in dermal lesions. In ancient Egypt, propolis was also employed in mummification [6].

The main mechanism of action of honey derives from its antioxidant and anti-inflammatory activities [1,2,10]. The antioxidant potential of honey is due to the presence of phenolic compounds, namely gallic acid and flavonoids, that promote the free radical scavenging [2,10]. Its anti-inflammatory activity is associated with the ability to increase cellular proliferation, autolytic debridement, and to stimulate the immune system, reducing edema and pain [10]. Propolis has a higher content in flavonoids and phenolic acids than honey, which suggests that propolis should have a higher antioxidant and anti-inflammatory activities [1,7].

Currently, an increasing interest in mixing propolis with honeys was verified, as products including propolis are already available on the market. The commercial achievement of these products is dependent of the level of acceptance and expectations of the consumers [1]. However, the potential benefits of adding propolis to honeys is still to be scientifically validated.

Wounds are responsible for the consumption of large amounts of healthcare resources to ameliorate the quality of life of patients, and so they represent a considerable health challenge. Several scientific experiments have been performed to find novel compounds that possess wound-healing properties, particularly from natural sources [13]. Honey was employed for the repair of battle wounds in both World War I and in modern history. The healing activity of honey in infected wounds was initially described in Europe and USA in mid-20th century [10]. Moreover, evidences suggest that propolis has therapeutic activity, through quantitative and qualitative analyses of collagen types I and III expression and degradation in wounds matrix, which suggests the favorable biochemical environment supporting re-epithelization of propolis [11].

The first aim of this study was to phytochemically characterize different samples of honey and propolis together with mixtures of propolis extracts with honey in different concentrations, determining the contents in total phenolic compounds and flavonoids, as well as evaluating the antioxidant and anti-inflammatory activities. The second purpose of this work was to evaluate the *in vitro* wound-healing activity of the samples using normal human dermal fibroblasts (NHDF) by means of the scratch assay.

2. Materials and Methods

2.1. Honey and Propolis Samples

Honey and propolis samples were collected from a *Langstroth* beehive in a mountainous forest in the north central Portugal, specifically in the region of Sabugal (Guarda) during September 2018 (GPS coordinates: 40°23′33.5″ N 7°03′45.4″ W).

Three different samples of honey and propolis were collected from the same beehive to maintain consistent the conditions of production. Each honey sample was collected from a different box inside the hive and was then filtered individually until there was no visible debris or particles. Propolis samples were scraped from the boxes and frames of the hive and then separated according to the box from which they were collected. All samples were stored individually in an appropriate sterilized container.

Dark-brown honey was considered as Honey 1 (H1), red honey was labeled as Honey 2 (H2), and light-yellow honey was identified as Honey 3 (H3). Propolis samples were named according to the box from which the honey samples were collected as Propolis 1 (P1), Propolis 2 (P2), and Propolis 3 (P3).

2.2. Propolis Extracts

The samples of propolis were grounded individually in a marble mortar at room temperature. To start the extraction process, 100 mL of ethanol (Scharlab, Spain) were added to 14 g of each pulverized propolis sample. The mixtures were left for 48 h under magnetic stirring at room temperature and in the absence of light. After 48 h, the mixtures were filtered and transferred to round bottom flasks to evaporate the ethanol using a rotary evaporation system at 45 °C. To ensure the complete evaporation, the round bottom flasks were kept in a vacuum oven for 24 h at 35 °C.

The propolis extracts were labelled as Propolis Extract 1 (PE1), Propolis Extract 2 (PE2), and Propolis Extract 3 (PE3) and were frozen at −20 °C until they were used.

2.3. Mixtures of Honey with Propolis

Each propolis extract was added in a concentration of 0.3% (w/w) and 0.5% (w/w) to the Honey 1 (the one that presented the best results), according to the previously published work dealing with the evaluation of the bioactive properties of honey with propolis [1]. The mixtures were prepared by weighting the corresponding amounts of honey and propolis and mixing them to complete 100 g of each mixture. To guarantee the complete homogenization, all mixtures were subjected to mechanical stirring, after which they were stored at −20 °C until further use.

The mixtures were identified as Honey 1 followed by the corresponding Propolis Extract and percentage added resulting in Honey 1-Propolis Extract 1 (H1PE1), Honey 1-Propolis Extract 2 (H1PE2) and Honey 1-Propolis Extract 3 (H1PE3) at 0.3% or 0.5%.

2.4. Fourier-Transform Infrared Spectroscopy (FTIR)

FTIR was used to obtain spectra of the samples of honey, propolis and propolis extracts. These spectra were obtained with 64 scans and a 4 cm^{-1} resolution, between 4000 and 600 cm^{-1} using a Nicolet iS10 smart iTRBasic (Thermo Fisher Scientific, Waltham, MA, USA) model.

2.5. Phytochemical Characterization

For the phytochemical characterization, all the samples were diluted with methanol (Scharlab, Spain).

2.5.1. Total Phenolic Compounds Determination

The phenolic compounds were determined by Folin–Ciocalteu's colorimetric method [14,15], using gallic acid as the standard. Initially, 450 μL of distilled water were mixed with 50 μL of each sample or gallic acid (Sigma-Aldrich, USA) solution. Then, 2.5 mL of Folin–Ciocalteu's reagent (Sigma-Aldrich, USA) (0.2 N) were added, being the mixtures left for 5 min before the addition of 2 mL of aqueous Na_2CO_3 (Sigma-Aldrich, USA) (75 g/L). The reaction mixtures were incubated for 90 min at 30 °C. After incubation, the content in total phenolic compounds was determined by colorimetry at 765 nm [14,15].

A standard curve was prepared using methanolic solutions of gallic acid at 500, 300, 250, 200, 150, 100, and 50 mg/L (y = 0.0010x; R^2 = 0.9612). The total phenolic compounds content was expressed as g of gallic acid equivalents (GAE)/100 g of sample (honey, propolis, and mixtures of honey with propolis) [14,15].

2.5.2. Flavonoid Determination

The aluminum chloride colorimetric method was used to determine the flavonoids content according to a previously implemented method [14,15]. To 500 μL of each solution, either the samples or the quercetin (Sigma-Aldrich, USA) (used as standard), 1.5 mL of methanol, 0.1 mL of aluminum chloride (Sigma-Aldrich, USA) 10% (w/v), 0.1 mL of 1 M potassium acetate (Sigma-Aldrich, USA) and 2.8 mL of distilled water were added. These solutions remained for 30 min at room temperature and then the absorbances were measured using a spectrophotometer (Helios–Omega, Thermo Scientific, USA) at 415 nm [14,15].

To construct the calibration curve, eight quercetin solutions were prepared in methanol with a concentration of 200, 175, 150, 100, 75, 50, 25, and 12.5 μg/mL (y = 0.0146 x; R^2 = 0.9887). The flavonoids content was expressed as g of quercetin equivalents (QE)/100 g of sample (honey, propolis, and mixtures of honey with propolis) [14,15].

2.6. Antioxidant Activity Evaluation

For the antioxidant activity evaluation all the samples were diluted with methanol.

2.6.1. DPPH Free Radical Scavenging Assay

The DPPH (2,2-diphenyl-1-picrylhydrazyl) free radical scavenging assay was used to evaluate the antioxidant activity of the samples [16]. Briefly, to 100 μL of each sample, 3.9 mL of a 0.1 mM DPPH (Sigma-Aldrich, USA) methanolic solution were added, being this mixture stirred until complete homogenization. The control solution consisted in 100 μL of methanol with 3.9 mL of the DPPH solution. The reaction mixtures were kept at room temperature in the absence of light for 30 min, time after which the absorbances were read at 517 nm using a spectrophotometer (Helios–Omega, Thermo Scientific, USA) [16].

The percentage of inhibition (%Inhibition) of DPPH free radical by the samples was determined using the equation %Inhibition = $[(Abs_{control} - Abs_{sample})/Abs_{control}] \times 100$, where $Abs_{control}$ corresponds to the absorbance of the control and Abs_{sample} is the absorbance of each sample [14,15]. The results were expressed as %Inhibition/100 g sample (honey, propolis, and mixtures of honey with propolis).

2.6.2. β-Carotene Bleaching Test

The β-carotene bleaching test was also employed to evaluate the antioxidant properties of the samples [16]. Firstly, a β-carotene (Sigma-Aldrich, USA) solution in chloroform with a concentration of 20 mg/mL was prepared. To 500 μL of this solution 40 μL of linoleic acid (TCI Europe N.V., Belgium), 400 μL of Tween 40 (Riedel-de H¨aen, Germany) and 1 mL of chloroform (Scharlab, Spain) were added. This mixture was transferred to a round bottom flask and subjected to a rotary evaporation system at 45 °C to ensure complete evaporation of the chloroform. After this, 100 mL of water saturated with oxygen were added to the mixture, forming an emulsion. Secondly, 300 μL of each sample were transferred to test tubes and 5 mL of the previously prepared β-carotene emulsion were added. The tubes were stirred until complete homogenization and were placed in a water bath at 50 °C for 1 h. Using a spectrophotometer (Helios–Omega, Thermo Scientific, USA), the absorbances of the samples were measured at 470 nm at the initial (t = 0 h) and final time (t = 1 h). The antioxidant activity was determined as percentage of inhibition of β-carotene's oxidation (%Inhibition) using the following equation, %Inhibition = $[(Abs_{sample}{}^{t=1\,h} - Abs_{control}{}^{t=1\,h})/(Abs_{control}{}^{t=0\,h} - Abs_{control}{}^{t=1\,h})] \times 100$, where $Abs_{control}$ corresponds to the absorbance of the control and Abs_{sample} is the absorbance of each sample [14,15]. The results were expressed as %Inhibition/100 g sample (honey, propolis, and mixtures of honey with propolis).

2.7. Assessment of In Vitro Anti-Inflammatory Activity

The anti-inflammatory activity was determined by evaluating the capacity of the samples to inhibit protein denaturation [17]. Initially, a solution of bovine serum albumin (BSA) (Sigma-Aldrich, USA) at 1% (w/v) in phosphate buffer saline (PBS) solution was prepared. The pH of this solution was adjusted to 6.8 using glacial acetic acid (Scharlab, Spain). Then, 100 μL of the samples diluted in methanol were mixed, in test tubes pre-heated at 37 °C, with 900 μL of the BSA solution previously prepared. The control was composed of distilled water. The tubes were then incubated for 10 min at 72 °C and after this period cooled in ice for another 10 min. Finally, measurements of the absorbances were performed using a spectrophotometer (Helios–Omega, Thermo Scientific, USA) at 620 nm. The percentage of inhibition of protein denaturation (%Inhibition) was determined applying the following equation, %Inhibition = $100 - [(Abs_{sample} \times 100)/Abs_{control}]$, where $Abs_{control}$ corresponds to the absorbance of the control and Abs_{sample} is the absorbance of each sample [17]. The results were expressed as %Inhibition/100 g sample (honey, propolis, and mixtures of honey with propolis).

2.8. Evaluation of the In Vitro Wound-Healing Activity

2.8.1. Cell Culture

Normal human dermal fibroblasts (NHDF) cell line was maintained in RPMI-1640 culture medium (Sigma-Aldrich, USA) supplemented with 10% fetal bovine serum (FBS) (Sigma-Aldrich, USA), 1% mixture of antibiotic/antimycotic (Sigma-Aldrich, USA), 0.01 M of HEPES (Sigma-Aldrich, USA), 0.02 M of *L*-glutamine (Sigma-Aldrich, USA) and 0.001 M of sodium pyruvate (Sigma-Aldrich, USA). Subsequently, the cells were incubated at 37 °C in an air incubator with a humidified atmosphere with 5% CO_2 [18].

2.8.2. Wound Scratch Assay

The samples were tested for wound-healing activity by using the wound scratch assay [13,19]. NHDF cells were seeded in 12-well plates (4×10^4 cells/well) and cultured until a monolayer confluence was reached. After the adhesion of the cells, the medium was removed from the wells and the cell monolayer was scraped in a straight central line using a p200 micropipette tip, creating a scratch, with reference points being marked in the plates. The wells were washed with PBS to remove floating cells and cell debris. Then, the PBS was removed, the samples were prepared in RPMI-1640 and sonicated, then they were added to the wells. Supplemented RPMI-1640 culture medium was added to the control wells. After this, the plates were placed under a phase-contrast microscope and images were acquired at the initial moment (t = 0 h). Then, the plates were incubated at 37 °C (5% CO_2) and examined once again under the microscope after 2, 24, and 36 h [13,19].

The size of the scratch zones was assessed manually using a digital image analysis tool (IC Measure software version 2.0.0.161) (The Imaging Source, Germany) that allowed the estimation of the distance between the injury margins. Using the IC Measure, the distance between the margins of the lesion in the control at 0 h was estimated, which was considered the initial one and was used to scale all other measurements to more easily compare the estimated distances of the injuries between the samples and the control.

2.9. Statistical Analysis

The results were presented as mean values ± standard deviation. To determine the reproducibility of the measurements, each assay was performed at least in triplicate. The calculated distance between the margins of the injury were analyzed using the statistical program IBM SPSS Statistics 25 (https://www.ibm.com/analytics/spss-statistics-software) (IBM, Armonk, NY, USA). The significant difference among means was analyzed by Student's *t*-test (assuming the normal distribution of the continuous variables). A level of *p*-value < 0.05 was considered significant.

3. Results and Discussion

3.1. FTIR Analysis of the Samples of Honey, Propolis, and Propolis Extracts

Regarding the honey samples and given the flora that exists at the place of harvest, it is expected that Honey 1 was produced mainly during the autumn from species such as *Arbutus unedo*, *Castanea sativa*, *Quercus faginea*, and *Pinus pinaster*. Honey 2 was produced during the summer from flowers such as *Rosa* spp., *Dahlia* spp., *Hydrangea* spp., and flowers from fruit trees like *Prunus avium*, *Malus* spp., *Pyrus communis*, and *Prunus spinosa*. Finally, Honey 3 was produced during the spring, when most wildflowers and aromatic plants bloom (*Papaver rhoeas*, *Chrysanthemum coronarium*, *Lavandula stoechas*, *Rosmarinus officinalis*, *Baccharis trimera*, and *Thymus mastichina*).

The propolis samples subsequently collected did not show a direct correlation with the honey samples, because bees repair their hive continuously throughout the year, whenever this need arises, so the conditions under which propolis is produced will not necessarily be the same conditions under which honey is produced.

The FTIR spectra of the samples of honey, propolis, and propolis extracts were recorded (Figure 1).

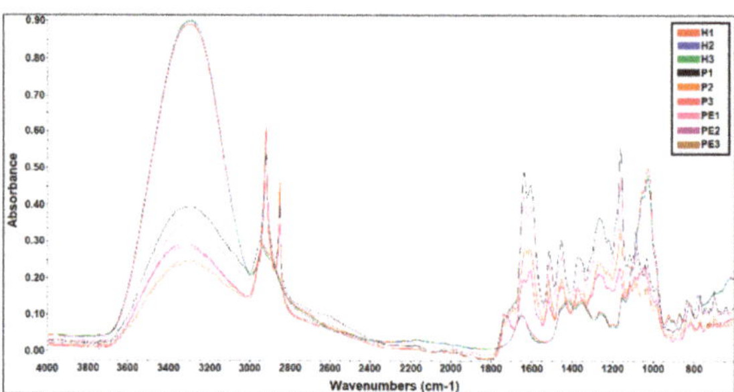

Figure 1. FTIR spectra of the samples.

The spectra of honeys show a water band at approximately 3300 cm^{-1} and 1650 cm^{-1}. Near 2900 cm^{-1} it is possible to observe a band that may be associated with groups present in amino acids. The bands between 1450 and 750 cm^{-1} may correspond to organic acids and to sugars commonly present in honey, such as sucrose, glucose, and fructose. Even though the honey samples have been produced in different seasons, they present quite similar spectra.

The spectra of propolis samples show approximately the same bands but with great dissimilarity in terms of the intensity of the signal. In these spectra, it can be observed a water band at around 3400 cm^{-1}, and very pronounced bands at nearly 2900 and 2850 cm^{-1} that may correspond to some aliphatic compounds. The bands observed between 1650 and 1600 cm^{-1}, as well as the bands between 1550 and 1400 cm^{-1} may be caused by the presence of flavonoids and other aromatic compounds. The band at 1150 cm^{-1} may be due to the presence of hydroxyflavonoids. Some of the bands were not considered and may be related to wax and other debris present in the samples. The spectra of propolis extracts show approximately the same bands, but with lower intensity than the propolis samples.

The water band observed in both propolis samples and extracts is less pronounced than the band registered in the honeys. In contrast, the bands identified as possible aromatic compounds and flavonoids in propolis samples and extracts spectra are more pronounced. Finally, the sugar bands in honey spectra are more evident and better outlined.

The results now obtained are in agreement with other previously published results for other honey and propolis samples [20,21].

3.2. Phytochemical Characterization

Plant polyphenols present at least 8000 distinct known structures, being the most important class of natural bioactive compounds, which exhibit various biological activities [22]. Honey presents three classes of flavonoids with analogous structure: flavonols, flavones, and flavanones. Flavonoids are responsible for the color, taste, and flavor of the honey and they also improve its beneficial health effects [22]. Furthermore, the floral sources used by bees to produce honey, whose predominance depends on seasonal and environmental issues, influences the phenolic composition and antioxidant activity of honey.

The results of the phytochemical characterization of the samples regarding total phenolics and flavonoids contents are presented in Table 1.

Table 1. Total phenolic compounds and flavonoids content of the samples.

Samples	Total Phenolic Compounds	Flavonoids
	(g GAE/100 g Sample) [1]	(g QE/100 g Sample) [1]
H1	0.107 ± 0.016	0.007 ± 0.001
H2	0.046 ± 0.005	Not detected
H3	0.029 ± 0.003	Not detected
PE1	28.947 ± 1.329	5.494 ±0.335
PE2	21.747 ± 1.062	1.786 ± 0.029
PE3	28.667 ± 0.774	4.280 ± 0.123
H1PE1 0.3%	2.394 ± 0.227	0.290 ± 0.007
H1PE1 0.5%	3.324 ± 0.044	0.452 ± 0.012
H1PE2 0.3%	1.219 ± 0.049	0.115 ± 0.005
H1PE2 0.5%	1.750 ± 0.076	0.199 ± 0.018
H1PE3 0.3%	1.969 ± 0.071	0.054 ± 0.011
H1PE3 0.5%	3.506 ± 0.257	0.308 ± 0.006

[1] Results expressed as mean ± standard deviation.

Total phenolic compounds content of honey samples ranged from 0.029 to 0.107 g GAE/100 g sample, the values observed for propolis extracts ranged from 21.747 to 28.947 g GAE/100 g sample, and finally for the mixtures of honey with propolis ranged from 1.219 to 3.506 g GAE/100 g sample. The honey that presented the highest content in phenolic compounds is Honey 1 and the lowest content can be found in Honey 3. These differences may be related with the different seasons in which the honeys were produced, as mentioned above.

Propolis extracts showed a much higher concentration of phenolic compounds than honey samples, with the highest content found in Propolis Extract 1 followed by Propolis Extract 3 and Propolis Extract 2. An increase in phenolic content was observed with the addition of higher concentrations of propolis extracts to honey, and the highest value was obtained with H1PE3 at 0.5%.

Flavonoids were almost absent from honey samples. The only one that presents flavonoids in its composition is Honey 1, but even this sample has a very low content. In contrast, the flavonoids determined in the propolis extracts ranged from 1.786 to 5.494 g QE/100 g sample, and in the mixtures ranged from 0.054 to 0.452 g QE/100 g sample. Propolis extracts showed a higher concentration of flavonoids than honey samples, with the highest content in Propolis Extract 1 followed by Propolis Extract 3 and Propolis Extract 2. An increase in flavonoid content was observed in all samples with the addition of higher concentrations of propolis extract to honey, as expected.

The values of total phenolic compounds and flavonoids determined in the present work are very similar to the ones obtained for selected Czech honeys [23].

3.3. Antioxidant and Anti-Inflammatory Activities

Honey is an important natural source of antioxidants and has potential therapeutic value in several inflammatory diseases and in the treatment of heart disease, cancer, and cataracts, in addition to its sweetening capacity and lower glycemic load [24]. The biological properties of honey comprise antioxidant, antimicrobial, anti-inflammatory, and wound-healing activities [24].

In this work, the antioxidant activity of the samples was evaluated by two different methods that measure distinct antioxidant properties (Table 2). The DPPH free radical scavenging assay is based on the capacity of the samples to scavenge free radicals, while the β-carotene bleaching test allows the indirect evaluation of the capacity of the samples to inhibit the lipid peroxidation [25].

Table 2. Antioxidant and anti-inflammatory activities of the samples.

Samples	DPPH	β-Carotene Bleaching Test	Anti-Inflammatory Activity
	% Inhibition/100 g Sample [1]	% Inhibition/100 g Sample [1]	% Inhibition/100 g Sample [1]
H1	0.431 ± 0.023	0.809 ± 0.042	20.625 ± 0.884
H2	0.133 ± 0.019	0.684 ± 0.032	23.438 ± 3.094
H3	Not detected	0.349 ± 0.028	Not detected
PE1	92.506 ± 1.249	51.441 ± 4.477	15.000 ± 3.536
PE2	92.012 ± 0.258	48.660 ± 1.876	31.250 ± 1.768
PE3	93.245 ± 0.687	53.909 ± 2.328	48.750 ± 1.768
H1PE1 0.3%	8.271 ± 0.044	3.027 ± 0.070	36.408 ± 6.865
H1PE1 0.5%	8.119 ± 0.040	3.929 ± 0.105	28.571 ± 0.001
H1PE2 0.3%	7.921 ± 0.097	2.878 ± 0.096	40.049 ± 1.716
H1PE2 0.5%	8.154 ± 0.158	3.108 ± 0.049	45.238 ± 0.001
H1PE3 0.3%	8.165 ± 0.026	2.919 ± 0.106	36.408 ± 3.433
H1PE3 0.5%	8.396 ± 0.321	3.596 ± 0.089	40.476 ± 6.734

[1] Results expressed as mean ± standard deviation.

The honey that presented the highest antioxidant activity measured by the DPPH assay was Honey 1, opposed to Honey 3 that showed no relevant activity measured by this method. Propolis extracts revealed extremely high levels of antioxidant activity across all samples, with Propolis Extract 3 presenting the highest value. An increase in the antioxidant activity was observed in all samples with the addition of propolis extract to honey, however adding a higher concentration of propolis did not result in a considerable rase in activity in most cases.

Concerning the results of β-carotene bleaching test, the honey that revealed the highest antioxidant activity was once again Honey 1. The antioxidant activity of this honey measured by both methods is related with the presence of great amounts of phenolic compounds, as previously mentioned. Propolis extracts revealed high levels of antioxidant activity measured by β-carotene bleaching test as it was also verified by DPPH assay.

Analyzing the data obtained throughout the different assays it was possible to verify that dark-brown honey (H1) presented a higher content in phenolic compounds and flavonoids, followed by red honey (H2) and finally by light-yellow honey (H3). These results were consistent with the bioactive activity of the different samples. Since Honey 1 presented better phytochemical results, it was used in all the mixtures of honey with propolis.

The anti-inflammatory activity was evaluated using an in vitro assay that studied the ability of the samples to inhibit protein denaturation using a BSA solution (Table 2). It was noted that propolis extracts reveal a higher anti-inflammatory activity than honeys. The honey that presented the highest activity was Honey 2, and among the Propolis Extracts, PE3 revealed the highest activity. Generally, an increase in the anti-inflammatory activity was observed in all samples when adding propolis extract to honey. In a previous work involving Malaysian honeys, the authors concluded that the anti-inflammatory activity may be attributed, at least in part, to the phenolic compounds [24].

3.4. Wound-Healing Activity

In the present study, NHDF cells were used in a scratch assay. Although all the preliminary characterization data showed that the propolis extracts always presented better results than the other samples under study, it was decided to also evaluate the wound-healing activity for all the samples including the mixtures of honey with propolis.

It is already known that honey is not toxic against normal cells but is extremely cytotoxic to the tumor or cancer cells, as it was previously described [26]. Similar results were found in propolis extracts, which demonstrated cytotoxicity in human fibrosarcoma and colon adenocarcinoma cells while presenting no cytotoxic action in normal human skin fibroblasts [27].

By using the microscopic images, it was possible to evaluate the evolution of the gap created in the confluent cell monolayer in the presence of the samples (Table 3, Table 4, and Table 5).

Analyzing the different images and comparing them to the control samples it is possible to say that the honey that shows better results after 36 h is Honey 2 (Table 3), while the propolis extract that presented better results was Propolis Extract 2 at 0.5% (Table 4); the best mixture is the honey with propolis—H1PE3 0.3% (Table 5).

Examining all the images, it is possible to observe that the cells continue alive when incubated with the samples. Moreover, it is clear that the samples promote cell migration, demonstrating the wound-healing potential of honey and propolis.

Furthermore, by estimating the distance between the margins of the scratch (Table 6) the conclusions were the same. For all the samples, except for Honey 3 at 2 h, a significant (p-value < 0.05) reduction of the scratch was observed when compared to the control at the same time of incubation. In general, the samples that showed the best results were the mixtures of honey with propolis. However, the sample that presented the maximum activity was the Propolis Extract 2 at 0.5%.

Table 3. Microscopic images obtained from the scratch wound-healing assay with the honeys (magnification: 100×).

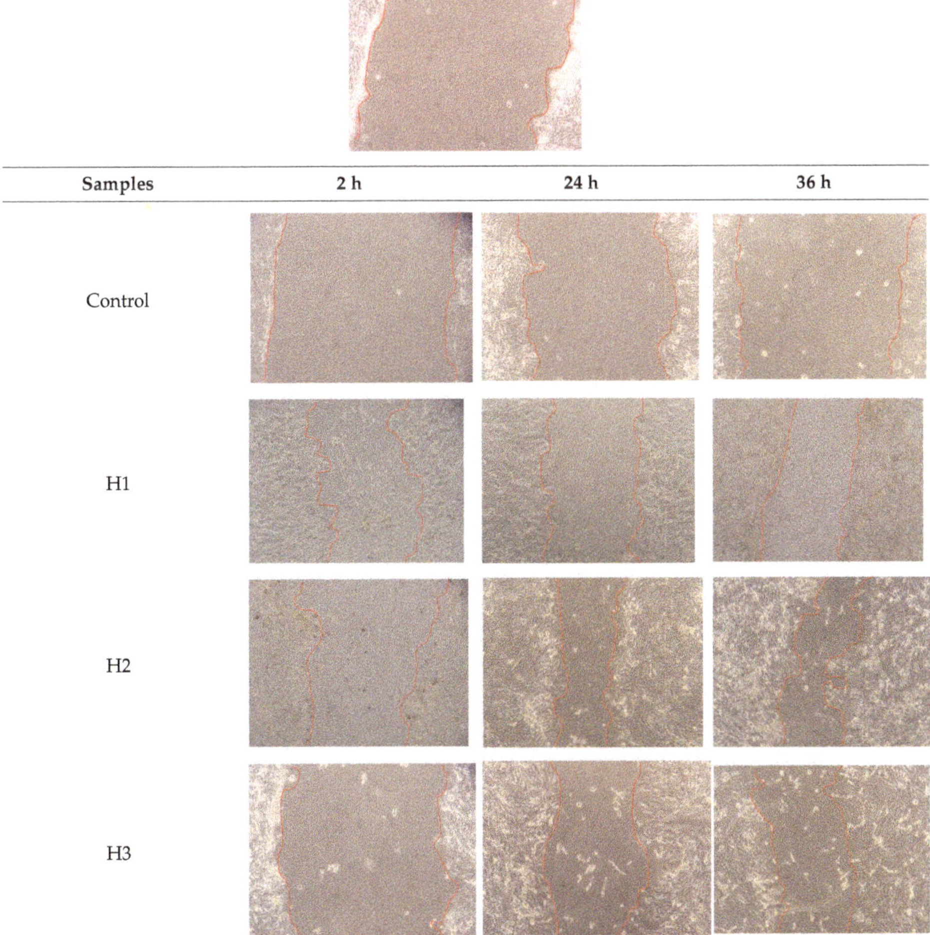

Table 4. Microscopic images obtained from the scratch wound-healing assay with the propolis extracts (magnification: 100×).

Samples	2 h	24 h	36 h
PE1 0.3%			
PE1 0.5%			
PE2 0.3%			
PE2 0.5%			
PE3 0.3%			
PE3 0.5%			

Table 5. Microscopic images obtained from the scratch wound-healing assay with the mixtures of honey with propolis (magnification: 100×).

Samples	2 h	24 h	36 h
H1PE1 0.3%			
H1PE1 0.5%		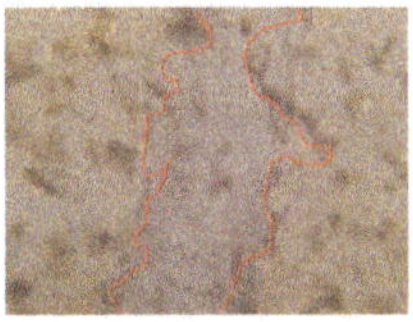	
H1PE2 0.3%			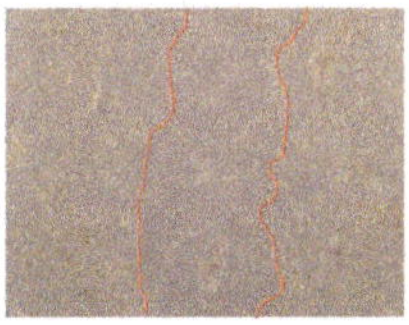
H1PE2 0.5%	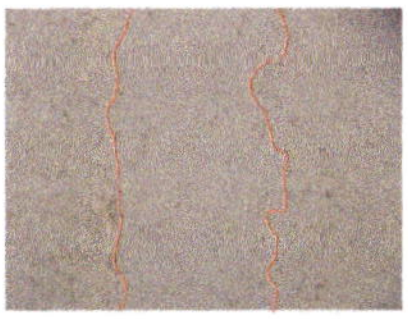		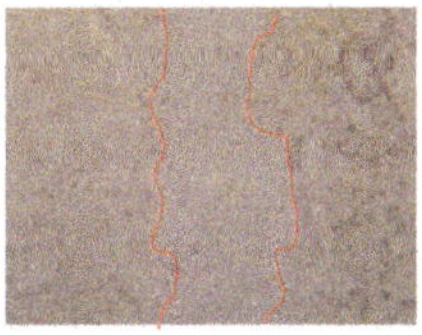
H1PE3 0.3%		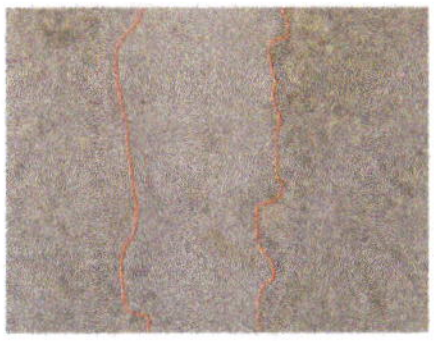	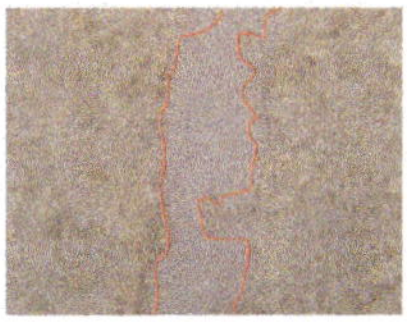
H1PE3 0.5%	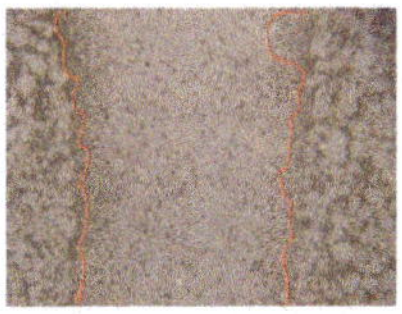		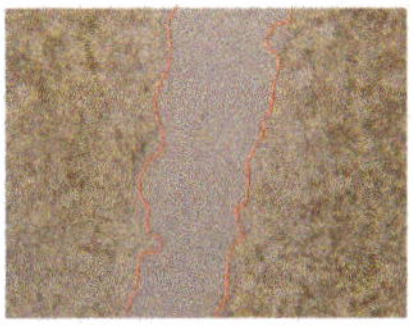

In opposition to what was previously observed [1], in the present work the obtained results suggest that the effect of combining propolis with honey is not synergistic but just the combined effect of honey and propolis. This may be due to the chemical composition of each particular honey that will directly influence its bioactivities. In the honey samples now studied, the concentration in total phenolic compounds is relatively lower than in other samples of honey [1]. Moreover, flavonoids were not detected in the honey samples. These observations may explain the additive results observed for the mixtures of honey with propolis, suggesting the contribution of the propolis compounds to the biological activities.

Considering all the obtained results, it is possible to verify that the samples that presented higher cell migration levels also presented higher bioactivity.

During the inflammation process, honey promotes the release of inflammatory cytokines (TNF-α IL-6, IL-1β, and NO) by monocytes, which might stimulate collagen synthesis by fibroblasts, playing important roles in the initiation and amplification of this process [10]. The modulation of the severity of inflammation can be associated with the anti-inflammatory properties of the polyphenols present in honey. Honey initiates an active but controlled inflammation but does not let the inflammation to develop in a chronic or exaggerated state, modulating the inflammatory phase of wound-healing [10]. The anti-ulcerous activity of honey and propolis can be attributed to flavonoids that can act alone or in combination with other compounds such as sterols, terpinens, saponins, gums, and mucilage [4].

A recently published paper, in which the potential wound-healing properties of propolis was evaluated, demonstrated that propolis promoted a marked increase in the wound repair capacity of keratinocytes [28]. It was also proved that the regenerative properties of propolis are mainly due to H_2O_2 (which is extracellularly released and passes across the plasma membrane) is able to modulate intracellular mechanisms [28].

Table 6. Calculated distance between the margins of the injury.

Samples	0 h [1]	2 h [1]	*p*-Value	24 h [1]	*p*-Value	36 h [1]	*p*-Value
Control		0.99 ± 0.05	0.818	0.99 ± 0.05	1.000	0.98 ± 0.05	0.817
H1		0.71 ± 0.04	0.002 *	0.59 ± 0.03	0.001 *	0.48 ± 0.02	0.001 *
H2		0.83 ± 0.04	0.011 *	0.37 ± 0.02	0.001 *	0.33 ± 0.02	0.001 *
H3		0.90 ± 0.05	0.062	0.57 ± 0.03	0.001 *	0.49 ± 0.02	0.001 *
PE1 0.3%		0.80 ± 0.04	0.006 *	0.78 ± 0.04	0.005 *	0.69 ± 0.03	0.002 *
PE1 0.5%		0.84 ± 0.04	0.014 *	0.81 ± 0.04	0.009 *	0.78 ± 0.04	0.005 *
PE2 0.3%		0.54 ± 0.03	0.001 *	0.39 ± 0.02	0.001 *	0.34 ± 0.02	0.001 *
PE2 0.5%	1.00 ± 0.05	0.66 ± 0.03	0.001 *	0.52 ± 0.03	0.001 *	0.27 ± 0.01	0.001 *
PE3 0.3%		0.74 ± 0.04	0.003 *	0.62 ± 0.03	0.001 *	0.53 ± 0.03	0.001 *
PE3 0.5%		0.45 ± 0.02	0.001 *	0.41 ± 0.02	0.001 *	0.39 ± 0.02	0.001 *
H1PE1 0.3%		0.74 ± 0.04	0.003 *	0.63 ± 0.03	0.001 *	0.55 ± 0.03	0.001 *
H1PE1 0.5%		0.43 ± 0.02	0.001 *	0.43 ± 0.02	0.001 *	0.42 ± 0.02	0.001 *
H1PE2 0.3%		0.54 ± 0.03	0.001 *	0.37 ± 0.02	0.001 *	0.49 ± 0.02	0.001 *
H1PE2 0.5%		0.64 ± 0.03	0.001 *	0.58 ± 0.03	0.001 *	0.46 ± 0.02	0.001 *
H1PE3 0.3%		0.80 ± 0.04	0.006 *	0.44 ± 0.02	0.001 *	0.31 ± 0.02	0.001 *
H1PE3 0.5%		0.74 ± 0.04	0.001 *	0.45 ± 0.02	0.001 *	0.34 ± 0.02	0.001 *

[1] Results expressed as mean ± standard deviation; * Indicates a significant result (*p*-value < 0.05).

4. Conclusions

This work demonstrated the biological potential of honey and propolis, particularly the wound-healing activity, which is related with their antioxidant and anti-inflammatory properties. Further studies should be performed to clarify the mechanism of action of honey and propolis by which cell migration is stimulated.

Author Contributions: Conceptualization, A.P.D.; methodology, A.M.A., J.G., and Â.L.; Formal analysis, E.G. and A.P.D.; Investigation, A.M.A., J.G., and Â.L.; Resources, E.G. and A.P.D.; Data curation, E.G. and A.P.D.; Writing—original draft preparation, A.M.A.; Writing—review and editing, J.G., Â.L., E.G., and A.P.D.; Supervision, A.P.D. All authors have read and agreed to the published version of the manuscript.

Funding: This work was partially supported by CICS-UBI that is financed by National Funds from Fundação para a Ciência e a Tecnologia (FCT) and Community Funds (UIDB/00709/2020). Joana Gonçalves acknowledges the PhD fellowship from FCT (reference: SFRH/BD/149360/2019). Ângelo Luís acknowledges the contract of Scientific Employment in the scientific area of Microbiology also financed by FCT.

References

1. Osés, S.M.; Pascual-Maté, A.; Fernández-Muiño, M.A.; López-Díaz, T.M.; Sancho, M.T. Bioactive properties of honey with propolis. *Food Chem.* **2016**, *196*, 1215–1223. [CrossRef]

2. Liu, J.R.; Ye, Y.L.; Lin, T.Y.; Wang, Y.W.; Peng, C.C. Effect of floral sources on the antioxidant, antimicrobial, and anti-inflammatory activities of honeys in Taiwan. *Food Chem.* **2013**, *139*, 938–943. [CrossRef] [PubMed]

3. Ferreira, I.C.F.R.; Aires, E.; Barreira, J.C.M.; Estevinho, L.M. Antioxidant activity of Portuguese honey samples: Different contributions of the entire honey and phenolic extract. *Food Chem.* **2009**, *114*, 1438–1443. [CrossRef]

4. Viuda-Martos, M.; Ruiz-Navajas, Y.; Fernández-López, J.; Pérez-Álvarez, J.A. Functional properties of honey, propolis, and royal jelly. *J. Food Sci.* **2008**, *73*, 117–124. [CrossRef] [PubMed]

5. Cianciosi, D.; Forbes-Hernández, T.Y.; Afrin, S.; Gasparrini, M.; Reboredo-Rodriguez, P.; Manna, P.P.; Zhang, J.; Lamas, L.B.; Flórez, S.M.; Toyos, P.A.; et al. Phenolic compounds in honey and their associated health benefits: A review. *Molecules* **2018**, *23*, 2322. [CrossRef]

6. Sforcin, J.M. Propolis and the immune system: A review. *J. Ethnopharmacol.* **2007**, *113*, 1–14. [CrossRef]

7. Anjum, S.I.; Ullah, A.; Khan, K.A.; Attaullah, M.; Khan, H.; Ali, H.; Bashir, M.A.; Tahir, M.; Ansari, M.J.; Ghram, H.A.; et al. Composition and functional properties of propolis (bee glue): A review. *Saudi J. Biol. Sci.* **2019**, *26*, 1695–1703. [CrossRef]

8. Zabaiou, N.; Fouache, A.; Trousson, A.; Baron, S.; Zellagui, A.; Lahouel, M.; Lobaccaro, J.A. Biological properties of propolis extracts: Something new from an ancient product. *Chem. Phys. Lipids* **2017**, *207*, 214–222. [CrossRef]

9. Silva-Carvalho, R.; Baltazar, F.; Almeida-Aguiar, C. Propolis: A complex natural product with a plethora of biological activities that can be explored for drug development. *Evid.-Based Complement. Altern. Med.* **2015**, *2015*, 206439. [CrossRef]

10. Oryan, A.; Alemzadeh, E.; Moshiri, A. Biological properties and therapeutic activities of honey in wound healing: A narrative review and meta-analysis. *J. Tissue Viability* **2016**, *25*, 98–118. [CrossRef]

11. Martinotti, S.; Ranzato, E. Propolis: A new frontier for wound healing? *Burn. Trauma* **2015**, *3*, 9. [CrossRef] [PubMed]

12. Kalogeropoulos, N.; Konteles, S.J.; Troullidou, E.; Mourtzinos, I.; Karathanos, V.T. Chemical composition, antioxidant activity and antimicrobial properties of propolis extracts from Greece and Cyprus. *Food Chem.* **2009**, *116*, 452–461. [CrossRef]

13. Felician, F.F.; Yu, R.; Li, M.; Li, C.; Chen, H.; Jiang, Y.; Tang, T.; Qi, W.; Xu, H. The wound healing potential of collagen peptides derived from the jellyfish *Rhopilema esculentum*. *Chin. J. Traumatol.* **2019**, *22*, 12–20. [CrossRef] [PubMed]

14. Luís, Â.; Sousa, S.; Duarte, A.P.; Pereira, L.; Domingues, F. Phytochemical characterization, and evaluation of rheological and antioxidant properties of commercially available juices of berries. *J. Berry Res.* **2018**, *8*, 11–23. [CrossRef]

15. Luís, Â.; Neiva, D.; Pereira, H.; Gominho, J.; Domingues, F.; Duarte, A.P. Stumps of *Eucalyptus globulus* as a source of antioxidant and antimicrobial polyphenols. *Molecules* **2014**, *19*, 16428–16446. [CrossRef]

16. Luís, Â.; Duarte, A.P.; Pereira, L.; Domingues, F. Interactions between the major bioactive polyphenols of berries: Effects on antioxidant properties. *Eur. Food Res. Technol.* **2018**, *244*, 175–185. [CrossRef]

17. Luís, Â.; Sousa, S.; Wackerlig, J.; Dobusch, D.; Duarte, A.P.; Pereira, L.; Domingues, F. Star anise (*Illicium verum* Hook. f.) essential oil: Antioxidant properties and antibacterial activity against *Acinetobacter baumannii*. *Flavour Fragr. J.* **2019**, *34*, 260–270. [CrossRef]

18. Luís, Â.; Breitenfeld, L.; Ferreira, S.; Duarte, A.P.; Domingues, F. Antimicrobial, antibiofilm and cytotoxic activities of *Hakea sericea* Schrader extracts. *Pharmacogn. Mag.* **2014**, *10*, S6–S13.

19. Liang, C.-C.; Park, A.Y.; Guan, J.-L. In vitro scratch assay: A convenient and inexpensive method for analysis of cell migration in vitro. *Nat. Protoc.* **2007**, *2*, 329–333. [CrossRef]

20. Oliveira, R.N.; Mancini, M.C.; Oliveira, F.C.S.; Passos, T.M.; Quilty, B.; Thiré, R.M.S.M.; McGuinness, G.B. FTIR analysis and quantification of phenols and flavonoids of five commercially available plants extracts used in wound healing. *Rev. Mater.* **2016**, *21*, 767–779. [CrossRef]

21. Anjos, O.; Campos, M.G.; Ruiz, P.C.; Antunes, P. Application of FTIR-ATR spectroscopy to the quantification of sugar in honey. *Food Chem.* **2015**, *169*, 218–223. [CrossRef] [PubMed]

22. Estevinho, L.; Pereira, A.P.; Moreira, L.; Dias, L.G.; Pereira, E. Antioxidant and antimicrobial effects of phenolic compounds extracts of Northeast Portugal honey. *Food Chem. Toxicol.* **2008**, *46*, 3774–3779. [CrossRef] [PubMed]

23. Lachman, J.; Orsák, M.; Hejtmánková, A.; Kovářová, E. Evaluation of antioxidant activity and total phenolics of selected Czech honeys. *LWT Food Sci. Technol.* **2010**, *43*, 52–58. [CrossRef]

24. Kassim, M.; Achoui, M.; Mustafa, M.R.; Mohd, M.A.; Yusoff, K.M. Ellagic acid, phenolic acids, and flavonoids in Malaysian honey extracts demonstrate in vitro anti-inflammatory activity. *Nutr. Res.* **2010**, *30*, 650–659. [CrossRef] [PubMed]

25. Luís, Â.; Domingues, F.; Gil, C.; Duarte, A.P. Antioxidant activity of extracts of Portuguese shrubs: *Pterospartum tridentatum*, *Cytisus scoparius* and *Erica* spp. *J. Med. Plants Res.* **2009**, *3*, 886–893.

26. Erejuwa, O.O.; Sulaiman, S.A.; Wahab, M.S. Effects of honey and its mechanisms of action on the development and progression of cancer. *Molecules* **2014**, *19*, 2497–2522. [CrossRef]

27. Watanabe, M.A.E.; Amarante, M.K.; Conti, B.J.; Sforcin, J.M. Cytotoxic constituents of propolis inducing anticancer effects: A review. *J. Pahrmacy Pharmacol.* **2011**, *63*, 1378–1386. [CrossRef]

28. Martinotti, S.; Pellavio, G.; Laforenza, U.; Ranzato, E. Propolis induces AQP3 expression: A possible way of action in wound healing. *Molecules* **2019**, *24*, 1544. [CrossRef]

 *applied sciences*

Review

Applications of Sesquiterpene Lactones: A Review of Some Potential Success Cases

Laila Moujir [1,*], Oliver Callies [2], Pedro M. C. Sousa [3], Farukh Sharopov [4] and Ana M. L. Seca [5,6,*]

[1]   Department of Biochemistry, Microbiology, Genetics and Cell Biology, Facultad de Farmacia, Universidad de La Laguna, 38206 San Cristóbal de La Laguna, Spain

[2]   AbbVie Deutschland GmbH & Co. KG, Knollstrasse, 67061 Ludwigshafen, Germany; oliver.callies@abbvie.com

[3]   Faculty of Sciences and Technology, University of Azores, 9500-321 Ponta Delgada, Portugal; sdoffich@gmail.com

[4]   Research Institution "Chinese-Tajik Innovation Center for Natural Products", Academy of Sciences of the Republic of Tajikistan, Ayni 299/2, Dushanbe 734063, Tajikistan; shfarukh@mail.ru

[5]   cE3c—Centre for Ecology, Evolution and Environmental Changes/Azorean Biodiversity Group & Faculty of Sciences and Technology, University of Azores, Rua Mãe de Deus, 9500-321 Ponta Delgada, Portugal

[6]   LAQV-REQUIMTE, University of Aveiro, 3810-193 Aveiro, Portugal

*   Correspondence: lmoujir@ull.es (L.M.); ana.ml.seca@uac.pt (A.M.L.S.); Tel.: +34-9-22-318513 (L.M.); +351-29-6650174 (A.M.L.S.)

Received: 25 March 2020; Accepted: 19 April 2020; Published: 25 April 2020

Abstract: Sesquiterpene lactones, a vast range of terpenoids isolated from Asteraceae species, exhibit a broad spectrum of biological effects and several of them are already commercially available, such as artemisinin. Here the most recent and impactful results of in vivo, preclinical and clinical studies involving a selection of ten sesquiterpene lactones (alantolactone, arglabin, costunolide, cynaropicrin, helenalin, inuviscolide, lactucin, parthenolide, thapsigargin and tomentosin) are presented and discussed, along with some of their derivatives. In the authors' opinion, these compounds have been neglected compared to others, although they could be of great use in developing important new pharmaceutical products. The selected sesquiterpenes show promising anticancer and anti-inflammatory effects, acting on various targets. Moreover, they exhibit antifungal, anxiolytic, analgesic, and antitrypanosomal activities. Several studies discussed here clearly show the potential that some of them have in combination therapy, as sensitizing agents to facilitate and enhance the action of drugs in clinical use. The derivatives show greater pharmacological value since they have better pharmacokinetics, stability, potency, and/or selectivity. All these natural terpenoids and their derivatives exhibit properties that invite further research by the scientific community.

Keywords: Asteraceae; sesquiterpene lactones; alantolactone; arglabin; parthenolide; thapsigargin; in vivo study; anticancer; anti-inflammatory

1. Introduction

Traditional or folk medicine relies heavily on the use of compounds-rich plants, like those of the Asteraceae family, of which many such species are commercially available in the form of herbal preparations. These are particularly rich in a wide range of natural terpenoids named sesquiterpene lactones [1] that, in some cases, are considered the active principles of such therapeutic plants [2]. Structurally speaking, sesquiterpene lactones are terpenes that have in common a basic structure of 15 carbons (thus the prefix sesqui-) resulting from biosynthesis involving three isoprene units with a cyclical structure along with a fused α-methylene-γ-lactone ring [3]. Sesquiterpene synthases catalyze a common biosynthesis route for sesquiterpene lactones, based on the cyclization of farnesyl phosphate resulting from the 2-C-methyl-D-erythritol-4-phosphate (MEP) and mevalonate (MVA) pathways of dimethylalyl diphosphate and isopentenyl diphosphate precursors in chloroplasts and cytosol, respectively [4]. However, sesquiterpene lactones have very different chemical structures regarding the type and position of the substituents, as well as the size of the non-lactone ring [5,6]. For this reason, in structural terms, sesquiterpene lactones are organized into several subclasses: eudesmanolide (a 6/6 bicyclic structure), guaianolide and pseudoguaianolide (both 5/7 bicyclic compounds), germacranolide (with a 10-membered ring) and xanthanolide (containing a non-cyclic carbon chain and a seven-membered ring) [7,8] (Figure 1).

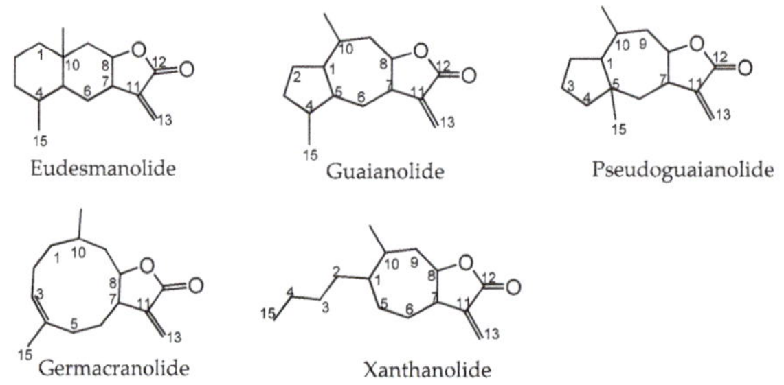

Figure 1. Basic chemical structure of each of the sesquiterpene lactone subclasses: eudesmanolide, guaianolide, pseudoguaianolide, germacranolide, and xanthanolide.

Some of these subclasses have compounds that exhibit a wide range of biological activities. These range from antitumor [9] to anti-inflammatory, including antimalarial, antimicrobial, antioxidant [10], neuroprotective [11], hepatoprotective, and immune-stimulant properties [12,13]. Regarding the structure/activity relationship of these compounds, it appears that the α-methylene-γ-lactone nucleus has a crucial role in almost all their observed biological effects, such as cytotoxic, antitrypanosomal, and anti-inflammatory actions [12,14]. Other specific structural moieties of sesquiterpene lactone seem to influence their activity. For example, the presence of electrophilic sites associated with medium/high lipophilicity increase antimycobacterial activity, while a double bond exo to the cyclopentenone ring seems to favor anti-inflammatory activity [7,8,14]. The interaction of the α,β-unsaturated cyclopentenone nucleus with the target depends largely on the geometry of the molecule, which is also a factor that influences the level of activity exhibited by sesquiterpene lactones [7,15]. Moreover, the number of alkylating groups in the structure of sesquiterpene lactones contributes to the level of activity they display, two groups being the optimal number [7,15]. The structure/activity relationship specific to each sesquiterpene lactone presented in this review will be discussed in detail throughout Section 2. Scientific evidence shows an action mechanism common to sesquiterpene lactones. The structural elements α-methylene-γ-lactone and α,β-unsaturated cyclopentenone act as alkylating groups on proteins found in cells through Michael addition, especially their thiol groups. They thereby affect cell functionality, i.e., gene regulation, protein synthesis, and cell metabolism [15–17].

In recent years, the scientific community has already shown an oustanding growth of interest in the sesquiterpene lactones, largely due to the success of artemisinin—one of the best known—and the broad spectrum of activities exhibited by this compound's chemical family [18–20]. Thus, the high number of studies published concerning their isolation from new natural sources, total and semi-synthetic syntheses, and evaluation of pharmacological potential, is not surprising. Natural sesquiterpene lactones exhibit poor pharmacokinetic properties due mainly to their low bioavailability, deriving from low solubility in water. As a result, in order to overcome these limitations, research interest in the synthesis of their derivatives has increased [21,22]. In addition, structural modification and synthesis has also allowed for an in-depth knowledge of their chemical properties, as well as the establishment of structure-activity relationships.

In terms of health promotion, the applications of sesquiterpene lactones and their derivatives are a key research area. From this point of view, in vivo, preclinical and clinical studies are those that allow a more realistic assessment of their medicinal potential [23].

The literature on this topic is extensive, especially for the most successful compounds; however, studies on less known sesquiterpenes and their derivatives show that these compounds deserve more attention, since they could play an important part in future human health maintenance. Therefore, this review aims to point out the results of the most impactful and recent in vivo and (pre)clinical studies of these insufficiently explored compounds and derivatives, which could be a valuable alternative in the development of new therapeutic drugs.

2. Sesquiterpene Lactones with Significant In Vivo Activity

2.1. Alantolactone

The eudesmanolide alantolactone (**1**) ($C_{15}H_{20}O_2$) (Figure 2) was first isolated from *Inula helenium* L. roots, and later was also found in other *Inula* species [24], as well as non-*Inula* spp. such as *Saussurea lappa* C.B. Clarke [25] and *Aucklandia lappa* DC. [26].

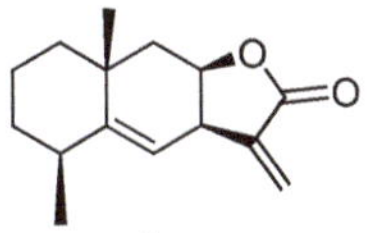

1

Figure 2. Structure of alantolactone (**1**).

Alantolactone (**1**) is obtained mainly by extraction and purification from natural sources, although its total synthesis is possible [27,28]. Modern laboratory techniques have resulted in a significant increase in its extraction yield [29–31]. For example, ~3% pure alantolactone (**1**) yields were obtained from *Inula helenium* roots by Zhao et al. [30], which meant a significant improvement over the ~1% yields from the roots of *Inula magnifica* Lipsky [32].

Although alantolactone is associated with allergic contact dermatitis triggered by *Inula* species [33–35], this compound was first described as anti-helminthic [36], with subsequent studies demonstrating its tremendous potential, mainly as antitumor [37–40], anti-inflammatory [26,41], and antioxidant [42] agent.

Regarding anticancer activity, the effect of alantolactone (**1**) on leukemia is well documented and was recently reviewed by Da Silva Castro et al. [43]. Xu et al. [44] reported a positive and significant effect (**1**) on B-cell acute lymphoblastic leukemia. In this study, 100 mg/kg (b.w.) doses were intravenously administered every two days in leukemia xenografted NOC/SCID mice. Results showed that treated mice lived an average of eight days more than non-treated mice (31.5 days after xenografting in treated compared to 23.5 days in non-treated mice), without significant weight loss. Before this study, Chun et al. [45] had also reported an in vivo antitumor effect on triple negative breast cancer in mice. The researchers used only a 2.5 mg/kg (b.w.) dose every two days to treat athymic nude mice xenografted with MDA-MB-231 cells. In this assay, alantolactone (**1**) reduced tumor size by over half, and significantly reduced tumor weight by about half after 24 days. Alantolactone (**1**) also exhibits in vivo activity against gastric cancer. Human gastric cancer cells (SGC-7901) were xenografted onto nude athymic mice, which were treated with 15 mg/kg (b.w.) of alantolactone (**1**) injected every two days [46]. Alantolactone (**1**) caused significant tumor growth inhibition and reduced Ki-67 and Bcl-2 expression (tumor associated proteins), without significant liver and kidney toxicity or impact on mouse weight [46].

Alantolactone (**1**) has also proven to be an especially potent sensitizing agent. In fact, it exhibits a synergistic effect with known chemotherapy drugs, such as oxaliplatin. Cao et al. [47] showed that a 10:2 mg/kg (b.w.) dose of alantolactone: oxaliplatin reduced tumor volume and weight by more than

50% in athymic mice xenografted with colorectal tumor cells (HCT116). The anticancer effect increased substantially when both compounds were used together, as opposed to alone. This result is in line with that obtained by He et al. [48], according to which, using a small weekly injected dose of 3 mg/kg (b.w.), tumor volume significantly decreased by around 75% in xenografted pancreatic cancer cells (PANC-1), while increasing cancer cell chemosensitivity to oxaliplatin, revealing a synergistic action.

Alantolactone's anticancer mechanism was also the main focus of several studies, some of them mentioned in recent reviews [49,50] The authors attribute its activity to the multiple pathways it activates. Alantolactone (**1**) acts as an alkylating agent leading to inhibition of key enzymes and proteins, and as an apoptosis inducer in cancer cells at mitochondrial level by interacting with cytochrome c. It promotes overproduction of reactive oxygen species (ROS) due to specific caspase activation, by inhibiting autophagic deregulation, among other processes.

Beyond its anticancer activity, compound **1** also exhibits interesting anti-inflammatory activity. Ren et al. [51] used the DSS-induced colitis mouse model to test alantolactone (**1**) anti-inflammatory activity. A 50 mg/kg dose reversed colitis symptoms (bloody diarrhea, colon shortening and weight loss), besides significantly reducing pro-inflammatory cytokine TNF-α expression (to about half that of positive control) and IL-6 (by over 2.5 times compared to positive control) [51]. Wang et al. [41], showed that a 10 mg/kg (b.w.) dose of alantolactone (**1**) significantly improved neurological function and reduced cerebral edema in a traumatic brain injury mouse model. This neuroprotective effect was attributed to alantolactone's capacity to inhibit the NF-κB inflammatory pathway and the cytochrome c/caspase-mediated apoptosis pathways [41]. To the best of our knowledge, no other authors have elaborated on this interesting double action: the fact that alantolactone (**1**) can simultaneously activate apoptotic pathways in cancer cells and inhibit these pathways in a cerebral edema model.

Seo et al. [52] described the neuroprotective effect of alantolactone (**1**) using amyloid β25–35-induced ex vivo neuronal cell death and scopolamine-induced amnesia in mouse models meant to emulate conditions common to neurodegenerative conditions like Alzheimer's disease. A 1 μM alantolactone (**1**) treatment increased cortical neuron viability to almost baseline control readings, and 1 mg/kg (b.w.) significantly decreased scopolamine-induced cognitive impairments. It is a interesting to note that this particular neuroprotective effect is attributed to a drop in ROS levels. High ROS concentrations are associated with the neurodegenerative damage of Alzheimer's disease. Again, alantolactone (**1**) exhibits a double action: it raises ROS levels to induce apoptosis in murine models of neurodegenerative damage.

The interesting advances with alantolactone (**1**) derivatives are also worth mentioning. Kumar et al. [53] assayed three of the 17 thiol derivatives synthesized, at 10 mg/kg (b.w.) doses, in mice showing that they have in vivo anti-inflammatory activity comparable to alantolactone (**1**). The novel compounds shared alantolactone's anti-inflammatory mechanisms. Another noteworthy study involving alantolactone (**1**) derivatives was published in 2018 by Li et al. [54], testing 44 derivatives for their ability to inhibit induced pulmonary fibrosis in mice. The results showed that 2 of these compounds are particularly active at 100 mg/kg (b.w.), reducing the fibrotic area by more than 60%. This is achieved by inhibition of the TGF-β1 pathway of myofibroblast differentiation. It should be noted that no toxic effects were observed for either compound in the chronic toxicity test (seven days), using a dose of 2 g/kg (b.w.) administered orally. Finally, a patent involving an alantolactone (**1**) spiro-isoxazoline derivative was filed in June 2019 (US patent N° 20190185487) for the development and production of these compounds that exhibit significant anti-inflammatory activity.

One factor which contributes to making alantolactone (**1**) very interesting as a future medicine is its low toxicity. In work by Khan et al. [55], Kunming mice treated with 100 mg/kg (b.w.) alantolactone (**1**) showed no significant signs of hepatotoxicity or nephrotoxicity, in line with the previously cited work by He et al. [38]. This is especially important for alantolactone (**1**) as an anticancer drug, because it is well known that the liver and kidneys are particularly susceptible to negative side effects from chemotherapy approaches [56,57].

Another engaging finding obtained by Khan et al. [55] is alantolactone's ability to cross the blood-brain barrier. It may thus become useful in the treatment of brain tumors or other conditions involving the central nervous system (CNS), since the blood-brain barrier is the greatest obstacle for drug delivery to those areas [58].

The metabolism and pharmacokinetics of alantolactone (**1**) have also been studied in vivo, with future pre-clinical testing in mind. Research shows that alantolactone (**1**) exhibits low absorption and is rapidly eliminated after intravenous and oral administration. Its metabolism involves conjugation with thiol, and α, β-unsaturated carbonyl is the preferential structural metabolic site. The low aqueous solubility of alantolactone (**1**) causes low oral bioavailability [59–61].

All these recent reports show there is great interest in alantolactone (**1**), and that it has potent proven in vivo activities, mainly several different types of anticancer activity. This broad-spectrum activity, combined with its synergistic action with known cancer therapy agents, shows alantolactone (**1**) has great potential for future drug development. However, further research is necessary, especially clinical trials, to identify its intracellular action sites and secondary targets and thereby elucidate its mode of action.

2.2. Arglabin

Arglabin (**2**) (Figure 3), is a guaianolide sesquiterpene lactone with the chemical formula $C_{15}H_{18}O_3$, isolated from several species including *Artemisia myriantha* Wall. ex Besser [62,63], *Artemisia jacutica* Drob [64], and *Artemisia glabella* Kar. and Kir. species [65] where the extraction yield was 0.27% [66]. Fortunately, several synthetic and hemisynthetic methods for preparing compound **2** have been reported [66–70], allowing it to be provided in the quantities necessary for research and medicinal applications.

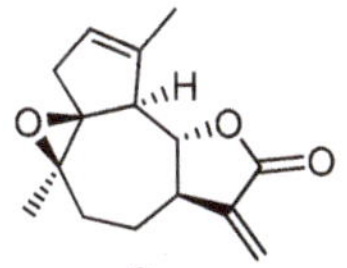

2

Figure 3. Structure of arglabin (**2**).

The arglabin (**2**) molecule contains a 5,7,5-tricyclic ring structure with contiguous stereo centers [70]. Furthermore, it contains both an epoxide and an α,β-unsaturated Michael acceptor, two functional groups that have important roles in its pharmacological activities [71]. There are many research lines advancing towards more effective drugs on the basis of the arglabin (**2**) molecule. New arglabin (**2**) derivatives have been obtained by chemical modification, the most successful being obtained via amination followed by treatment with gaseous hydrochloric acid. This converts arglabin (**2**) into the hydrochloride salt of the dimethyl amino adduct, which is very soluble in water [66,72]. This arglabin (**2**) salt form was used in the therapy of several cancer types such as breast, lung, liver and esophageal tumors in oncological clinics of Kazakhstan [73–75]. Treatment of an esophageal carcinoma patient with compound **2** contributed to a significant reduction in tumor volume, favoring its regression and a lower incidence of leucopenia [75].

In terms of mechanism of action, arglabin (**2**) inhibits farnesyl protein transferase enzyme [65,72]. It influences DNA synthesis in murine P388 lymphocytic leukemia cells [74]. Recently, Schepetkin and co-authors [76] have reported that compound **2** inhibits T-cell receptor activation and anti-CD3-induced movement of intercellular Ca^{2+} ions ($[Ca^{2+}]_i$), blocks ERK1/2 phosphorylation and depletes [GSH] in Jurkat T cells.

The anticancer effect of Arglabin (**2**) was also demonstrated by He and colleagues [77] on xenografted oral squamous cell carcinomas. They elucidated that tumor growth is inhibited via downregulation of relevant protein expression in the mTOR/PI3K/Akt signaling pathway and impairment of mitochondrial membrane potential, leading to apoptosis.

Recently, the pharmacokinetic properties of arglabin (**2**) have been reviewed [72], highlighting that it mainly accumulates in the liver, quickly reaches peripheral tissues and penetrates the blood-brain barrier. Furthermore, according to the literature, arglabin (**2**) has no described adverse effects, since it does not affect normal liver and kidney function or cause local irritative/allergenic reactions, nor mutagenic or embryotoxic effects [66,72,78].

Besides anticancer activity, arglabin (**2**) demonstrated in vivo anti-arthritis activity in a rat model. Arglabin (**2**) lowers the levels of inflammatory mediators and cytokines, and reduces expression of NF-κB (nuclear factor kappa-light-chain-enhancer of activated B cells), COX-2 (cyclooxygenase-2) and iNOS (inducible nitric oxide synthase) [79]. Abderrazak and co-authors [80] suggest that arglabin (**2**) is a compound with great potential in inflammation and atherosclerosis therapy. Using a high-fat diet ApoE2.Ki mouse model, experimental results [80] showed that arglabin (2) reduces inflammation by decreasing IL-1β and IL-18 levels and increases autophagy apoptosis [81].

Studies carried out with arglabin (**2**) show its potential for developing new anticancer and/or anti-inflammatory drugs, the main limiting factor being its low bioavailability. The preparation of soluble active versions of compound 2, and/or new delivery systems may attract further interest in this sesquiterpene lactone.

2.3. Costunolide

Costunolide (**3**) (Figure 4), has the same chemical formula as alantolactone (**1**) ($C_{15}H_{20}O_2$). It is a member of the germacranolide subclass and was first isolated from *Saussurea costus* (Falc.) Lipsch. roots in 1960 [82]. It is present in many Asteraceae genera such as *Inula* [24], *Lactuca* and *Helianthus* [83], but also those from other families like Magnoliaceae [84].

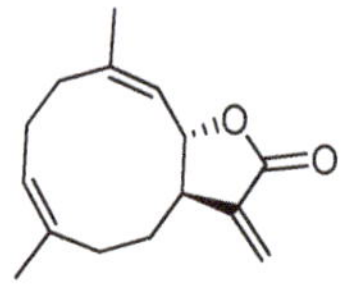

3

Figure 4. Structure of costunolide (**3**).

Biosynthesis of costunolide (**3**) is well documented and occurs through the mevalonate pathway [85]. Briefly, the process starts with the cyclization of farnesyl pyrophosphate, forming germacrene A. Next, the isoprenyl side chain of germacrene A undergoes hydroxylation by germacrene A hydroxylase, followed by oxidation to germacrene acid. It is finally synthesized after oxidation and cyclization of germacrene acid [85]. Some species are rich sources of costunolide (**3**), such as essential oil of *Saussurea lappa* roots (yield ~3%), of which 52% was identified as costunolide (**3**) [86]. The total synthesis of costunolide (**3**) has been described using different methods/strategies [87–89].

In a recent review [90], Kim and Choi detail the activities of custonolide (**3**) and discuss its therapeutic potential. Like alantolactone (**1**), compound **3** also exhibits anticancer bioactivity against different cancer cells via various routes. In fact, it acts as an apoptosis inducer, cell cycle regulator, angiogenesis and metastasis inhibitor and can also reverse the drug resistance mechanism [91].

There are several recent publications describing in vivo costunolide (**3**) studies, such as the work by Jin et al. [92] on costunolide's anticancer effect on osteosarcoma xenografted mice. Results showed a daily 20 mg/kg (b.w.) dose was enough to significantly reduce tumor weight by about half, as well as reduce the number of lung metastases to around one third of that in the control group. Western blot analysis of the tissues revealed this effect could be attributed to costunolide (**3**) inhibition of STAT3 transcription, a factor that is widely known to be linked to oncogenesis and cancer proliferation [93].

It is interesting to note the similarity between previously mentioned STAT3 inhibitory activity by alantolactone (**1**) [45]. To highlight another similarity, costunolide (**3**) also exhibits activity against gastric adenocarcinoma in xenografted mice [94]. Results from this study showed costunolide (**3**) induced caspase-mediated mitochondrial apoptosis in cancer cells and a 50 mg/kg (b.w.) dose on alternate days significantly reduced tumor size to about half. It achieved very similar results to the positive control cisplatin, a chemotherapeutic agent used in clinical treatment [94]. Interesting to note is costunolide's anticancer effect through telomerase reverse transcriptase inhibition [95], a mechanism not known in alantolactone (**1**). As reported by the researchers, a 5 mg/kg (b.w.) dose injected on alternate days in glioma xenografted mice significantly reduced tumor size by over 50%. This inhibition was associated with reduced telomerase activity, which leads to ROS-associated apoptosis [95]. A subsequent work by the same research group also showed the same dose affects lipid metabolism in glioma xenograft tumors, by lowering expression of FASN, SREBP-1, and PGC-1α [96], which are key genes targeted for cancer treatment [97–99].

Beyond costunolide's anticancer effect, it has also proven to be a potent anti-inflammatory agent [90], as well as an antidiabetic, antihelminth, antimicrobial, antiulcer and antioxidant [91]. Its anti-inflammatory action has been well documented in vitro [84,100] and in vivo, appearing to be linked to NF-κB pathway inhibition [101]. For example, costunolide (**3**) is able to suppress inflammatory angiogenesis [102], alleviating gastric ulcers [103], and acute lung and liver damage [104–107].

Recently, studies have shown that costunolide (**3**) also exhibited anti-osteoarthritic effects [108]. In fact, treatment in rats with osteoarthritis causes attenuation of cartilage degeneration compared to the control osteoarthritic group. The observed effect has been attributed to the inhibitory action of this compound on the Wnt/β-catenin and NF-κB signaling pathways, and on the expression of matrix metalloproteinases [108].

Costunolide (**3**) seems to be also a powerful antiasthmatic [109]. In this work, the researchers treated asthma-induced mice with 10 mg/kg (b.w.) before an immune challenge. Results showed a 61.8% inhibition of asthma-associated eosinophil increase, as well as significantly reducing lung inflammation scores and mucin production [109].

Recently, it has been shown that costunolide (**3**) is an effective inducer of hair growth in mice [110]. For this assay, the researchers implanted mouse dermal cells, treated with 3 mM costunolide (**3**) for two days. Results showed a 2.5-fold increase in induced hair follicles in the implanted treated cells, and topically applied costunolide (**3**) significantly and visibly improved hair growth. The authors claim this might be due to activation by compound **3** of key follicle-cell cycle pathways, including the Wnt/b-catenin, Shh/Gli, and TGF-β/Smad pathways. It is worth mentioning the TGF-β/Smad pathway was also mentioned in previous studies related to anti-inflammatory action of costunolide (**3**) [106].

Costunolide (**3**) derivatives have also proven to be highly interesting, with remarkable in vivo effects. In recent work by Cala et al. in an agro-research context [111], many different functional groups were added to the sesquiterpene backbone, yielding, among others, two amino and two methyl ether derivatives with strong herbicidal activity. The etiolated wheat coleoptile assay indicated treatment with these costunolide (**3**) derivatives had dose-dependent growth inhibition responses equivalent to a widely used synthetic herbicide used as positive control. These results show there is potential for some of these derivatives as bio-herbicides, but further studies are necessary. There was also another interesting result of this work which the authors did not address, but is worth pointing out: two tested thiol derivatives boosted coleoptile growth instead of inhibiting it, with one compound increasing coleoptile length by ~30% with a 30 μM treatment. The general implication of these results is that costunolide (**3**) modification can be powerful and versatile, capable of yielding diametrically opposed bioactive compounds depending on the functional group added.

In conclusion, costunolide (**3**) is a highly multitalented and potent bioactive compound. Similar to other sesquiterpene lactones such as the previously discussed alantolactone (**1**), much of its potential for future drug development lies in its anticancer and anti-inflammatory effects. It is interesting that there appears to be more work done with costunolide (**3**) derivatives than alantolactone (**1**) derivatives, which in general suggests greater interest of the scientific community in costunolide (**3**) as opposed to alantolactone (**1**). With so many relevant publications in 2019 alone, we expect to see costunolide (**3**) research ramping up to a pre-clinical stage very soon.

2.4. Cynaropicrin

Cynaropicrin (**4**) (Figure 5) is a guaianolide type sesquiterpene lactone with a chemical formula of $C_{19}H_{22}O_6$ and a 5,7,5 fused tricyclic skeleton [112].

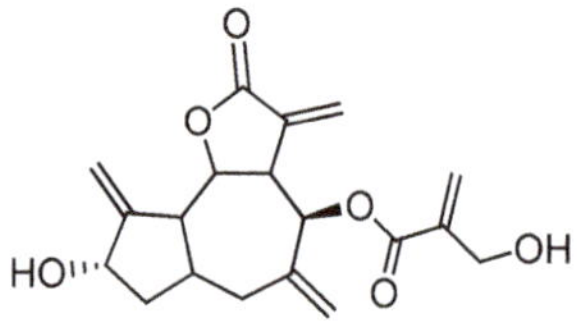

Figure 5. Structure of cynaropicrin (**4**).

The sesquiterpene lactone cynaropicrin was isolated from artichoke (*Cynara scolymus* L.) in 1960 for the first time [113] and later it was found in *Cynara cardunculus* L. [114,115] and *Cynara scolymus* L. species [116], being considered as a chemotaxonomic marker for artichoke plant species [112]. Cynaropicrin (**4**) was also found in many species of Asteraceae family such as *Centaurea drabifolia* subsp. *floccosa* (Boiss.) Wagenitz and Greuter [117], *Psephellus sibiricus* (L.) Wagenitz [118], *Rhaponticum pulchrum* Fisch. and C.A.Mey. [119], *Moquinia kingii* (H.Rob.) Gamerro [120], *Saussurea calcicola* Nakai [121], *Saussurea costus* (Falc.) Lipsch. [122,123], *Tricholepis glaberrima* DC. [124], and many others [112]. Yields of cynaropicrin (**4**) extraction from *Cynara cardunculus* L. leaves using ethanol, ethyl acetate, dichloromethane and water were 56.9, 37.5, 40.3 and 13.6 mg/g dry weight, respectively [125].

The biological activities of cynaropicrin (**4**), as with other sesquiterpenic lactones, are related to its pharmacophore γ-butyrolactone ring [112]. There are many studies reporting on the important pharmacological activities of cynaropicrin (**4**), and plants rich in cynaropicrin (**4**), such as antitumor, anti-inflammatory, antitrypanosomal, and antihepatitis C virus, among many others. Due to these notable effects, cynaropicrin (**4**) will be suitable for the development of medicinal compounds [123].

Cynaropicrin (**4**) is the first natural product that in vivo potently inhibits the African trypanosome diseases [126]. Using the acute model of mice infected with *Trypanosoma brucei rhodesiense* STIB 900, when treated with two doses of 10 mg/kg (b.p.) per day, on the 7th day after infection, there was a 92% reduction in parasitemia when compared to the untreated group. Additionally, selectivity indices of 7.8 were obtained for cynaropicrin (**4**) against L6 cells of rat myoblasts [127]. The action mechanism is still under study to date, but is thought to be related to the interaction of compound **4** with the trypanothione redox system in *Trypanosoma brucei* [127]. However, Da Silva et al. [128] demonstrated that cynaropicrin (**4**) at a dose of up to 50 mg/kg (b.w.) per day has no effect in mice infected with *Trypanosoma cruzi*, in either Y or Colombiana strains. The synthesis and semi-synthesis of several cynaropicrin (**4**) derivatives allowed the structure/antitrypanosomal activity of these compounds to be evaluated. It was concluded that the α-methylene-γ-lactone structure is indispensable to maintain the biological effect, whereas 3-OH and 19-OH derivatization does not change the activity and some types of side-chain promote the selectivity of the compound [129,130]. The in vivo evaluation of some derivatives, using the *Trypanosoma brucei rhodesiense* acute mouse model, indicated that the dimethylamino derivative exhibits much less toxicity than cynaropicrin (**4**), but also less activity [129].

Cynaropicrin (**4**) also showed the ability in vivo to delay the effects of skin photoaging, promoting the proliferation of melanocytes and keratinocytes, by acting as inhibitor of NF-kB transcription activity [131].

2.5. Helenalin

Helenalin (**5**) ($C_{15}H_{18}O_4$) (Figure 6), a 5/7-fused bicyclic sesquiterpene lactone that belongs to the pseudoguaianolide subclass. It is very abundant in *Arnica montana* L. but is also identified in other species from *Arnica* and *Helenium* genera and from *Centipeda minima* (L.) A.Braun and Asch. [132–136].

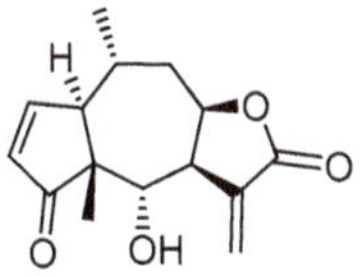

Figure 6. Structure of helenalin (**5**).

Alcohol extracts containing helenalin (**5**) and its derivatives have been used as a staple of traditional herbal anti-inflammatory medicine for many decades [134]. For instance *Arnica montana* L. solutions are used to treat rheumatism, arthritis, inflammation, hematoma, soreness, sprains, swelling and muscle spasms from athletic activity etc., seasonally triggered arthritis, arteriosclerosis, angina pectoris, postoperative conditions, and joint pain. The plant is used externally in creams, alcoholic tincture, and ointment form but also taken highly diluted in homeopathic remedies [134]. Many such preparations, even hair oil and shampoo, are commercially available from a range of suppliers in healthfood shops and pharmacies almost worldwide [137]. Likewise, clinical trials have aimed to assess topical *Arnica* applications, regarding possible reduction of laser-induced minor hematomas and osteoarthritis-type symptoms. Orally administered homeopathic formulations are also widely employed in the clinical setting to treat and manage conditions such as carpal tunnel, slow knee-surgery recovery, tonsil and wisdom tooth extractions, facelifts, neuralgia, hysterectomy, venous surgery, hallux valgus, heart-valve surgery, hemarthrosis, prolonged intravenous perfusion, joint sprains and strains, muscle pain, etc. [133].

Helenalin (**5**) inhibits NF-κB transcription of inflammatory cytokines, which have an essential role in both inflammation and cancer [138]. It also efficiently inhibits cancer cell proliferation through a variety of action modes, e.g., telomerase inhibition [139], DNA and protein synthesis attenuation, apoptosis induction and promoting reactive oxygen species generation [140]. It is noteworthy that inhibition of NF-κB activation associated with another sesquiterpene, helenin, occurs in T-cells, B-cells and epithelial cells in response to four different stimuli, nullifying kB-driven gene expression. Since this activity does not affect transcription factors Oct-1, TBP, Sp1 or STAT 5, this NF-κB activation is probably inhibited selectively [141,142]. Lyss et al. [141] described how helenalin modifies the NF-κB/IκB complex, preventing IκB release. They proposed a molecular action mode for the anti-inflammatory influence, which is different from other nonsteroidal anti-inflammatory drugs (NSAIDs) such as aspirin and indomethacin. Furthermore, it targets Cys38 on p65, ablating DNA-binding [143], inhibiting neutrophil chemotaxis and migration as well as 5-lipoxygenase and leukotriene C4 synthase activities [144].

Experiments performed by Schröder et al. [145] demonstrate that compound **5** inhibits platelet aggregation induced by collagen, 5-hydroxytryptamine secretion and thromboxane formation, depending its on concentration (between 3–300 μM). They concluded that helenalin inhibits platelet function via interaction with platelet sulfhydryl groups in a way most likely associated with reduced phospholipase A2 activity.

Helenalin (**5**) inhibited complete Freund's adjuvant-induced arthritis and carrageenan-induced paw edema in the rat [146]. Topical application of Arnica 3D gel (10%), combined with a 10 mA microcurrent for 2 min also significantly improved wound healing in a linear incision wound model in the rat back [147]. The evidence was consistent with a higher percentage of mature collagen fibers and a significantly larger total number of cells in the wound, as assessed by structural and morphometric analysis. According to this, different proportions of *Arnica montana* extracts alone or combined with other plants have been patented for their therapeutic potential [137]. Widrig et al. [148] performed a randomized, double-blind study in 204 patients with osteoarthritis (OA) to compare the effects of ibuprofen (5%) and *Arnica montana* gel (50 g tincture/100 g) containing helenalin (**5**), 11α,13-dihydrohelenalin and its ester. The results show that short-term use, up to three weeks, of Arnica gel improves pain and hand function in OA, indistinguishably from ibuprofen gel. Substantial antiosteoarthritic activity by blocking transcription factors NF-κB and NF-AT is attributable to helenalin and derivatives.

Boulanger et al. [149] demonstrated that helenalin (**5**), intraperitoneally delivered at 20 mg/kg in lactating-Balb/C mice 9 and 3 h prior to infection, reduced intracellular growth of *Staphylococcus aureus* in mouse mammary glands This suggests the compound interferes with host molecular mechanisms and not directly with *Staphylococcus aureus* growth. The authors conclude helenalin might be worth investigating as a potential treatment for *Staphylococcus aureus*-induced mastitis in bovine species. There is however some concern regarding this treatment, mainly regarding helenin persistence in the animal's milk, and whether or not the therapeutic doses pose short- and long-term toxicity risks. It is therefore imperative to further study and characterize its safety and pharmacological properties.

Valan et al. [150] showed that helenalin (**5**) has a significant biphasic positive inotropic effect on the myocardium of guinea pigs at doses of 10^{-5}–10^{-3} mol. Nevertheless, concentrations above 10^{-3} mol. cause an irreversible negative inotropic action leading to a blocking of muscle contraction.

The skin is susceptible to environmental damage by multiple agents, particularly solar ultraviolet (UV), which induces skin hyperpigmentation disorders. Expression of heat shock proteins (HSPs, particularly HSP70) is receiving consideration in the field of cosmetics, to reduce skin damage and signs of aging. Usui et al. [151] isolated AM-2 (helenalin 2-methylbutyrate) from *Arnica montana* as a good inductor of HSP70, with low cytotoxicity. Treatment of cultured mouse melanoma cells with AM-2 or *Arnica montana* extract up-regulated the expression of HSP70 in a dose-dependent manner. It also activated the transcription factor for hsp genes, i.e., heat shock factor-1. They concluded that both *Arnica montana* extract and AM-2 are likely to show beneficial effects if incorporated in hypopigmenting cosmetics.

Acute liver injury is a life-threatening syndrome frequently associated with hepatocyte damage and characterized by oxidative and inflammatory responses. Li et al. [136] recently observed that intragastric administration of helenalin (**5**) for 10 days significantly ameliorated hepatic injury induced previously in mice with LPS/D-GalN. These results were evidenced by the attenuation of histopathological changes and the decrease in serum aminotransferase and total bilirubin activities. Therefore, helenalin (**5**) shows a hepatoprotective effect against damage induced by LPS/D-GalN. This in turn may be associated with reduced hepatocyte apoptosis, by protection of mitochondrial function and oxidative stress inhibition by Nrf2 pathway activation, as well as attenuating inflammation by inhibiting NF-κB activation. The co-authors of that study [152] submitted Patent No. CN 110283151 in 2019 as a method for isolation of helenalin from *Centipeda minima* and its application for treating hepatic fibrosis and inflammation. Furthermore, the same authors [153] patented it for inhibiting hepatic stellate cell activation, showing the advantages of reducing collagen deposition and synthesis of inflammation-related proteins, promoting death of stellate cells, and its application in liver fibrosis treatment.

2.6. Inuviscolide

The guaianoline-type sesquiterpene lactone inuviscolide (**6**) ($C_{15}H_{20}O_3$) (Figure 7) was identified in *Ferula communis* (Apiaceae) and described to be the main metabolite in *Inula viscosa* (L.) Ait. [154].

Inuviscolide (**6**) inhibited in vivo inflammation in mice, as shown in the investigation performed by Hernández et al. [155]. The authors used inuviscolide (**6**) and ilicic acid from *Inula viscosa* and found that compound **6** influenced cell degranulation and leukotriene biosynthesis, as well as neurogenic drive and glucocorticoid-like interactions. During the testing, ear and paw edema were induced in Swiss female mice using phorbol esters or ethyl phenylpropiolate (EPP), and phospholipase A2 (PLA2) or serotonin, respectively. The drug dose was applied topically in the ear models but as subcutaneous or intraperitoneal injections in the paw models. For quantitation of leukotriene B4 (LTB4) formation, high-performance liquid chromatography (HPLC) was performed on rat peritoneal neutrophils. The results showed that compound **6** reduced PLA2-induced edema with an ID_{50} of 98 µmol/kg. The results did not indicate that glucocorticoid response modifiers had an influence on the edema induced by serotonin. In intact cells, inuviscolide (**6**) resulted in reduced generation of LTB4 (IC_{50} value of 94 µM). The overall results indicated that compound **6** has an important role in the anti-inflammatory activity of *Inula viscosa*, being more active than ilicic acid. The action mechanism was suggested to be related to an interference with leukotriene synthesis and to PLA2-induced mastocyte release of inflammatory mediators [156].

Figure 7. Structure of inuviscolide (**6**).

2.7. Lactucin and Its Derivatives Lactupicrin, and Lactucopicrin

Other compounds from the guaianolide subclass are lactucin (**7**) ($C_{15}H_{16}O_5$) and its 8- and 15-(4-hydroxyphenylacetate) derivatives lactupicrin (**8**) and lactucopicrin (**9**), respectively (Figure 8). Dolejš et al. [156] and Ruban et al. [157] had previously described the chemical structure of lactucin (**7**). These compounds are distributed within Asteraceae like *Lactuca serriola* [158], especially plants commonly used in salads. *Lactuca sativa* (lettuce), *Cichorium intybus* (chicory and radicchio), and *Cichorium endivia* (endive) have been reported to contain lactucin (**7**) and lactucin-related substances [159–164]. The content depends largely on the species, variety and the part of the plant analyzed. In fact, lettuce guaianolide content was high and reached concentrations between 61.7 mg/mL and 147.1 mg/mL in its latex [160] and 2.9 mg/g to 36.1 mg/g in the overall plant, expressed as dry weight [162].

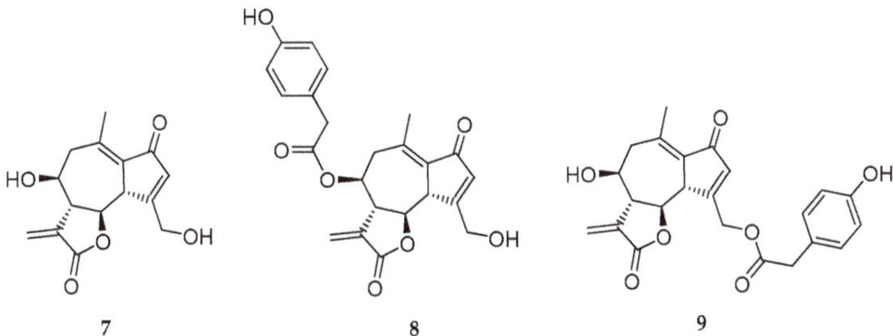

Figure 8. Structures of lactucin (**7**) and its derivatives lactupicrin (**8**) and lactucopicrin (**9**).

The biological activities of lactucin (**7**) and lactucin-related compounds had already attracted interest at the end of the 19th century, as they were credited to be responsible for the bitter taste and pharmacological effect of lactucarium or lettuce opium [165]. This is the condensed latex of wild lettuce *Lactuca virosa*. This dried exudate was used in Europe for centuries with similar applications to opium as an analgesic, antitussive, and sedative [163]. Due to its widespread use, it was described and standardized in the United States Pharmacopoeia (USP) and the British Pharmaceutical Codex as a sedative for irritable cough and as a mild hypnotic for insomnia [166–168]. Lactucopicrin (**9**) and lactucin (**7**) were identified as its major active compounds, although compound **7** was suggested to be one of the main metabolites related to the sedative and the sleep-promoting effects [163]. In an in vivo study in mice, both substances were confirmed to be analgesics, with an activity equal to or greater than ibuprofen. In addition, when lactucin (**7**) and lactucopicrin (**9**) were administered to mice on the hot plate, both compounds were revealed to have analgesic and sedative effects at concentrations of 15 and 30 mg/kg, respectively. Furthermore, in a spontaneous locomotor activity test, these compounds were active at concentrations of 30 mg/kg [168].

Although the pure compounds currently are not easily available on the market, several lettuce extracts and seed oils containing these sesquiterpene lactones as main active compounds are commercialized, e.g., Sedan® (Pharco Pharmaceutical Company, Egypt). The sedative effects were addressed in the pilot study by Yakoot et al. [169]. The authors investigated if lettuce seed oil was efficient and safe to treat patients with sleep disorders. The results showed that the seed oil of *L. sativa* was a potentially hazard-free agent, able to reduce sleeping difficulties and alleviate mild to moderate forms of anxiety in geriatric patients, without showing side effects [169,170]. Additionally, the study performed by Kim et al. [163] examined the sleep-inducing and sleep-prolonging effect of four lettuce varieties. The seed and leaf extracts were evaluated using a four-week-old ICR mouse model to analyze their effects on pentobarbital-induced sleep. The results showed that both extract types lengthened sleep duration and significantly reduced sleep latency at both evaluated doses of 80 mg/kg and of 160 mg/kg [163]. Although these studies in patients were performed with plant extracts reported to contain lactucin (**7**) and its derivatives as principal active pharmacological compounds, they confirm that their application for treatment of anxiety and sleep disorders is worthy of further evaluation. Additionally, the results showed that these sesquiterpenes might be assessed as potential alternatives to currently used sleep-promoting, sedative, and anxiolytic agents, with their varied negative side-effects and addiction-potential.

2.8. Parthenolide

Parthenolide (**10**) (Figure 9) has the same chemical formula as inuviscolide (**6**) ($C_{15}H_{20}O_3$) and is the best known sesquiterpene lactone in the germacranolide class. It was isolated for the first time from feverfew leaves and flowerheads (*Tanacetum parthenium* (L.) (Sch.Bip.), a plant known in traditional Chinese medicine for centuries to treat various ailments. Among them, it is used to relieve fever, pain of different etiologies such as migraine and rheumatoid arthritis and even to treat insect bites [171].

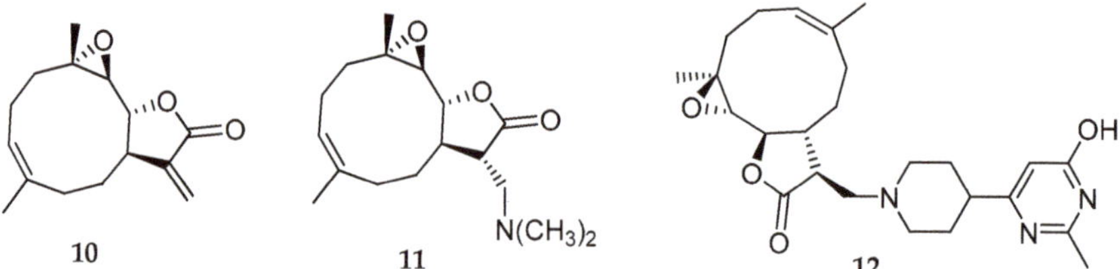

Figure 9. Structure of parthenolide (**10**) and its derivatives DMAPT (**11**) and HMPPPT (**12**).

Compound **10** has been applied as an anti-inflammatory through specific inhibition of the signal proteins IKK2, STAT3, and MAPK, along with the activity and expression of many inflammatory mediators including COX, which is involved in the NF-κB mediated proinflammatory signal transduction pathway [172]. It also exhibits antitumoral activity by proliferation inhibition in various cancer cell types, including prostate, pancreas, cervical, breast, lung, colorectal, glioblastoma, multiple myeloma, and leukemia [8,173]. There are numerous reviews describing the outstanding anti-inflammatory and antitumor activities of parthenolide, its analogs, and derivatives upon different pathways in human cancer cells [8,174–178].

Parthenolide (**10**) has a multitarget action mechanism. It triggers EGF receptor phosphorylation, interferes with AP-1 [179] and the signal transducer and activator of transcription 3 (STAT3) [180] and induces activation of c-Jun N-terminal kinase (JNK) [181]. Furthermore, the molecular mechanisms of parthenolide action are strongly associated with DNA-binding inhibition of two transcription factors

NF-κB [182,183], as well as the proapoptotic activation of p53, along with reduced glutathione (GSH) depletion [184]. Parthenolide rapidly induces ROS generation [185,186], and lowers histone deacetylase activity (HDAC) [187] and DNA methyltranspherase 1 (DNMT1) [188]. Additionally, parthenolide can interfere with microtubule function through tubulin binding [189].

Recent advances regarding in vivo therapeutic applications of parthenolide (**10**), mainly focused but not limited to anticancer and anti-inflammatory activities, are discussed below, together with the synergistic effects and toxicity of compound **10** and some derivatives.

One of the advantages of parthenolide (**10**) is its cancer stem-cell selectivity while remaining non-cytotoxic to non-tumor cells [190,191]. In this case, the molecular mechanism of parthenolide involves induction of apoptosis through mitochondrial and caspase signaling pathways, and also an increase in the cytosolic concentration of calcium, cell cycle arrest, and inhibition of metastasis [178,192].

Cholangiocarcinoma (CC) is an intrahepatic bile duct carcinoma with a poor prognosis due to being chemoresistant. Parthenolide induced oxidative stress-mediated apoptosis in CC cells and the Bcl-2-related family of proteins modulated that susceptibility [193]. Moreover, intraperitoneal injection of parthenolide at 4 mg/kg caused significant inhibition of tumor growth and angiogenesis in the xenograft model [194]. Recently, Yun et al. [195] demonstrated that low concentrations of parthenolide (5–10 μM) suppressed HO-1 expression, enhancing oxidative stress by the PKC-α inhibitor Ro317549 (Ro) through inhibition of Nrf2 expression and its nuclear translocation. The effects of parthenolide (**10**) and Ro at 2.5 mg/kg on tumor growth were tested using a xenograft nude mouse tumor model with subcutaneously implanted ChoiCK and SCK cells. This assay indicated that their combined application more effectively inhibited cancer cell growth inhibition as compared to treatment with either compound by itself. Furthermore, the effect of parthenolide on the development of colitis-associated colon cancer (CAC) was investigated using a murine model of azoxymethane (AOM)/dextran sulfate sodium (DSS) induced CAC. This study showed that parthenolide administration (**10**) at 2 and 4 mg/kg can significantly inhibit the inflammation-carcinoma sequence and be crucial in experimental CAC regulation. The mechanism of action involves decreased NF-κB p65 expression levels blocking phosphorylation and subsequent degradation of κB-α inhibitor (IκBα) [196]. The authors conclude that parthenolide (**10**) could be a novel chemopreventive agent for CAC treatment [196].

Oral cancer is one of the five most common cancers worldwide. Chemoprevention is a new approach to cancer research, focusing more on the prevention, suppression, and reversal of the carcinogenic process by the use of natural plant products and/or synthetic chemical compounds. Thus, Baskaran et al. [197] tested the chemopreventive potential of parthenolide in DMBA-induced hamster buccal pouch carcinogenesis (DMBA, 7,12-dimethylbenz[a]anthracene). Oral administration of parthenolide (**10**) at 2 mg/kg b.w. completely prevented tumor formation and significantly reduced the nefarious histopathological changes. In addition, the parthenolide treated group showed significant improvement regarding antioxidants, detoxification enzymes and lipid peroxidation.

Glioblastoma, or glioblastoma multiforme (GBM), is the most aggressive type of brain cancer and is very difficult to treat. Nakabayashi and Shimizu [198] examined the effect of compound **10** on tumor growth using a xenograft mouse model of glioblastoma, administering it intraperitoneally (10 mg/kg/day) for 22 days. It significantly inhibited the growth of transplanted glioblastoma cells with respect to the control group.

Zhang et al. [199] demonstrated that a high parthenolide dose (8 mg/kg/day) impedes initiation of experimental autoimmune neuritis (EAN), an animal model for peripheral nervous system acute inflammatory disease. This is achieved by parthenolide suppressing TNF-α, IFN-γ, IL-1β and IL-17 pathways and quickly decreasing Th1 and Th17 cells in the early stages. Although this anti-inflammatory effect is short-lived, compound **10** also suppresses late-stage recovery of EAN models, along with inhibiting the apoptosis of inflammatory cells. Such results suggest that parthenolide is not suitable for nervous system autoimmune disease treatment.

Nitric oxide (NO) plays a key role in the etiopathology of central nervous system (CNS) diseases like multiple sclerosis (MS). It has been proposed that inhibition of NO synthesis could prove a relevant mechanism of action in treating multiple sclerosis and migraine. Accordingly, Fiebich et al. [200] investigated the effect of parthenolide (**10**) on iNOS synthesis and NO release using primary rat microglia. The results indicated that compound **10** prevents iNOS/NO synthesis and inhibits the activation of p42/44 mitogen-activated protein kinase (MAPK), but not IkBα degradation or NF-kB p65 activation. These results show parthenolide may be a potential therapeutic agent in the treatment of CNS diseases.

Mechanisms of axon regeneration and optimal functional recovery after nerve injury are key to in higher animals. However, insufficient growth rates of injured axons often lead to incomplete peripheral nerve regeneration. Gobrecht et al. [201], demonstrated that a single parthenolide injection at 5 nM into the injured sciatic nerve or its systemic intraperitoneal application was already enough to significantly increase the number and length of regenerating axons in the distal nerve at three days post-lesion. This application of parthenolide (**10**) appears to act on the great instability of microtubules in promoting axonal growth, at least in the CNS [174]. For this reason, the efficacy of parthenolide is very promising for a therapeutic promotion of nerve regeneration, since compound application and recurrent treatments are facilitated, compared to invasive local nerve injections [202].

Pulmonary fibrosis (PF) in general and idiopathic pulmonary fibrosis (IPF) in particular, is a disease for which there is no effective therapy. In vivo studies have shown that parthenolide (**10**), via intragastric administration, inhibited the NF-κB/Snail pathway, attenuating bleomycin-induced pulmonary fibrosis. Moreover, there were significant improvements in body weight and other pathological changes associated with this disease [203].

NF-κB has been associated with the cardiovascular system; in fact, its function is related to the protection of cardiovascular tissues against injuries. However, its activation can also contribute to tissue pathogenesis, depending on the type of cells in which it is activated [204]. It is known that myocardial infarct size could be reduced up to 60% by antagonizing NF-κB activity [205]. To achieve this, parthenolide (**10**) at 250 or 500 mg/kg (b.w.) was administered intraperitoneally before reperfusion in rats, and caused a significant improvement in myocardial injury, with a reduction in the oxidative and inflammation state, consequently reducing infarct size [206]. However, Tsai et al. [207] reported that a prolonged treatment in bEND.3 cells affected Ca2þ signaling in the endothelial cells that regulate vascular tone; therefore, care should be taken on using this compound in experimental designs and clinically.

Parthenolide (**10**) has relatively poor pharmacological properties, derived from its low solubility in water and consequently reduced bioavailability, which limit its potential clinical use as anticancer drug. To increase its solubility, a series of parthenolide derivatives were obtained by diastereoselective addition of several primary and secondary amines to the exocyclic double bond [208,209]. *N,N*-Dimethylaminoparthenolide (DMAPT) (**11**) (Figure 9) was selected as a leader compound according to its pharmacokinetic, pharmacodynamic and bioavailability properties [209,210]. When formulated as a fumarate salt, DMPAPT is 1000-fold more soluble in water than parthenolide and maintained the anticancer activity because, DMAPT (**11**) is rapidly converted back to parthenolide in body fluids (**10**). Recently, molecular studies indicate that DMAPT has a similar action to parthenolide [192,211–213].

DMAPT has approximately 70% oral bioavailability and induces rapid death of primary human leukemia stem cells (LSCs) from both myeloid and lymphoid leukemias and is highly cytotoxic to bulk leukemic cell populations. Pharmacological studies carried out by Guzman et al. [210] using both mouse xenograft models and spontaneous acute canine leukemias demonstrate in vivo bioactivity. Indeed, DMAPT eliminates human AML stem and progenitor cells without harm to normal hematopoietic stem and progenitor cells, and eradicates phenotypically primitive blast-crisis chronic myeloid leukemia (bcCML) and acute lymphoblastic leukemia (ALL) cells. Moreover, it inhibited metastasis in a mouse xenograft model of breast cancer and enhanced the survival of treated mice [210].

DMAPT was assayed in a phase I trial against acute myeloid leukemia (AML), acute lymphoblastic leukemia (ALL) and other blood and lymph node cancers [12,176]. However, a year later the clinical trials were suspended [214].

Radiotherapy is widely used in cancer treatment; however, the benefits can be reduced by radiation-induced damage to neighboring healthy tissues. Morel et al. [215] demonstrated in mice that DMAPT (**11**) selectively induces radio-sensitivity in prostate cancer cell-lines, while protecting primary prostate epithelial cell lines from radiation-induced damage. Compound 11 has the advantage of being well-tolerated orally without the need to adjust the administration time to radiation exposure. Radiation-induced lung fibrosis is considered a critical determinant for late normal tissue complications. Therefore, the same group [216] examined the radioprotective effect of DMAPT (**11**) on fibrosis in normal tissues, according to the image-guided fractionated radiotherapy protocols used clinically. The results obtained show that DMAPT reduced radiation-induced fibrosis in the corpus cavernosum of the rat penis (98.1%) and in the muscle layer around the bladder (80.1%), and also the tendency towards reduced collagen infiltration into the submucosal and muscle layers in the rectum. They concluded that DMAPT could be useful in providing protection from the radiation-induced side effects such as impotence and infertility, urinary incontinence and fecal urgency resulting from prostate cancer radiotherapy [216].

On the other hand, radiation-resistant prostate cancer cells often overexpress the transcription factor NFκB. Mendonca et al. [217] suggest that DMAPT might have a potential clinical role as radio-sensitizing agent in prostate cancer treatment. This conclusion is based on the finding that combined treatment of PC-3 prostate tumor xenografts with oral DMAPT and radiation therapy significantly reduced tumor growth, when compared to those treated with either DMAPT or radiation therapy alone.

Li et al. [218], obtained a novel parthenolide derivative, HMPPPT, a 13-substituted derivative ((3*R*,3a*S*,9a*R*,10a*R*,10b*S*,*E*)-3-((4-(6-hydroxy-2-methylpyrimidin-4-yl)piperidin-1-yl)methyl)-6,9a-dimethyl-3a,4,5,8,9,9a,10a,10b-octahydrooxireno [2′,3′:9,10]cyclodeca[1,2-b]furan-2(3*H*)-one) (**12**) (Figure 9), with better bioavailability and pharmacological properties than DMAPT. In vitro studies pointed to compound **12** as the most promising derivative, from safety profile and ADME property standpoints. The newly identified compound was shown to have pro-oxidant activity and in silico molecular docking studies with components of the NF-κB pathway also supporting a pro-drug mode of action. This mechanism included release of parthenolide and covalent interaction with one or more proteins involved in that pathway [218]. The in vivo oral bioavailability study of compound 12 in murine PK at 10 mg/kg indicated that it has advantageous pharmacological properties and therefore can be considered an agent to be considered in therapy against drug resistant chronic lymphocytic leukemia [218].

In the last few years, nanotechnology has provided many selective strategies for the detection and treatment of cancer, overcoming the problems associated with conservative cancer diagnosis and therapy. In this context, the development and testing of parthenolide (**10**) nanoencapsulation and derivatives is a way to enhance its potential to provide effective pharmaceutical products for clinical use and resolve drawbacks such as low bioavailability [219–222]. Accordingly, several patents for elaboration of parthenolide and its derivatives in nanocarriers, and various pharmaceutical preparations combined with other products, have been recently registered in China: patents N° CN 110292640, 2019; N° CN108721276, 2018; N° CN1087211330, 2018; N° CN106366068, 2017; N° CN 109276553, 2017.

In recent years, parthenolide has been suggested for use in combination therapy with other anticancer agents, to overcome obstacles in the treatment of cancer, such as a) different types of cancer cells, b) resistance to chemotherapy, and c) drug toxicity to normal cells.

TRAIL (tumor necrosis factor (TNF)-related apoptosis inducing ligand) is now being developed as a promising natural immunity-stimulating molecule for clinical trials in cancer patients. However, various malignant tumors are currently resistant. Kim et al. [223] investigated how parthenolide (**10**) sensitizes colorectal cancer (CRC) cells to TRAIL-induced apoptosis. For this, HT-29 (TRAIL-resistant) and HCT116 (TRAIL-sensitive) cells were treated with compound **10** and/or TRAIL. The results revealed that parthenolide (**10**) increases induced apoptosis and upregulates DR5 protein level and surface expression in both cell lines, suggesting that combined therapy with TRAIL is a good strategy to overcome the resistance of certain CRC cells.

Pancreatic cancer is a common malignancy with high occurrence worldwide, and a poor survival rate. Recent research indicates that combination therapy with DMAPT (**11**) can enhance the antiproliferative effects of gemcitabine in pancreatic cancer cells in vitro and in vivo [224]. Yip-Schneider et al. [225] showed that celecoxib (a cyclooxygenase 2 (COX-2) inhibitor) at 50 mg/kg/day combined with DMPTA at 40 mg/kg/da has a significant inhibitory effect on tumor invasion of adjacent organs and metastasis in pancreatic cancer induced in Syrian golden hamsters. It reduced NFκB activity, and expression of prostaglandin E2 and its metabolite. They also demonstrated that compound **11** (40 mg/kg/day) in combination with sulindac (20 mg/kg/day) and gemcitabine (50 mg/kg twice weekly) can delay or prevent progression of premalignant pancreatic lesions) in a genetically engineered mouse model of pancreatic cancer [226]. Likewise, they demonstrated that DMAPT (**11**) (40 mg/kg/day) with gemcitabine (50 mg/kg/day) considerably improved average survival, lowering the frequency and multiplicity of pancreatic adenocarcinomas. This combination also significantly reduced tumor size and the incidence of metastasis into the liver [227].

Parthenolide exerted a cytotoxic effect on MDA-MB231 cells, a triple-negative breast cancer (TNBC) cell-line, however its success is scarce at low doses. In order to overcome this difficulty, Carlisi et al. [228] tested a histone deacetylase inhibitor, SAHA (suberoylanilide hydroxamic acid), which synergistically sensitized MDA-MB231 cells to the cytotoxic effect of parthenolide (**10**). It is noteworthy that treatment with parthenolide alone increased the survival of cell pathway Akt/mTOR and the consequent nuclear translocation of Nrf2, a protein regulating the expression of antioxidant proteins that protect against the oxidative damage triggered by injury and inflammation, while treatment with SAHA alone activated autophagy.

A phase 2 clinical trial indicated that actinomycin-D (ActD), a polypeptide antibiotic that intercalates to DNA and inhibits mRNA transcription in mammalian cells, could be a potent drug against pancreatic cancer. However, it is not a good candidate due to toxicity issues. Thus, given the modes of action of DMPTA and Actinomycin-D (ActD), Lamture et al. [229] postulated that combining both drugs would result in synergistic inhibition of Panc-1 pancreatic cancer cell growth, since the inhibitory activity of DMAPT on FkB would enhance apoptosis induction by ActD, through phosphorylation of c-Jun. Indeed, the combination of these two drugs produced a higher cell-death percentage than each drug alone.

2.9. Thapsigargin

In 1978, the structure of thapsigargin (**13**) ($C_{34}H_{50}O_{12}$) (Figure 10) was elucidated [230] for the first time. It was found as the main component of *Thapsia garganica* L. [230,231], an umbelliferous Apiaceae species distributed in the Mediterranean area. This plant has long been used in folk medicine for treating common lung diseases (acute bronchitis and pneumonia) and dental pains [232]. The resin from *Thapsia garganica* root was described in the French Pharmacopoeia [232].

Thapsigargin (**13**) yield ranged between 0.2–4.91% of the dry weight of leaves and roots of *Thapsia garganica* and depends on the plant tissue, collection site (locality) and extraction methods [231,233]. For example, classical maceration was more efficient than other methods such as microwave-assisted extraction or simple extraction with liquid nitrogen [233].

Due to the importance of thapsigargin's biological activities, there is great interest in its synthetic and semi-synthetic preparation, thus several studies have been published on this subject [234,235].

Thapsigargin (**13**) is an inhibitor of the sarco/endoplasmic reticulum (ER) calcium ATPase (SERCA) pump. The blockage of the SERCA pump results in malfunction of cellular calcium homeostasis and exerts a critical role in normal cell metabolism, leading to apoptosis [236,237]. Moreover, this sesquiterpene lactone causes apoptosis at all stages of the cell cycle.

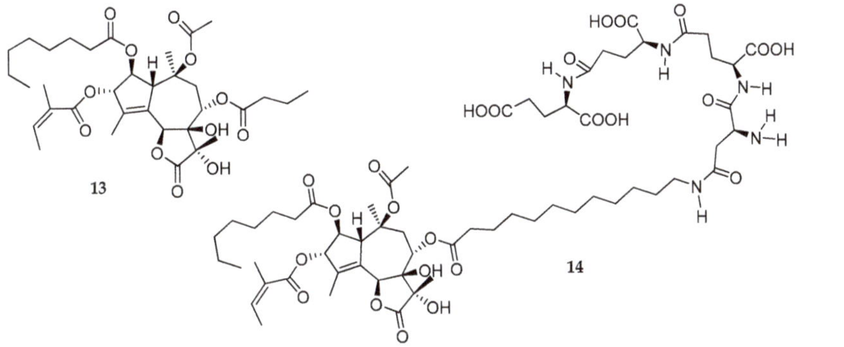

Figure 10. Structure of thapsigargin (**13**) and its derivative mipsagargin (**14**).

Compound **13** strongly inhibited all the subtypes of SERCA. The inhibitory constants (Ki) of thapsigargin (**13**) were 0.21, 1.3, and 12 nm for SERCA1b, SERCA2b, and SERCA3a, respectively [238,239]. Its affinity with the SERCA1a pump is significantly reduced by removal of the acyl groups at O-3, O-8 and O-10 [240].

Thapsigargin (**13**) possesses interesting pharmacological properties; for instance, it induces expression of the L-histidine decarboxylase enzyme responsible for converting L-histidine to histamine in cells [241]. It also evokes ROS generation in cells through calcium-ion mediated changes (mitochondrial depolarization), which can result in cell dysfunction and damage [242,243]. In addition, compound **13** is known for its cytotoxic action against different cancer cell-lines, for instance melanoma [244], insulinoma [245,246], neuronal [247], and breast [248].

The thapsigargin LD_{100} value is 0.8 mg/kg in mice [249]. Zhong et al. [250] proposed that thapsigargin causes vomiting by triggering the phosphorylation of CaMKIIα (Ca^{2+}/calmodulin kinase IIα) and ERK1/2 (extracellular signal-regulated protein kinase 1/2) cascade in the brainstem. Pharmacological preconditioning with the cell-stress inducer thapsigargin (0.3 mg/kg) protects against experimental sepsis in male KM (Kunming) mice [237].

Thapsigargin (**13**) is widely used as an experimental tool for endoplasmic reticulum stress inhibition and the discovery of new active therapeutic derivatives [232,251]. A new thapsigargin (**13**) derivative named mipsagargin, (8-*O*-(12-aminododecanoyl)-8-*O*-debutanoyl thapsigargin)-Asp-γ-Glu-γ-Glu-γ-GluGluOH (**14**) (Figure 10), has been developed for anticancer therapy, notably for the treatment of prostate cancer. Mipsagargin (**14**) has the ability to link the C-8 with prostate-specific membrane antigen (PSMA) peptide. It is a brand new example of a thapsigargin prodrug, which is currently undergoing preclinical evaluation as a targeted chemotherapeutic agent with selective toxicity against cancer cells [232,249]. Mahalingam and coauthors reported that mipsagargin (**14**) has an acceptable pharmacokinetic profile in patients with solid tumors [252]. It is relatively well tolerated, promoting prolonged disease stabilization in patients with advanced liver cancer [253].

2.10. Tomentosin

Tomentosin (**15**) (Figure 11), known also as xanthalongin, is a xanthanolide sesquiterpene lactone with the same chemical formula as inuviscolide (**6**) and parthenolide (**10**) ($C_{15}H_{20}O_3$). It has been isolated from plants such as *Dittrichia viscosa* (L.) Greuter, *Carpesium faberi* C. Winkl. and *Carpesium macrocephalum* Franch. and Sav. [254], *Leucophyta brownii* Cass. [255], the sunflower (*Helianthus annuus* L.) [256] and *Inula* species [254,257]. *Inula viscosa* was widely used as a medicinal plant. A steam distilled extract of its leaves and flowers yielding 2% of tomentosin (**15**) [258] exhibited anti-inflammatory, antibacterial, and anticancer activities [257].

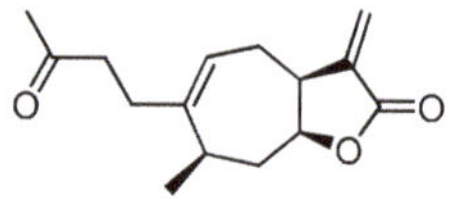

15

Figure 11. Structure of tomentosin (**15**).

The biological activities of tomentosin (**15**) have been addressed in several studies focused on anticancer, antifungal, and insecticidal activities.

Anticancer activity observed in vitro tests was confirmed by in vitro and in vivo studies performed by Bar-Shalom et al. [259]. The authors found that application of an aqueous extract of *Inula viscosa* leaves at a concentration of 300 μg/mL to colorectal cancer cells induced apoptosis, through activation of caspases. During the subsequent in vivo study, mice transplanted with MC38 cells were treated intraperitoneally with the extract at concentrations of 150 and 300 mg/kg. The results indicated that tumor weight and volume were reduced significantly by this treatment after comparison with the untreated control group. Furthermore, staining the paraffin section of the tumors showed inhibition of cell proliferation, as well as induction of apoptosis. Interestingly, side effects, which often include weight loss, impact on fur, and changes in behavior or kidney or liver functions, were not observed. This suggests that the extract was non-toxic at the concentrations applied. However, this remains to be confirmed [259].

Fungal infection of grapevines causes severe economic damage to grape producers. As several fungicide-resistant strains of the pathogens have developed, efective fungicide application is becoming more and more difficult. Therefore, natural products such as sesquiterpene lactones and plant extracts may provide alternatives to current synthetic fungicides. The antifungal activity of six *Inula viscosa* leaf extracts on downy mildew under field conditions was addressed in the study performed by Cohen et al. [260]. This mildew is caused by the fungus *Plasmopara viticola* (Berk. and Curt.). The results established that the effective concentration of oily paste extract in water required to control 90% of the treated shoots in the field was lower than 0.125%. For whole vines, the required concentration ranged between 0.30% and 0.37%. Additionally, the results did not appear to depend on seasonal fluctuations. The evaluated leaf extracts contained 10.6% of tomentosin (**15**) and 10.6% of costic acid, which were identified as the active principles with antifungal activity. The overall results indicated that *Inula viscosa* extracts can be of great value as an alternative source for non-synthetic fungicidal preparations to treat downy mildew infestations of grapes [260].

In an in vivo investigation performed by Ahern et al. [261] on insecticidal activity, the effect of the stereochemistry of the lactone ring on feeding behavior of the herbivorous polyphagous grasshopper *Schistocerca americana* (Drury) was addressed. *Schistocerca americana* is locally related in the south of the U.S. to significant damage to crops when temporary mass populations occur. To evaluate the antifeedant activity of diastereomeric sesquiterpene lactones, the study focused on tomentosin (**15**) with a *cis*-configuration and the corresponding isomer xanthinosin with a *trans*-configuration. The results showed both compounds were able to reduce plant consumption by the insects. However, the *trans*-fused compounds were consumed less than the corresponding *cis*-fused compounds. The results of this study not only permitted conclusions on the importance of the stereochemistry-dependence of diastereomeric sesquiterpene lactones for insecticidal activity, but also might help to understand the geographical distribution and evolution of different clades among the Asteraceae family. The distribution of *cis*- and *trans*-fused compounds within plants of this family is still not fully understood, in this study, 12.5% of the sampled Asteraceae genera contained only *cis*-fused sesquiterpene lactones, 64% only *trans*-fused sesquiterpene lactones, and 23% both types [261].

3. Conclusions

Taking into account the research described above, the sesquiterpene lactones dealt with in this review have proved to be very useful compounds with promising anticancer and anti-inflammatory bioactivities in particular. For instance, alantolactone (**1**) was shown in vivo to be able to reduce tumor size by over 50%, while increasing life expectancy and reducing pathological indicators across several different cancer types. In addition, synergistic effects with well-known cancer therapeutics were also reported. Arglabin (**2**) in salt form could be used to treat several cancers and possesses valuable pharmacokinetic characteristics highly sought for in therapeutic drugs. It has no adverse effects described in the literature. Antitrypanosomal, antitumor and anti-inflammatory activities are reported for cynaropicrin (**4**). Another compound, helenalin (**5**), inhibits essential factors in both cancer and inflammation, i.e., the NF-κB pathway and the transcription of inflammatory cytokines. Inuviscolide (**6**) demonstrates potential as an anti-inflammatory therapeutic compound, being able to reduce rat ear and paw edema and LTB4 generation in intact cells. Lactucin (**7**) and derivatives lactupicrin (**8**) and lactucopicrin (**9**) have a strong potential as phytopharmaceutical agents for treatment of anxiety and sleep disorders, being the main active components of long commercialized substances, e.g., Sedan®, without observing side-effects.

Of all the sesquiterpene lactones reviewed in this work, parthenolide (**10**) and its derivatives DMAPT (**11**) and HMPPPT (**12**) are those most mentioned in the literature, with several studies particularly reporting their excellent anti-inflammatory and antitumor activities. Some work reveals synergistic effects of these compounds with current anticancer drugs, favoring more potent and selective antitumor action. Parthenolide derivatives have been designed and DMAPT (**11**) is 1000-fold more soluble in water than the natural product, while maintaining a similar mechanism of action with improved anticancer and anti-inflammatory activities.

Thapsigargin (**13**) has the particular capacity to cause endoplasmic reticulum stress and thus cell apoptosis, an interesting pharmacological property to be retained or potentiated in derivatives aimed at anticancer therapy. An example is mipsagargin (**14**), which is currently undergoing preclinical evaluation for selective toxicity against cancer cells in tumor sites, with minimal side-effects for the host. Another active compound is tomentosin (**15**), additionally with potential applications in cancer but also as antifungal and antifeedant.

In summary, given the activities described in this review these sesquiterpene lactones exhibit properties that justify more research by the scientific community, to drive preclinical and clinical studies leading to development of new drugs.

Author Contributions: Conceptualization, A.M.L.S. and L.M.; writing—original draft preparation, A.M.L.S., L.M., O.C., P.M.C.S., F.S.; writing—review and editing, L.M., O.C., A.M.L.S. All authors have read and agreed to the published version of the manuscript.

Funding: This research was funded by FCT–Fundação para a Ciência e a Tecnologia, the European Union, QREN, FEDER, COMPETE, by funding the cE3c centre (UIDB/00329/2020), the LAQV-REQUIMTE (UIDB/50006/2020) and QOPNA (UID/QUI/00062/2019) research units, and by the Spanish Ministry Science and Research (MINECO RTI2018-094356-B-C21).

Acknowledgments: Thanks are due to the University of Azores and University of La Laguna. AbbVie participated in the revision of the manuscript.

Conflicts of Interest: O.C. is an employee of AbbVie and owns AbbVie stock. The authors L.M., P.M.C.S., F.S., and A.M.L.S. declare they have no conflict of interest.

References

1. Li, Q.; Wang, Z.; Xie, Y.; Hu, H. Antitumor activity and mechanism of costunolide and dehydrocostus lactone: Two natural sesquiterpene lactones from the Asteraceae family. *Biomed. Pharmacother.* **2020**, *125*, 109955. [CrossRef] [PubMed]

2. Stavrianidi, A. A classification of liquid chromatography mass spectrometry techniques for evaluation of chemical composition and quality control of traditional medicines. *J. Chromatog. A* **2020**, *1609*, 460501. [CrossRef] [PubMed]

3. Gou, J.; Hao, F.; Huang, C.; Kwon, M.; Chen, F.; Li, C.; Liu, C.; Ro, D.-K.; Tang, H.; Zhang, Y. Discovery of a non-stereoselective cytochrome P450 catalyzing either 8α- or 8β-hydroxylation of germacrene A acid from the Chinese medicinal plant, *Inula hupehensis*. *Plant J.* **2018**, *93*, 92–106. [CrossRef] [PubMed]

4. Perassolo, M.; Cardillo, A.B.; Busto, V.D.; Giulietti, A.M.; Talou, J.R. Biosynthesis of sesquiterpene lactones in plants and metabolic engineering for their biotechnological production. In *Sesquiterpene Lactones: Advances in Their Chemistry and Biological Aspects*; Sülsen, V.P., Martino, V.S., Eds.; Springer International Publishing: Cham, Switzerland, 2018; pp. 47–91. [CrossRef]

5. Sülsen, V.P.; Martino, V.S. Overview. In *Sesquiterpene Lactones: Advances in Their Chemistry and Biological Aspects*; Sülsen, V.P., Martino, V.S., Eds.; Springer International Publishing: Cham, Switzerland, 2018; pp. 3–17. [CrossRef]

6. Choodej, S.; Pudhom, K.; Yamauchi, K.; Mitsunaga, T. Inhibition of melanin production by sesquiterpene lactones from *Saussurea lappa* and their analogues. *Med. Chem. Res.* **2019**, *28*, 857–862. [CrossRef]

7. Seca, A.M.L.; Grigore, A.; Pinto, D.C.G.A.; Silva, A.M.S. The genus *Inula* and their metabolites: From ethnopharmacological to medicinal uses. *J. Ethnopharmacol.* **2014**, *154*, 286–310. [CrossRef]

8. Bosco, A.; Golsteyn, R.M. Emerging anti-mitotic activities and other bioactivities of sesquiterpene compounds upon human cells. *Molecules* **2017**, *22*, 459. [CrossRef]

9. Beer, M.F.; Bivona, A.E.; Sánchez Alberti, A.; Cerny, N.; Reta, G.F.; Martín, V.S.; Padrón, J.M.; Malchiodi, E.L.; Sülsen, V.P.; Donadel, O.J. Preparation of sesquiterpene lactone derivatives: Cytotoxic activity and selectivity of action. *Molecules* **2019**, *24*, 1113. [CrossRef]

10. Sokovic, M.; Ciric, A.; Glamoclija, J.; Skaltsa, H. Biological activities of sesquiterpene lactones isolated from the genus *Centaurea* L. (Asteraceae). *Curr. Pharm. Des.* **2017**, *23*, 2767–2786. [CrossRef]

11. Ma, C.; Meng, C.-W.; Zhou, Q.-M.; Peng, C.; Liu, F.; Zhang, J.-W.; Zhou, F.; Xiong, L. New sesquiterpenoids from the stems of *Dendrobium nobile* and their neuroprotective activities. *Fitoterapia* **2019**, *138*, 104351. [CrossRef]

12. Ghantous, A.; Gali-Muhtasib, H.; Vuorela, H.; Saliba, N.A.; Darwiche, N. What made sesquiterpene lactones reach cancer clinical trials? *Drug Discov. Today* **2010**, *15*, 668–678. [CrossRef]

13. Cheriti, A.; Belboukhari, N. Terpenoids of the Saharan medicinal plants *Launaea* Cass. genus (Asteraceae) and their biological activities. In *Terpenoids and Squalene: Biosynthesis, Functions and Health Implications*; Bates, A.R., Ed.; Nova Science Publishers, Inc.: New York, NY, USA, 2015; pp. 51–70. ISBN 978-1-63463-656-8.

14. Wang, J.; Su, S.; Zhang, S.; Zhai, S.; Sheng, R.; Wu, W.; Guo, R. Structure-activity relationship and synthetic methodologies of α-santonin derivatives with diverse bioactivities: A mini-review. *Eur. J. Med. Chem.* **2019**, *175*, 215–233. [CrossRef] [PubMed]

15. Schmidt, T.J. Structure-activity and activity-activity relationships of sesquiterpene lactones. In *Sesquiterpene Lactones: Advances in Their Chemistry and Biological Aspects*; Sülsen, V., Martino, V., Eds.; Springer International Publishing AG: Cham, Switzerland, 2018; pp. 349–371. [CrossRef]

16. Wang, G.-W.; Qin, J.-J.; Cheng, X.-R.; Shen, Y.-H.; Shan, L.; Jin, H.-Z.; Zhang, W.-D. *Inula* sesquiterpenoids: Structural diversity, cytotoxicity and anti-tumor activity. *Expert Opin. Investig. Drugs* **2014**, *23*, 317–345. [CrossRef] [PubMed]

17. Reddy, D.M.; Qazi, N.A.; Sawant, S.D.; Bandey, A.H.; Srinivas, J.; Shankar, M.; Singh, S.K.; Verma, M.; Chashoo, G.; Saxena, A.; et al. Design and synthesis of spiro derivatives of parthenin as novel anti-cancer agents. *Eur. J. Med. Chem.* **2011**, *46*, 3210–3217. [CrossRef] [PubMed]

18. Liu, X.; Cao, J.; Huang, G.; Zhao, Q.; Shen, J. Biological activities of artemisinin derivatives beyond malaria. *Curr. Top. Med. Chem.* **2019**, *19*, 205–222. [CrossRef]

19. Zhang, C.; Zhu, Y.; Yin, X.-P.; Wei, Q.-H.; Zhang, N.-N.; Li, C.-X.; Xie, T.; Chen, R. [Advances in synthesis of artemisinin based on plant genetic engineering]. *Zhongguo Zhong Yao Za Zhi* **2019**, *44*, 4285–4292. [CrossRef]

20. Gao, F.; Sun, Z.; Kong, F.; Xiao, J. Artemisinin-derived hybrids and their anticancer activity. *Eur. J. Med. Chem.* **2020**, *188*, 112044. [CrossRef]

21. Xu, R.; Peng, Y.; Wang, M.; Li, X. Intestinal absorption of isoalantolactone and alantolactone, two sesquiterpene lactones from radix inulae, using Caco-2 cells. *Eur. J. Drug Metab. Pharmacokinet.* **2019**, *44*, 295–303. [CrossRef]

22. Circioban, D.; Pavel, I.Z.; Ledeti, A.; Ledeti, I.; Danciu, C.; Dehelean, C. Cytotoxic activity evaluation on breast cells of guest-host complexes containing artemisinin. *Rev. Chim.* **2019**, *70*, 2843–2846. [CrossRef]

23. Shakeri, A.; Amini, E.; Asili, J.; Masullo, M.; Piacente, S.; Iranshahi, M. Screening of several biological activities induced by different sesquiterpene lactones isolated from *Centaurea behen* L. and *Rhaponticum repens* (L.) Hidalgo. *Nat. Prod. Res.* **2018**, *32*, 1436–1440. [CrossRef]

24. Seca, A.M.L.; Pinto, D.C.G.A.; Silva, A.M.S. Metabolomic profile of the genus *Inula*. *Chem. Biodivers.* **2015**, *12*, 859–906. [CrossRef]

25. Lim, H.S.; Jin, S.E.; Kim, O.S.; Shin, H.K.; Jeong, S.J. Alantolactone from *Saussurea lappa* exerts antiinflammatory effects by inhibiting chemokine production and STAT1 phosphorylation in TNF-α and IFN-γ-induced in HaCaT cells. *Phytother. Res.* **2015**, *29*, 1088–1096. [CrossRef] [PubMed]

26. Chun, J.; Choi, R.J.; Khan, S.; Lee, D.-S.; Kim, Y.-C.; Nam, Y.-J.; Lee, D.-U.; Kim, Y.S. Alantolactone suppresses inducible nitric oxide synthase and cyclooxygenase-2 expression by down-regulating NF-κB, MAPK and AP-1 via the MyD88 signaling pathway in LPS-activated RAW 264.7 cells. *Int. Immunopharmacol.* **2012**, *14*, 375–383. [CrossRef] [PubMed]

27. Marshall, J.A.; Cohen, N. The stereoselective total synthesis of alantolactone. *J. Am. Chem. Soc.* **1965**, *87*, 2773–2774. [CrossRef]

28. Schultz, A.G.; Godfrey, J.D. An annelation approach to the synthesis of eudesmane and elemane sesquiterpene lactones. Total synthesis of dl-dihydrocallitrisin, dl-7,8-epialantolactone, dl-7,8-epiisoalantolactone, and dl-atractylon. *J. Am. Chem. Soc.* **1980**, *102*, 2414–2428. [CrossRef]

29. Trendafilova, A.; Chanev, C.; Todorova, M. Ultrasound-assisted extraction of alantolactone and isoalantolactone from *Inula helenium* roots. *Pharmacogn. Mag.* **2010**, *6*, 234–237. [CrossRef] [PubMed]

30. Zhao, Y.M.; Wang, J.; Liu, H.B.; Guo, C.Y.; Zhang, W.M. Microwave-assisted extraction of alantolactone and isoalantolactone from *Inula helenium*. *Indian J. Pharm. Sci.* **2015**, *77*, 116–120. [CrossRef]

31. Chi, X.-F.; Yue, H.-L.; Zhao, X.-H.; Hu, F.-Z. Obtaining alantolactone and isoalantolactone from *Inula racemose* Hook.f. by optimized supercritical fluid extraction. *Ind. Crops Prod.* **2016**, *79*, 63–69. [CrossRef]

32. Nikonova, L.P. Lactones of *Inula magnifica*. *Chem. Nat. Comp.* **1973**, *9*, 528. [CrossRef]

33. Stampf, J.L.; Benezra, C.; Klecak, G.; Geleick, H.; Schulz, K.H.; Hausen, B. The sensitizing capacity of helenin and of two of its main constituents, the sesquiterpene lactones alantolactone and isoalantolactone: A comparison of epicutaneous and intradermal sensitizing methods in different strains of guinea pig. *Contact Derm.* **1982**, *8*, 16–24. [CrossRef]

34. Kuno, Y.; Kawabe, Y.; Sakakibara, S. Allergic contact dermatitis associated with photosensitivity, from alantolactone in a chrysanthemum farmer. *Contact Derm.* **1999**, *40*, 224–225. [CrossRef]

35. Marc, E.B.; Nelly, A.; Annick, D.-D.; Frederic, D. Plants used as remedies antirheumatic and antineuralgic in the traditional medicine of Lebanon. *J. Ethnopharmacol.* **2008**, *120*, 315–334. [CrossRef] [PubMed]

36. Sekera, A.; Rahm, J. Natural antihelminthics. III. Helenin. *Cesk. Farm.* **1953**, *2*, 22–24. [PubMed]

37. Kang, X.; Wang, H.; Li, Y.; Xiao, Y.; Zhao, L.; Zhang, T.; Zhou, S.; Zhou, X.; Li, Y.; Shou, Z.; et al. Alantolactone induces apoptosis through ROS-mediated AKT pathway and inhibition of PINK1-mediated mitophagy in human HepG2 cells. *Artif. Cells Nanomed. Biotechnol.* **2019**, *47*, 1961–1970. [CrossRef] [PubMed]

38. He, Y.; Cao, X.; Kong, Y.; Wang, S.; Xia, Y.; Bi, R.; Liu, J. Apoptosis-promoting and migration-suppressing effect of alantolactone on gastric cancer cell lines BGC-823 and SGC-7901 via regulating p38 MAPK and NF-κB pathways. *Hum. Exp. Toxicol.* **2019**, *38*, 1132–1144. [CrossRef] [PubMed]

39. Liu, J.; Yang, Z.; Kong, Y.; He, Y.; Xu, Y.; Cao, X. Antitumor activity of alantolactone in lung cancer cell lines NCI-H1299 and Anip973. *J. Food Biochem.* **2019**, *43*, 12972. [CrossRef] [PubMed]

40. Zhang, X.; Zhang, H.-M. Alantolactone induces gastric cancer BGC-823 cell apoptosis by regulating reactive oxygen species generation and the AKT signaling pathway. *Oncol. Lett.* **2019**, *17*, 4795–4802. [CrossRef] [PubMed]

41. Wang, X.; Lan, Y.L.; Xing, J.S.; Lan, X.Q.; Wang, L.T.; Zhang, B. Alantolactone plays neuroprotective roles in traumatic brain injury in rats via anti-inflammatory, anti-oxidative and anti-apoptosis pathways. *Am. J. Transl. Res.* **2018**, *10*, 368–380.

42. Tavares, W.R.; Seca, A.M.L. *Inula* L. secondary metabolites against oxidative stress-related human diseases. *Antioxidants* **2019**, *8*, 122. [CrossRef]

43. Da Silva Castro, E.; Azeredo Alves Antunes, L.; Felipe Revoredo Lobo, J.; Arthur Ratcliffe, N.; Moreira Borges, R.; Rocha, L.; Burth, P.; Maria Fonte Amorim, L. Antileukemic properties of sesquiterpene lactones: A systematic review. *Anti-Cancer Agents Med. Chem.* **2018**, *18*, 323–334. [CrossRef]

44. Xu, X.; Huang, L.; Zhang, Z.; Tong, J.; Mi, J.; Wu, Y.; Zhang, C.; Yan, H. Targeting non-oncogene ROS pathway by alantolactone in B cell acute lymphoblastic leukemia cells. *Life Sci.* **2019**, *227*, 153–165. [CrossRef]

45. Chun, J.; Li, R.-J.; Cheng, M.-S.; Kim, Y.S. Alantolactone selectively suppresses STAT3 activation and exhibits potent anticancer activity in MDA-MB-231 cells. *Cancer Lett.* **2015**, *357*, 393–403. [CrossRef] [PubMed]

46. He, W.; Cao, P.; Xia, Y.; Hong, L.; Zhang, T.; Shen, X.; Zheng, P.; Shen, H.; Zhao, Y.; Zou, P. Potent inhibition of gastric cancer cells by a natural compound via inhibiting TrxR1 activity and activating ROS-mediated p38 MAPK pathway. *Free Radic. Res.* **2019**, *53*, 104–114. [CrossRef] [PubMed]

47. Cao, P.; Xia, Y.; He, W.; Zhang, T.; Hong, L.; Zheng, P.; Shen, X.; Liang, G.; Cui, R.; Zou, P. Enhancement of oxaliplatin-induced colon cancer cell apoptosis by alantolactone, a natural product inducer of ROS. *Int. J. Biol. Sci.* **2019**, *15*, 1676–1684. [CrossRef] [PubMed]

48. He, R.; Shi, X.; Zhou, M.; Zhao, Y.; Pan, S.; Zhao, C.; Guo, X.; Wang, M.; Li, X.; Qin, R. Alantolactone induces apoptosis and improves chemosensitivity of pancreatic cancer cells by impairment of autophagy-lysosome pathway via targeting TFEB. *Toxicol. Appl. Pharmacol.* **2018**, *356*, 159–171. [CrossRef]

49. Rasul, A.; Khan, M.; Ali, M.; Li, J.; Li, X. Targeting apoptosis pathways in cancer with alantolactone and isoalantolactone. *Sci. World J.* **2013**, *2013*, 248532. [CrossRef]

50. Quintana, J.; Estévez, F. Recent advances on cytotoxic sesquiterpene lactones. *Curr. Pharm. Des.* **2018**, *24*, 4355–4361. [CrossRef]

51. Ren, Y.; Yue, B.; Ren, G.; Yu, Z.; Luo, X.; Sun, A.; Zhang, J.; Han, M.; Wang, Z.; Dou, W. Activation of PXR by alantolactone ameliorates DSS-induced experimental colitis via suppressing NF-κB signaling pathway. *Sci. Rep.* **2019**, *9*, 1–12. [CrossRef]

52. Seo, J.Y.; Lim, S.S.; Kim, J.; Lee, K.W.; Kim, J.-S. Alantolactone and isoalantolactone prevent amyloid β25-35 -induced toxicity in mouse cortical neurons and scopolamine-induced cognitive impairment in mice. *Phytother. Res.* **2017**, *31*, 801–811. [CrossRef]

53. Kumar, C.; Kumar, A.; Nalli, Y.; Lone, W.I.; Satti, N.K.; Verma, M.K.; Ahmed, Z.; Ali, A. Design, synthesis and biological evaluation of alantolactone derivatives as potential anti-inflammatory agents. *Med. Chem. Res.* **2019**, *28*, 849–856. [CrossRef]

54. Li, X.; Lu, C.; Liu, S.; Liu, S.; Liu, S.; Su, C.; Xiao, T.; Bi, Z.; Sheng, P.; Huang, M.; et al. Synthesis and discovery of a drug candidate for treatment of idiopathic pulmonary fibrosis through inhibition of TGF-β1 pathway. *Eur. J. Med. Chem.* **2018**, *157*, 229–247. [CrossRef]

55. Khan, M.; Yi, F.; Rasul, A.; Li, T.; Wang, N.; Gao, H.; Gao, R.; Ma, T. Alantolactone induces apoptosis in glioblastoma cells via GSH depletion, ROS generation, and mitochondrial dysfunction. *IUBMB Life* **2012**, *64*, 783–794. [CrossRef] [PubMed]

56. Markowiak, T.; Kerner, N.; Neu, R.; Potzger, T.; Großer, C.; Zeman, F.; Hofmann, H.-S.; Ried, M. Adequate nephroprotection reduces renal complications after hyperthermic intrathoracic chemotherapy. *J. Surg. Oncol.* **2019**, *120*, 1220–1226. [CrossRef] [PubMed]

57. Leach, C. Complications of systemic anti-cancer treatment. *Medicine* **2020**, *48*, 48–51. [CrossRef]

58. Jain, K.K. An overview of drug delivery systems. *Meth. Mol. Biol.* **2020**, *2059*, 1–54. [CrossRef]

59. Guo, C.; Zhang, S.; Teng, S.; Niu, K. Simultaneous determination of sesquiterpene lactones isoalantolactone and alantolactone isomers in rat plasma by liquid chromatography with tandem mass spectrometry: Application to a pharmacokinetic study. *J. Sep. Sci.* **2014**, *37*, 950–956. [CrossRef] [PubMed]

60. Xu, R.; Zhou, G.; Peng, Y.; Wang, M.; Li, X. Pharmacokinetics, tissue distribution and excretion of isoalantolactone and alantolactone in rats after oral administration of radix inulae extract. *Molecules* **2015**, *20*, 7719–7736. [CrossRef]

61. Zhou, B.; Ye, J.; Yang, N.; Chen, L.; Zhuo, Z.; Mao, L.; Liu, Q.; Lan, G.; Ning, J.; Ge, G.; et al. Metabolism and pharmacokinetics of alantolactone and isoalantolactone in rats: Thiol conjugation as a potential metabolic pathway. *J. Chromatogr. B Analyt. Technol. Biomed. Life Sci.* **2018**, *1072*, 370–378. [CrossRef]

62. Bottex-Gauthier, C.; Vidal, D.; Picot, F.; Potier, P.; Menichini, F.; Appendino, G. *In vitro* biological activities of arglabin, a sesquiterpene lactone from the Chinese herb *Artemisia myriantha* Wall. (Asteraceae). *Biotechnol. Ther.* **1993**, *4*, 77–98.

63. Zan, K.; Chen, X.Q.; Tu, P.F. Guaianolides from aerial parts of *Artemisia myriantha*. *Zhongguo Zhongyao Zazhi.* **2018**, *43*, 2295–2299. [CrossRef]

64. Dylenova, E.P.; Randalova, T.E.; Tykheev, Z.A.; Zhigzhitzhapova, S.V.; Radnaeva, L.D. *Artemisia jacutica* Drob. as the source of terpenoids. *IOP Conf. Ser. Earth Environ. Sci.* **2019**, *320*, 012054. [CrossRef]

65. Shaikenov, T.E.; Adekenov, S.M.; Williams, R.M.; Prashad, N.; Baker, F.L.; Madden, T.L.; Newman, R. Arglabin-DMA, a plant derived sesquiterpene, inhibits farnesyltransferase. *Oncol. Rep.* **2001**, *8*, 173–179. [CrossRef] [PubMed]

66. Lone, S.H.; Bhat, K.A.; Khuroo, M.A. Arglabin: From isolation to antitumor evaluation. *Chem.-Biol. Interact.* **2015**, *240*, 180–198. [CrossRef] [PubMed]

67. Seitz, M.; Reiser, O. Synthetic approaches towards structurally diverse gamma-butyrolactone natural-product-like compounds. *Curr. Opin. Chem. Biol.* **2005**, *9*, 285–292. [CrossRef] [PubMed]

68. Kalidindi, S.; Jeong, S.W.B.; Schall, A.; Bandichhor, R.; Nosse, B.; Reiser, O. Enantioselective synthesis of arglabin. *Angew. Chem. Int. Ed.* **2007**, *46*, 6361–6363. [CrossRef] [PubMed]

69. Gharpure, S.J.; Nanda, L.N. Application of oxygen/nitrogen substituted donor-acceptor cyclopropanes in the total synthesis of natural products. *Tetrahedron Lett.* **2017**, *58*, 711–720. [CrossRef]

70. Lone, S.H.; Bhat, K.A. Hemisynthesis of a naturally occurring clinically significant antitumor arglabin from ludartin. *Tetrahedron Lett.* **2015**, *56*, 1908–1910. [CrossRef]

71. Ganesan, A. The impact of natural products upon modern drug discovery. *Curr. Opin. Chem. Biol.* **2008**, *12*, 306–317. [CrossRef]

72. Adekenov, S.M. Chemical modification of arglabin and biological activity of its new derivatives. *Fitoterapia* **2016**, *110*, 196–205. [CrossRef]

73. Musulmanbekov, K.Z. Clinical use of the antitumor drug arglabin. In Proceedings of the International Scientific-Practical Conference, Karaganda, Kazakhstan, 16–18 October 2002; p. 46.

74. Zhangabylov, N.S.; Dederer, L.Y.; Gorbacheva, L.B.; Vasil'eva, S.V.; Terekhov, A.S.; Adekenov, S.M. Sesquiterpene lactone arglabin influences DNA synthesis in P388 leukemia cells *in vivo*. *Pharm. Chem. J.* **2004**, *38*, 651–653. [CrossRef]

75. Sirota, V.B.; Bochkova, N.V.; Kostrova, E.V.; Tselikova, N.L.; Kabildina, N.A. Application of the phytopreparate arglabin in combined treatment of patients with cancer. In Proceedings of the 1st Russian Phytotherapeutic Congress, Moscow, Russia, 14–16 March 2008; p. 148.

76. Schepetkin, I.A.; Kirpotina, L.N.; Mitchell, P.T.; Kishkentaeva, A.C.; Shaimerdenova, Z.R.; Atazhanova, G.A.; Adekenov, S.M.; Quinn, M.T. The natural sesquiterpene lactones arglabin, grosheimin, agracin, parthenolide, and estafiatin inhibit T cell receptor (TCR) activation. *Phytochemistry* **2018**, *146*, 36–46. [CrossRef]

77. He, W.; Lai, R.F.; Lin, Q.; Huang, Y.M.; Wang, L. Arglabin is a plant sesquiterpene lactone that exerts potent anticancer effects on human oral squamous cancer cells via mitochondrial apoptosis and downregulation of the mTOR/PI3K/Akt signaling pathway to inhibit tumor growth *in vivo*. *J. Buon* **2018**, *23*, 1679–1685. [PubMed]

78. Tewari, D.; Rawat, P.; Singh, P.K. Adverse drug reactions of anticancer drugs derived from natural sources. *Food Chem. Toxicol.* **2019**, *123*, 522–535. [CrossRef] [PubMed]

79. Liu, X.; Jia, H.; Xia, H. Arglabin as a potential drug in the treatment of Freund's complete adjuvant-induced arthritis in rats. *Trop. J. Pharm. Res.* **2018**, *17*, 1585–1590. [CrossRef]

80. Abderrazak, A.; Couchie, D.; Mahmood, D.F.; Elhage, R.; Vindis, C.; Laffargue, M.; Mateo, V.; Buchele, B.; Ayala, M.R.; El Gaafary, M.; et al. Anti-inflammatory and antiatherogenic effects of the NLRP3 inflammasome inhibitor arglabin in ApoE2.Ki mice fed a high-fat diet. *Circulation* **2015**, *131*, 1061–1070. [CrossRef] [PubMed]

81. Baldrighi, M.; Mallat, Z.; Li, X. NLRP3 inflammasome pathways in atherosclerosis. *Atherosclerosis* **2017**, *267*, 127–138. [CrossRef] [PubMed]

82. Rao, A.S.; Kelkar, G.R.; Bhattacharyya, S.C. Terpenoids—XXI: The structure of costunolide, a new sesquiterpene lactone from costus root oil. *Tetrahedron* **1960**, *9*, 275–283. [CrossRef]

83. Ikezawa, N.; Göpfert, J.C.; Nguyen, D.T.; Kim, S.-U.; O'Maille, P.E.; Spring, O.; Ro, D.-K. Lettuce costunolide synthase (CYP71BL2) and its homolog (CYP71BL1) from sunflower catalyze distinct regio- and stereoselective hydroxylations in sesquiterpene lactone metabolism. *J. Biol. Chem.* **2011**, *286*, 21601–21611. [CrossRef]

84. Kassuya, C.A.L.; Cremoneze, A.; Barros, L.F.L.; Simas, A.S.; Lapa, F.d.R.; Mello-Silva, R.; Stefanello, M.E.A.; Zampronio, A.R. Antipyretic and anti-inflammatory properties of the ethanolic extract, dichloromethane fraction and costunolide from *Magnolia ovata* (Magnoliaceae). *J. Ethnopharmacol.* **2009**, *124*, 369–376. [CrossRef]

85. De Kraker, J.-W.; Franssen, M.C.R.; Joerink, M.; de Groot, A.; Bouwmeester, H.J. Biosynthesis of costunolide, dihydrocostunolide, and leucodin. Demonstration of cytochrome p450-catalyzed formation of the lactone ring present in sesquiterpene lactones of chicory. *Plant Physiol.* **2002**, *129*, 257–268. [CrossRef]

86. Abdelwahab, S.I.; Taha, M.M.E.; Alhazmi, H.A.; Ahsan, W.; Rehman, Z.U.; Bratty, M.A.; Makeen, H. Phytochemical profiling of costus (*Saussurea lappa* Clarke) root essential oil, and its antimicrobial and toxicological effects. *Trop. J. Pharm. Res.* **2019**, *18*, 2155–2160. [CrossRef]

87. Yang, Z.-J.; Ge, W.-Z.; Li, Q.-Y.; Lu, Y.; Gong, J.-M.; Kuang, B.-J.; Xi, X.; Wu, H.; Zhang, Q.; Chen, Y. Synthesis and biological evaluation of costunolide, parthenolide, and their fluorinated analogues. *J. Med. Chem.* **2015**, *58*, 7007–7020. [CrossRef] [PubMed]

88. Pavan Kumar, C.; Devi, A.; Ashok Yadav, P.; Rao Vadaparthi, R.; Shankaraiah, G.; Sowjanya, P.; Jain, N.; Suresh Babu, K. "Click" reaction mediated synthesis of costunolide and dehydrocostuslactone derivatives and evaluation of their cytotoxic activity. *Asian Nat. Prod. Res.* **2016**, *18*, 1063–1078. [CrossRef] [PubMed]

89. Poon, P.S.; Banerjee, A.K.; Vera, W.J.; Bedoya, L. Reagents for bromination; Application in the synthesis of diterpenes, sesquiterpenes and bioactive compounds. *Curr. Org. Chem.* **2017**, *21*, 889–907. [CrossRef]

90. Kim, D.Y.; Choi, B.Y. Costunolide - A bioactive sesquiterpene lactone with diverse therapeutic potential. *Int. J. Mol. Sci.* **2019**, *20*, 2926. [CrossRef]

91. Lin, X.; Peng, Z.; Su, C. Potential anti-cancer activities and mechanisms of costunolide and dehydrocostuslactone. *Int. J. Mol. Sci.* **2015**, *16*, 10888–10906. [CrossRef]

92. Jin, X.; Wang, C.; Wang, L. Costunolide inhibits osteosarcoma growth and metastasis via suppressing STAT3 signal pathway. *Biomed. Pharmacother.* **2020**, *121*, 109659. [CrossRef]

93. Kolosenko, I.; Yu, Y.; Busker, S.; Dyczynski, M.; Liu, J.; Haraldsson, M.; Palm Apergi, C.; Helleday, T.; Tamm, K.P.; Page, B.D.G.; et al. Identification of novel small molecules that inhibit STAT3-dependent transcription and function. *PLoS ONE* **2017**, *12*, 0178844. [CrossRef]

94. Yan, Z.; Xu, T.; An, Z.; Hu, Y.; Chen, W.; Ma, J.; Shao, C.; Zhu, F. Costunolide induces mitochondria-mediated apoptosis in human gastric adenocarcinoma BGC-823 cells. *BMC Complement. Altern. Med.* **2019**, *19*, 151. [CrossRef]

95. Ahmad, F.; Dixit, D.; Sharma, V.; Kumar, A.; Joshi, S.D.; Sarkar, C.; Sen, E. Nrf2-driven TERT regulates pentose phosphate pathway in glioblastoma. *Cell Death Dis.* **2016**, *7*, 2213. [CrossRef]

96. Ahmad, F.; Patrick, S.; Sheikh, T.; Sharma, V.; Pathak, P.; Malgulwar, P.B.; Kumar, A.; Joshi, S.D.; Sarkar, C.; Sen, E. Telomerase reverse transcriptase (TERT) - enhancer of zeste homolog 2 (EZH2) network regulates lipid metabolism and DNA damage responses in glioblastoma. *J. Neurochem.* **2017**, *143*, 671–683. [CrossRef]

97. Menendez, J.A.; Lupu, R. Fatty acid synthase (FASN) as a therapeutic target in breast cancer. *Expert Opin. Ther. Targets* **2017**, *21*, 1001–1016. [CrossRef] [PubMed]

98. Guo, D.; Bell, E.H.; Mischel, P.; Chakravarti, A. Targeting SREBP-1-driven lipid metabolism to treat cancer. *Curr. Pharm. Des.* **2014**, *20*, 2619–2626. [CrossRef] [PubMed]

99. Bost, F.; Kaminski, L. The metabolic modulator PGC-1α in cancer. *Am. J. Cancer Res.* **2019**, *9*, 198–211. [PubMed]

100. Park, E.; Song, J.H.; Kim, M.S.; Park, S.-H.; Kim, T.S. Costunolide, a sesquiterpene lactone, inhibits the differentiation of pro-inflammatory CD4+ T cells through the modulation of mitogen-activated protein kinases. *Int. Immunopharmacol.* **2016**, *40*, 508–516. [CrossRef] [PubMed]

101. Turk, A.; Ahn, J.H.; Jo, Y.H.; Song, J.Y.; Khalife, H.K.; Gali-Muhtasib, H.; Kim, Y.; Hwang, B.Y.; Lee, M.K. NF-κB inhibitory sesquiterpene lactones from Lebanese *Laurus nobilis*. *Phytochem. Lett.* **2019**, *30*, 120–123. [CrossRef]

102. Saraswati, S.; Alhaider, A.A.; Abdelgadir, A.M. Costunolide suppresses an inflammatory angiogenic response in a subcutaneous murine sponge model. *APMIS* **2018**, *126*, 257–266. [CrossRef]

103. Xie, Y.; Li, Q.-J.; Wang, Z.-G.; Hu, H.-L. Effects of active components from vladimiriae radix and their combinations on ethanol-induced acute gastric ulcer in mice. *Chin. J. New Drugs* **2019**, *28*, 2754–2760.

104. Chen, Z.; Zhang, D.; Li, M.; Wang, B. Costunolide ameliorates lipoteichoic acid-induced acute lung injury via attenuating MAPK signaling pathway. *Int. Immunopharm.* **2018**, *61*, 283–289. [CrossRef]

105. Chen, Y.-T.; Du, Y.; Zhao, B.; Gan, L.-X.; Yu, K.-K.; Sun, L.; Wang, J.; Qian, F. Costunolide alleviates HKSA-induced acute lung injury via inhibition of macrophage activation. *Acta Pharmacol. Sin.* **2019**, *40*, 1040–1048. [CrossRef]

106. Liu, B.; Rong, Y.; Sun, D.; Li, W.; Chen, H.; Cao, B.; Wang, T. Costunolide inhibits pulmonary fibrosis via regulating NF-κB and TGF-β1/Smad2/Nrf2-NOX4 signaling pathways. *Biochem. Biophys. Res. Commun.* **2019**, *510*, 329–333. [CrossRef]

107. Wang, Y.; Zhang, X.; Zhao, L.; Shi, M.; Wei, Z.; Yang, Z.; Guo, C.; Fu, Y. Costunolide protects lipopolysaccharide/D-galactosamine–induced acute liver injury in mice by inhibiting NF-κB signaling pathway. *J. Surg. Res.* **2017**, *220*, 40–45. [CrossRef] [PubMed]

108. He, Y.; Moqbel, S.A.A.; Xu, L.; Ran, J.; Ma, C.; Xu, K.; Bao, J.; Jiang, L.; Chen, W.; Xiong, Y.; et al. Costunolide inhibits matrix metalloproteinases expression and osteoarthritis via the NF-κB and Wnt/β-catenin signaling pathways. *Mol. Med. Rep.* **2019**, *20*, 312–322. [CrossRef] [PubMed]

109. Lee, B.-K.; Park, S.-J.; Nam, S.-Y.; Kang, S.; Hwang, J.; Lee, S.-J.; Im, D.-S. Anti-allergic effects of sesquiterpene lactones from *Saussurea costus* (Falc.) Lipsch. determined using *in vivo* and *in vitro* experiments. *J. Ethnopharmacol.* **2018**, *213*, 256–261. [CrossRef] [PubMed]

110. Kim, Y.E.; Choi, H.C.; Nam, G.; Choi, B.Y. Costunolide promotes the proliferation of human hair follicle dermal papilla cells and induces hair growth in C57 BL/6 mice. *J. Cosmet. Dermatol.* **2019**, *18*, 414–421. [CrossRef]

111. Cala, A.; Zorrilla, J.G.; Rial, C.; Molinillo, J.M.G.; Varela, R.M.; Macías, F.A. Easy access to alkoxy, amino, carbamoyl, hydroxy, and thiol derivatives of sesquiterpene lactones and evaluation of their bioactivity on parasitic weeds. *J. Agric. Food Chem.* **2019**, *67*, 10764–10773. [CrossRef]

112. Elsebai, M.F.; Mocan, A.; Atanasov, A.G. Cynaropicrin: A comprehensive research review and therapeutic potential as an anti-hepatitis C virus agent. *Front. Pharmacol.* **2016**, *7*, 472. [CrossRef]

113. Suchy, M.; Herout, V.; Šorm, F. On terpenes. CXVI. Structure of cynaropicrin. *Collect. Czech. Chem. Commun.* **1960**, *25*, 2777–2782. [CrossRef]

114. Eljounaidi, K.; Comino, C.; Moglia, A.; Cankar, K.; Genre, A.; Hehn, A.; Bourgaud, F.; Beekwilder, J.; Lanteri, S. Accumulation of cynaropicrin in globe artichoke and localization of enzymes involved in its biosynthesis. *Plant Sci.* **2015**, *239*, 128–136. [CrossRef]

115. Colantuono, A.; Ferracane, R.; Vitaglione, P. Potential bioaccessibility and functionality of polyphenols and cynaropicrin from breads enriched with artichoke stem. *Food Chem.* **2018**, *245*, 838–844. [CrossRef]

116. Liu, T.; Zhang, J.; Han, X.; Xu, J.; Wu, Y.; Fang, J. Promotion of HeLa cells apoptosis by cynaropicrin involving inhibition of thioredoxin reductase and induction of oxidative stress. *Free Rad. Biol. Med.* **2019**, *135*, 216–226. [CrossRef]

117. Formisano, C.; Sirignano, C.; Rigano, D.; Chianese, G.; Zengin, G.; Seo, E.J.; Efferth, T.; Taglialatela-Scafati, O. Antiproliferative activity against leukemia cells of sesquiterpene lactones from the Turkish endemic plant *Centaurea drabifolia* subsp. *detonsa*. *Fitoterapia* **2017**, *120*, 98–102. [CrossRef] [PubMed]

118. Nawrot, J.; Budzianowski, J.; Nowak, G. Phytochemical profiles of the leaves of *Stizolophus balsamita* and *Psephellus sibiricus* and their chemotaxonomic implications. *Phytochemistry* **2019**, *159*, 172–178. [CrossRef] [PubMed]

119. Cis, J.; Nowak, G.; Kisiel, W. Antifeedant properties and chemotaxonomic implications of sesquiterpene lactones and syringin from *Rhaponticum pulchrum*. *Biochem. Syst. Ecol.* **2006**, *34*, 862–867. [CrossRef]

120. Schinor, E.C.; Salvador, M.J.; Ito, I.Y.; de Albuquerque, S.; Dias, D.A. Trypanocidal and antimicrobial activities of *Moquinia kingii*. *Phytomedicine* **2004**, *11*, 224–229. [CrossRef] [PubMed]

121. Choi, S.Z.; Choi, S.U.; Lee, K.R. Cytotoxic sesquiterpene lactones from *Saussurea calcicola*. *Arch. Pharmacol. Res.* **2005**, *28*, 1142–1146. [CrossRef]

122. Cho, J.Y.; Kim, A.R.; Jung, J.H.; Chun, T.; Rhee, M.H.; Yoo, E.S. Cytotoxic and pro-apoptotic activities of cynaropicrin, a sesquiterpene lactone, on the viability of leukocyte cancer cell lines. *Eur. J. Pharmacol.* **2004**, *492*, 85–94. [CrossRef]

123. Pandey, M.M.; Rastogi, S.; Rawat, A.K.S. *Saussurea costus*: Botanical, chemical and pharmacological review of an ayurvedic medicinal plant. *J. Ethnopharmacol.* **2007**, *110*, 379–390. [CrossRef]

124. Bhattacharyya, P.R.; Barua, N.C.; Ghosh, A.C. Cynaropicrin from *Tricholepis glaberrima*: A potential insect feeding deterrent compound. *Ind. Crops Prod.* **1995**, *4*, 291–294. [CrossRef]

125. Brás, T.; Neves, L.A.; Crespo, J.G.; Duarte, M.F. Effect of extraction methodologies and solvent selection upon cynaropicrin extraction from *Cynara cardunculus* leaves. *Sep. Purif. Technol.* **2019**, *236*, 116283. [CrossRef]

126. Zimmermann, S.; Kaiser, M.; Brun, R.; Hamburger, M.; Adams, M. Cynaropicrin: The first plant natural product with *in vivo* activity against *Trypanosoma brucei*. *Planta Med.* **2012**, *78*, 553–556. [CrossRef]

127. Zimmermann, S.; Oufir, M.; Leroux, A.; Krauth-Siegel, R.L.; Becker, K.; Kaiser, M.; Brun, R.; Hamburger, M.; Adams, M. Cynaropicrin targets the trypanothione redox system in *Trypanosoma brucei*. *Bioorg. Med. Chem.* **2013**, *21*, 7202–7209. [CrossRef] [PubMed]

128. Da Silva, C.F.; Batista, D.d.G.; De Araújo, J.S.; Batista, M.M.; Lionel, J.; de Souza, E.M.; Hammera, E.R.; Silva, P.B.; De Mieri, M.; Adams, M.; et al. Activities of psilostachyin A and cynaropicrin against *Trypanosoma cruzi in vitro* and *in vivo*. *Antimicrob. Agents Chemother.* **2013**, *57*, 5307–5314. [CrossRef] [PubMed]

129. Zimmermann, S.; Fouché, G.; De Mieri, M.; Yoshimoto, Y.; Usuki, T.; Nthambeleni, R.; Parkinson, C.J.; van der Westhuyzen, C.; Kaiser, M.; Hamburger, M.; et al. Structure-activity relationship study of sesquiterpene lactones and their semi-synthetic amino derivatives as potential antitrypanosomal products. *Molecules* **2014**, *19*, 3523–3538. [CrossRef] [PubMed]

130. Usuki, T.; Sato, M.; Hara, S.; Yoshimoto, Y.; Kondo, R.; Zimmermann, S.; Kaiser, M.; Brun, R.; Hamburger, M.; Adams, M. Antitrypanosomal structure-activity-relationship study of synthetic cynaropicrin derivatives. *Bioorg. Med. Chem. Lett.* **2014**, *24*, 794–798. [CrossRef]

131. Tanaka, Y.T.; Tanaka, K.; Kojima, H.; Hamada, T.; Masutani, T.; Tsuboi, M.; Akao, Y. Cynaropicrin from *Cynara scolymus* L. suppresses photoaging of skin by inhibiting the transcription activity of nuclear factor-kappa B. *Bioorg. Med. Chem. Lett.* **2013**, *23*, 518–523. [CrossRef]

132. Willuhn, G. Arnica flowers: Pharmacology, toxicolgy, and analysis of the sesquiterpene lactones—Their main active substances. *ACS Symp. Ser.* **1998**, *691*, 118–132.

133. Iannitti, T.; Morales-Medina, J.C.; Bellavite, P.; Rottigni, V.; Palmieri, B. Effectiveness and safety of *Arnica montana* in post-surgical setting, pain and inflammation. *Am. J. Ther.* **2016**, *23*, 184–197. [CrossRef]

134. Todorova, M.; Trendafilova, A.; Vitkova, A.; Petrova, M.; Zayova, E.; Antonova, D. Developmental and environmental effects on sesquiterpene lactones in cultivated *Arnica montana* L. *Chem. Biodiv.* **2016**, *13*, 976–981. [CrossRef]

135. Widen, J.C.; Kempema, A.M.; Villalta, P.W.; Harki, D.A. Targeting NF-κB p65 with a helenalin inspired bis-electrophile. *Chem. Biol.* **2017**, *12*, 102–113. [CrossRef]

136. Li, Y.; Zeng, Y.; Huang, Q.; Wen, S.; Wei, Y.; Chen, Y.; Zhang, X.; Bai, F.; Lu, Z.; Wei, J.; et al. Helenalin from *Centipeda minima* ameliorates acute hepatic injury by protecting mitochondria function, activating Nrf2 pathway and inhibiting NF-κB activation. *Biomed. Pharmacother.* **2019**, *119*, 109435. [CrossRef]

137. Kriplani, P.; Guarve, K.; Baghael, U.S. *Arnica montana* L.—A plant of healing: Review. *J. Pharm. Pharmacol.* **2017**, *69*, 925–945. [CrossRef] [PubMed]

138. Chadwick, M.; Trewin, H.; Gawthrop, F.; Wagstaff, C. Sesquiterpenoids lactones: Benefits to plants and people. *Int. J. Mol. Sci.* **2013**, *14*, 12780–12805. [CrossRef] [PubMed]

139. Huang, P.R.; Yeh, Y.M.; Wang, T.C. Potent inhibition of human telomerase by helenalin. *Cancer Lett.* **2005**, *227*, 169–174. [CrossRef] [PubMed]

140. Berges, C.; Fuchs, D.; Opelz, G.; Daniel, V.; Naujokat, C. Helenalin suppresses essential immune functions of activated CD4+ T cells by multiple mechanisms *Mol. Immunol.* **2009**, *46*, 2892–2901. [CrossRef]

141. Lyss, G.; Knorre, A.; Schmidt, T.J.; Pahl, H.L.; Merfort, I. The anti-inflammatory sesquiterpene lactone helenalin inhibits the transcription factor NF-κB by directly targeting p65. *J. Biol. Chem.* **1998**, *273*, 33508–33516. [CrossRef]

142. Zwicker, P.; Schultze, N.; Niehs, S.; Albrecht, D.; Methling, K.; Wurster, M.; Wachlin, G.; Lalk, M.; Lindequist, U.; Haertel, B. Differential effects of helenalin, an anti-inflammatory sesquiterpene lactone, on the proteome, metabolome and the oxidative stress response in several immune cell types. *Toxicol. In Vitro* **2017**, *40*, 45–54. [CrossRef]

143. Büchele, B.; Zugmaier, W.; Lunov, O.; Syrovets, T.; Merfort, I.; Simmet, T. Surface plasmon resonance analysis of nuclear factor-κB protein interactions with the sesquiterpene lactone helenalin. *Anal. Biochem.* **2010**, *401*, 30–37. [CrossRef]

144. Tornhamre, S.; Schmidt, T.J.; Nasman-Glaser, B. Inhibitory effects of helenalin and related compounds on 5-lipoxygenase and leukotriene C(4) synthase in human blood cells. *Biochem Pharmacol.* **2001**, *62*, 903–911. [CrossRef]

145. Schröder, H.; Lösche, W.; Strobach, H.; Leven, W.; Willuhn, G.; Till, U.; Schrör, K. Helenalin and 11α,13-dihydrohelenalin, two constituents from *Arnica montana* L., inhibit human platelet function via thiol-dependent pathway. *Thromb. Res.* **1990**, *57*, 839–845. [CrossRef]

146. Macêdo, S.B.; Ferreira, L.R.; Perazzo, F.F.; Tavares Carvalho, J.C. Anti-inflammatory activity of *Arnica montana* 6cH: Preclinical study in animals. *Homeopathy* **2004**, *93*, 84–87. [CrossRef]

147. Castro, F.C.; Magre, A.; Cherpinski, R.; Zelante, P.M.; Neves, L.M.; Esquisatto, M.A.; Mendonça, F.A.; Santos, G.M. Effects of microcurrent application alone or in combination with topical *Hypericum perforatum* L. and *Arnica montana* L. on surgically induced wound healing in Wistar rats. *Homeopathy* **2012**, *101*, 147–153. [CrossRef] [PubMed]

148. Widrig, R.; Suter, A.; Saller, R.; Melzer, J. Choosing between NSAID and Arnica for topical treatment of hand osteoarthritis in a randomised, double-blind study. *Rheumatol. Int.* **2007**, *27*, 585–591. [CrossRef] [PubMed]

149. Boulanger, D.; Brouillette, E.; Jaspar, F.; Malouin, F.; Mainil, J.; Bureau, F.; Lekeux, P. Helenalin reduces *Staphylococcus aureus* infection *in vitro* and *in vivo*. *Vet. Microbiol.* **2007**, *119*, 330–338. [CrossRef] [PubMed]

150. Valan, M.F.; Britto, J.B.; Venkataraman, R. Phytoconstituents with hepatoprotective activity. *Int. J. Chem. Sci.* **2010**, *8*, 1421–1432.

151. Usui, K.; Ikeda, T.; Horibe, Y.; Nakao, M.; Hoshino, T.; Mizushima, T. Identification of HSP70- inducing activity in *Arnica montana* extract and purification and characterization of HSP70-inducers. *J. Dermatol. Sci.* **2015**, *78*, 67–75. [CrossRef]

152. Lin, X.; Huang, Q.; Bai, F.; Wei, J.; Huang, R.; Wen, S.; Wei, Y.; Zhang, X. Preparation Method of Traditional Chinese Medicine Centipeda minima Sesquiterpene Monomer Helenalin, and Its Application in Preparing Drug for Treating Hepatic Fibrosis and. Inflammation. Patent No. CN 110283151, 27 September 2019.

153. Lin, X.; Huang, Q.; Bai, F.; Wei, J.; Huang, R.; Wen, S.; Wei, Y.; Zhang, X. Application of Helenalin for Inhibiting Hepatic Stellate Cell Activation from Centipeda. minima. Patent No. CN 110179790, 30 August 2019.

154. Messaoudi, M.; Chahmi, N.; El-Mzibri, M.; Gmouh, S.; Amzazi, S.; Benbacer, L.; El-Hassouni, M. Cytotoxic effect and chemical composition of *Inula viscosa* from three different regions of Morocco. *Eur. J. Med. Plants* **2016**, *16*, 1–9. [CrossRef]

155. Hernández, V.; Recio, M.D.C.; Máñz, S.; Prieto, J.M.; Giner, R.M.; Ríos, J.L. A mechanistic approach to the *in vivo* anti-inflammatory activity of sesquiterpenoid compounds isolated from *Inula viscosa*. *Planta Med.* **2001**, *67*, 726–731. [CrossRef]

156. Dolejš, L.; Souček, M.; Horák, M.; Herout, V.; Šorm, F. On terpenes. XCIV. The structure of lactucin. *Collect. Czech. Chem. Commun.* **1958**, *23*, 2195–2200. [CrossRef]

157. Ruban, G.; Zabel, V.; Gensch, K.H.; Smalla, H. The crystal structure and absolute configuration of lactucin. *Acta Cryst.* **1978**, *34*, 1163–1167. [CrossRef]

158. Besharat, S.; Besharat, M.; Jabbari, A. Wild lettuce (*Lactuca virosa*) toxicity. *BMJ Case Rep.* **2009**, *2009*, 10–13. [CrossRef]

159. De Kraker, J.W.; Franssen, M.C.R.; Dalm, M.C.F.; De Groot, A.; Bouwmeester, H.J. Biosynthesis of germacrene a carboxylic acid in chicory roots. Demonstration of a cytochrome p450 (+)-germacrene a hydroxylase and NADP$^+$-dependent sesquiterpenoid dehydrogenase(s) involved in sesquiterpene lactone biosynthesis. *Plant Physiol.* **2001**, *125*, 1930–1940. [CrossRef] [PubMed]

160. Sessa, R.A.; Bennett, M.H.; Lewis, M.J.; Mansfield, J.W.; Beale, M.H. Metabolite profiling of sesquiterpene lactones from *Lactuca* Species. *J. Biol. Chem.* **2000**, *275*, 26877–26884. [CrossRef] [PubMed]

161. Testone, G.; Mele, G.; di Giacomo, E.; Tenore, G.C.; Gonnella, M.; Nicolodi, C.; Frugis, G.; Iannelli, M.A.; Arnesi, G.; Schiappa, A.; et al. Transcriptome driven characterization of curly- and smooth-leafed endives reveals molecular differences in the sesquiterpenoid pathway. *Hortic. Res.* **2019**, *6*, 1–19. [CrossRef] [PubMed]

162. Seo, M.W.; Yang, D.S.; Kays, S.J.; Lee, G.P.; Park, K.W. Sesquiterpene lactones and bitterness in korean leaf lettuce cultivars. *HortScience* **2009**, *44*, 246–249. [CrossRef]

163. Kim, H.D.; Hong, K.B.; Noh, D.O.; Suh, H.J. Sleep-inducing effect of lettuce (*Lactuca sativa*) varieties on pentobarbital-induced sleep. *Food Sci. Biotechnol.* **2017**, *26*, 807–814. [CrossRef]

164. Bahmani, M.; Shahinfard, N.; Rafieian-Kopaei, M.; Saki, K.; Shahsavari, S.; Taherikalani, M.; Ghafourian, S.; Baharvand-Ahmadi, B. Chicory: A review on ethnobotanical effects of *Cichorium intybus* L. *J. Chem. Pharm. Sci.* **2015**, *8*, 672–682.

165. Ludwig, H. Ueber die Bestandtheile des Lactucariums. *Arch. Pharm. Pharm. Med. Chem.* **1847**, *100*, 1–19. [CrossRef]

166. Christison, R. *A Dispensatory, or Commentary on the Pharmacopoeias of Great Britain*; Adam and Charles Black Ed.: Edinburgh, UK, 1842.

167. Thomson, A.T. *A Conspectus of the Pharmacopoeias of the London, Edinburgh and Dublin Collegues of Physicians and of the United States Pharmacopoeia Being a Practical Compendium of Materia Medica and Pharmacy*; Lee, C.A., Ed.; Henry, G. Langley, 8 Astor House: New York, NY, USA, 1844.

168. Wesołowska, A.; Nikiforuk, A.; Michalska, K.; Kisiel, W.; Chojnacka-Wójcik, E. Analgesic and sedative activities of lactucin and some lactucin-like guaianolides in mice. *J. Ethnopharmacol.* **2006**, *107*, 254–258. [CrossRef]

169. Yakoot, M.; Helmy; Fawal, K. Pilot study of the efficacy and safety of lettuce seed oil in patients with sleep disorders. *Int. J. Gen. Med.* **2011**, *4*, 451–456. [CrossRef]

170. Schmidt, B.; Ilic, N.; Poulev, A.; Raskin, I. Toxicological evaluation of chicory extract (humans). *Food Chem. Toxicol.* **2007**, *45*, 1–18. [CrossRef]

171. Ghantous, A.; Sinjab, A.; Herceg, Z.; Darwiche, N. Parthenolide: From plant shoots to cancer roots. *Drug Discov. Today* **2013**, *18*, 894–905. [CrossRef] [PubMed]

172. Kreuger, M.R.O.; Grootjans, S.; Biavatti, M.; Vandanabeele, P.; Dherde, K. Sesquiterpene lactones as drugs with multiple targets in cancer treatment: Focus on parthenolide. *Anticancer Drugs* **2011**, *23*, 883–896. [CrossRef]

173. Jafari, N.; Nazeri, S.; Enferadi, S.T. Parthenolide reduces metastasis by inhibition of vimentin expression and induces apoptosis by suppression elongation factor α–1 expression. *Phytomedicine* **2018**, *41*, 67–73. [CrossRef] [PubMed]

174. Freund, R.R.A.; Gobrecht, P.; Fischer, D.; Arndt, H.-D. Advances in chemistry and bioactivity of parthenolide. *Nat. Prod. Rep* **2020**. [CrossRef] [PubMed]

175. Seca, A.M.L.; Silva, A.M.S.; Pinto, D.C.G.A. Parthenolide and parthenolide-like sesquiterpene lactones as multiple targets drugs: Current knowledges and new developments. In *Studies in Natural Products Chemistry*; Atta-Ur-Rahman, Ed.; Elsevier Science Publishers: Amsterdam, The Netherlands, 2017; Volume 52, pp. 337–372. [CrossRef]

176. Ren, Y.; Yu, J.; Kinghorn, A.D. Development of anticancer agents from plant-derived sesquiterpene lactones. *Curr. Med. Chem.* **2016**, *23*, 2397–2420. [CrossRef]

177. Babaei, G.; Aliarab, A.; Abroon, S.; Rasmi, Y.; Aziz, S.G. Application of sesquiterpene lactone: A new promising way for cancer therapy based on anticancer activity. *Biomed. Pharmacother.* **2018**, *106*, 239–246. [CrossRef]

178. Dey, S.; Sarkar, M.; Giri, B. Anti-inflammatory and anti-tumor activities of parthenolide: An update. *Chem. Biol. Ther.* **2016**, *1*, 107. [CrossRef]

179. Saadane, A.; Eastman, J.; Berger, M.; Bonfield, T.M. Parthenolide inhibits ERK and AP-1 which are dysregulated and contribute to excessive IL-8 expression and secretion in cystic fibrosis cells. *J. Inflamm.* **2011**, *8*, 26. [CrossRef]

180. Carlisi, D.; D'Anneo, A.; Angileri, L.; Lauricella, M.; Emanuele, S.; Santulli, A.; Vento, R.; Tesoriere, G. Parthenolide sensitizes hepatocellular carcinoma cells to TRAIL by inducing the expression of death receptors through inhibition of STAT3 activation. *J. Cell Physiol.* **2011**, *226*, 1632–1641. [CrossRef]

181. Nakshatri, P.; Rice, S.E.; Bhat-Nakshatri, P. Antitumor agent parthenolide reverses resistance of breast cancer cells to tumor necrosis factor-related apoptosis-inducing ligand through sustained activation of c-Jun N-terminal kinase. *Oncogene* **2004**, *23*, 7330–7344. [CrossRef]

182. Rüngeler, P.; Castro, V.; Mora, G.; Gören, N.; Vichnewski, W.; Pah, H.L.; Merfort, I.; Schmidt, T.J. Inhibition of transcription factor NFkappaB by sesquiterpene lactones: A proposed molecular mechanism of action. *Bioorg. Med. Chem.* **1999**, *7*, 2343–2352. [CrossRef]

183. Steele, A.J.; Jones, D.T.; Ganeshaguru, K.; Duke, V.; Yogashangary, B.C.; North, J.M.; Lowdell, M.W.; Kottaridis, P.D.; Mehta, A.; Prentice, A.G.; et al. The sesquiterpene lactone parthenolide induces selective apoptosis of B-chronic lymphocytic leukemia cells *in vitro*. *Leukemia* **2006**, *20*, 1073–1079. [CrossRef] [PubMed]

184. D'Anneo, A.; Carlisi, D.; Lauricella, M.; Puleio, R.; Martinez, R.; Di Bella, S.; Di Marco, P.; Emanuele, S.; Di Fiore, R.; Guercio, A.; et al. Parthenolide generates reactive oxygen species and autophagy inMDA-MB231 cells. A soluble parthenolide analogue inhibits tumour growth and metastasis in a xenograft model of breast cancer. *Cell Death Dis.* **2013**, *4*, 89. [CrossRef]

185. Guzman, M.L.; Rossi, R.M.; Karnischky, L.; Li, X.; Peterson, D.R.; Howard, D.S.; Jordan, C.T. The sesquiterpene lactone parthenolide induces apoptosis of human acute myelogenous leukemia stem and progenitor cells. *Blood* **2005**, *105*, 4163–4169. [CrossRef] [PubMed]

186. Zunino, S.J.; Ducore, J.M.; Storms, D.H. Parthenolide induces significant apoptosis and production of reactive oxygen species in high-risk pre-B leukemia cells. *Cancer Lett.* **2007**, *254*, 119–127. [CrossRef] [PubMed]

187. Gopal, Y.N.; Arora, T.S.; Van Dyke, M.W. Parthenolide specifically depletes histone deacetylase 1 protein and induces cell death through ataxia telangiectasia mutated. *Chem. Biol.* **2007**, *14*, 813–823. [CrossRef]

188. Liu, Z.; Liu, S.; Xie, Z.; Pavlovicz, R.E.; Wu, J.; Chen, P.; Aimiuwu, J.; Pang, J.; Bhasin, D.; Neviani, P.; et al. Modulation of DNA methylation by a sesquiterpene lactone parthenolide. *J. Pharmacol. Exp. Ther.* **2009**, *329*, 505–514. [CrossRef]

189. Fonrose, X.; Ausseil, F.; Soleilhac, E.; Masson, V.; David, B.; Pouny, I.; Cintrat, J.C.; Rousseau, B.; Barette, C.; Massiot, G.; et al. Parthenolide inhibits tubulin carboxypeptidase activity. *Cancer Res.* **2007**, *67*, 3371–3378. [CrossRef]

190. Mathema, V.B.; Koh, Y.S.; Thakuri, B.C.; Sillanpa, M. Parthenolide, a sesquiterpene lactone, expresses multiple anti-cancer and anti-inflammatory activities. *Inflammation* **2012**, *35*, 560–565. [CrossRef]

191. Pei, S.; Jordan, C.T. How close are we to targeting the leukemia stem cell? *Best Pract. Res. Clin. Haematol.* **2012**, *25*, 415–418. [CrossRef]

192. Gach, K.; Długosz, A.; Janecka, A. The role of oxidative stress in anticancer activity of sesquiterpene lactones. *Naunyn-Schmiedeberg's Arch. Pharmacol.* **2015**, *388*, 477–486. [CrossRef] [PubMed]

193. Kim, J.H.; Liu, L.; Lee, S.O.; Kim, Y.T.; You, K.R.; Kim, D.G. Susceptibility of cholangiocarcinoma cells to parthenolide-induced apoptosis. *Cancer Res.* **2005**, *65*, 6312–6320. [CrossRef] [PubMed]

194. Kim, S.L.; Trang, K.T.; Kim, A.H.; Kim, I.H.; Lee, S.O.; Lee, S.T.; Kim, D.G.; Kim, S.W. Parthenolide suppresses tumor growth in a xenograft model of colorectal cancer cells by inducing mitochondrial dysfunction and apoptosis. *Int. J. Oncol.* **2012**, *41*, 1547–1553. [CrossRef] [PubMed]

195. Yun, B.R.; Lee, M.J.; Kim, J.H.; Kim, I.H.; Yu, G.R.; Kim, D.G. Enhancement of parthenolide-induced apoptosis by a PKC-alpha inhibition through heme oxygenase-1 blockage in cholangiocarcinoma cells. *Exp. Mol. Med.* **2010**, *42*, 787–797. [CrossRef] [PubMed]

196. Kim, S.L.; Liu, Y.C.; Seo, S.Y.; Kim, S.H.; Kim, I.H.; Lee, S.O.; Lee, S.T.; Kim, D.; Kim, S.W. Parthenolide induces apoptosis in colitis-associated colon cancer, inhibiting NF-κB signaling. *Oncol. Lett.* **2015**, *9*, 2315–2342. [CrossRef] [PubMed]

197. Baskaran, N.; Selvam, G.S.; Yuvaraj, S.; Abhishek, A. Parthenolide attenuates 7,12-dimethylbenz[a]anthracene induced hamster buccal pouch carcinogenesis. *Mol. Cell. Biochem.* **2018**, *440*, 11–22. [CrossRef] [PubMed]

198. Nakabayashi, H.; Shimizu, K. Involvement of Akt/NF-κB pathway in antitumor effects of parthenolide on glioblastoma cells *in vitro* and *in vivo*. *BMC Cancer.* **2012**, *12*, 453. [CrossRef]

199. Zhang, M.; Liu, R.T.; Zhang, P.; Zhang, N.; Yang, C.L.; Yue, L.T.; Li, X.L.; Liu, Y.; Li, H.; Du, J.; et al. Parthenolide inhibits the initiation of experimental autoimmune neuritis. *J. Neuroimmunol.* **2017**, *15*, 154–161. [CrossRef]

200. Fiebich, B.L.; Lieb, K.; Engels, S.; Heinrich, M. Inhibition of LPS-induced p42/44 MAP kinase activation and iNOS/NO synthesis by parthenolide in rat primary microglial cells. *J. Neuroimmunol.* **2002**, *132*, 18–24. [CrossRef]

201. Gobrecht, P.; Andreadaki, A.; Diekmann, H.; Heskamp, A.; Leibinger, M.; Fischer, D. Promotion of functional nerve regeneration by inhibition of microtubule detyrosination. *J. Neurosci.* **2016**, *36*, 3890–3902. [CrossRef]

202. Diekmann, H.; Fischer, D. Parthenolide: A novel pharmacological approach to promote nerve regeneration. *Neural. Regen. Res.* **2016**, *11*, 1566–1567. [CrossRef] [PubMed]

203. Li, X.H.; Xiao, T.; Yang, J.H.; Qin, Y.; Gao, J.J.; Liu, H.J.; Zhou, H.G. Parthenolide attenuated bleomycin-induced pulmonary fibrosis via the NF-κB/Snail signaling pathway. *Respir. Res.* **2018**, *19*, 111. [CrossRef] [PubMed]

204. Van der Heiden, K.; Cuhlmann, S.; Luong, A.; Zakkar, M.; Evans, P.C. Role of nuclear factor kappaB in cardiovascular health and disease. *Clin. Sci. (Lond.).* **2010**, *23*, 593–605. [CrossRef] [PubMed]

205. Latanich, C.A.; Toledo-Pereyra, L.H. Searching for NF-kappaB-based treatments of ischemia reperfusion injury. *J. Investig. Surg.* **2009**, *22*, 301–315. [CrossRef] [PubMed]

206. Zingarelli, B.; Hake, P.W.; Denenberg, A.; Wong, H.R. Sesquiterpene lactone parthenolide, an inhibitor of IkappaB kinase complex and nuclear factor-kappaB, exerts beneficial effects in myocardial reperfusion injury. *Shock* **2002**, *17*, 127–134. [CrossRef]

207. Tsai, T.Y.; Lou, S.L.; Cheng, K.S.; Wong, K.L.; Wang, M.L.; Su, T.H.; Chan, P.; Leu, Y.M. Repressed Ca^{2+} clearance in parthenolide-treated murine brain bEND.3 endothelial cells. *Eur. J. Pharmacol.* **2015**, *769*, 280–286. [CrossRef]

208. Nasim, S.; Crooks, P.A. Antileukemic activity of aminoparthenolide analogs. *Bioorg. Med. Chem. Lett.* **2008**, *18*, 3870–3873. [CrossRef]

209. Neelakantan, S.; Nasim, S.; Guzman, M.L.; Jordan, C.T.; Crooks, P.A. Aminoparthenolides as novel anti-leukemic agents: Discovery of the NF-kappaB inhibitor, DMAPT (LC-1). *Bioorg. Med. Chem. Lett.* **2009**, *19*, 4346–4349. [CrossRef]

210. Guzman, M.L.; Rossi, R.M.; Neelakantan, S.; Li, X.; Corbett, C.A.; Hassane, D.C.; Becker, M.W.; Bennett, J.M.; Sullivan, E.; Lachowicz, J.L.; et al. An orally bioavailable parthenolide analog selectively eradicates acute myelogenous leukemia stem and progenitor cells. *Blood* **2007**, *110*, 4427–4431. [CrossRef]

211. Carlisi, D.G.; Di Fiore, B.R.; Scerri, C.; Drago-Ferrante, R.; Tesoriere, V.G. Parthenolide and DMAPT exert cytotoxic effects on breast cancer stem-like cells by inducing oxidative stress, mitochondrial dysfunction and necrosis. *Cell Death Dis.* **2016**, *7*, 2194. [CrossRef]

212. Song, J.M.; Qian, X.; Upadhyayya, P.; Hong, K.H.M.; Kassie, F. Dimethylaminoparthenolide, a water soluble parthenolide, suppresses lung tumorigenesis through down-regulating the STAT3 signaling pathway. *Curr. Cancer Drug Targets* **2014**, *14*, 59–69. [CrossRef] [PubMed]

213. Nakshatri, H.; Appaiah, H.N.; Anjanappa, M.; Gilley, D.; Tanaka, H.; Badve, S.; Crooks, P.A.; Mathews, W.; Sweeney, C.; Bhat-Nakshatri, P. NF-κB-dependent and -independent epigenetic modulation using the novel anti-cancer agent DMAPT. *Cell Death Dis.* **2015**, *22*, 1608. [CrossRef] [PubMed]

214. Adis Insight. Available online: http://adisinsight.springer.com/drugs/800029612 (accessed on 15 January 2020).

215. Morel, K.L.; Ormsby, R.J.; Klebe, S.; Sweeney, C.J.; Sykes, P.J. Parthenolide selectively sensitizes prostate tumor tissue to radiotherapy while protecting healthy tissues *in vivo*. *Radiat. Res.* **2017**, *187*, 501–512. [CrossRef] [PubMed]

216. Morel, K.L.; Ormsby, R.J.; Klebe, S.; Sweeney, C.J.; Sykes, P.J. DMAPT is an effective radioprotector from long-term radiation-induced damage to normal mouse tissues *in vivo*. *Radiat. Res.* **2019**, *192*, 231–239. [CrossRef]

217. Mendonca, M.; Turchan, W.; Alpucha, M.; Watson, C.N.; Estabrook, N.; Chin-Sinex, H.; Shapiro, J.B.; Imasuen-Williams, I.E.; Rangela, G.; Gilley, D.; et al. DMAPT inhibits NF-κB activity and increases sensitivity of prostate cancer cells to X-rays *in vitro* and in tumor xenografts *in vivo*. *Free Radic. Biol. Med.* **2017**, *112*, 318–326. [CrossRef]

218. Li, X.; Payne, D.T.; Ampolu, B.; Bland, N.; Brown, J.T.; Dutton, M.J.; Fitton, C.A.; Gulliver, A.; Hale, L.; Hamza, D.; et al. Derivatisation of parthenolide to address chemoresistant chronic lymphocytic leukaemia. *Med. Chem. Commun.* **2019**, *10*, 1379. [CrossRef]

219. Baranello, M.P.; Bauer, L.; Danielle, S.W.; Benoit, D. Poly (styrene-alt-maleic anhydride)-based diblock copolymer micelles exhibit versatile hydrophobic drug loading, drug-dependent release, and internalization by multidrug resistant ovarian cancer cells. *Biomacromolecules* **2014**, *15*, 2629–2641. [CrossRef]

220. Karmakar, A.; Xu, Y.; Mustafa, T.; Kannarpady, G.; Bratton, S.M.; Radominska-Pandya, A.; Crooks, P.A.; Biris, A.S. Nanodelivery of parthenolide using functionalized nanographene enhances its anticancer activity. *RSC Adv.* **2015**, *5*, 2411. [CrossRef]

221. Jin, X.; Zhou, J.; Zhang, Z.; Lv, H. The combined administration of parthenolide and ginsenoside CK in long circulation liposomes with targeted tLyp-1 ligand induce mitochondria-mediated lung cancer apoptosis. artificial cells. *Artif. Cells Nanomed. Biotechnol.* **2018**, *46*, S931–S942. [CrossRef]

222. Darwish, N.H.E.; Sudha, T.; Godugu, K.; Bharali, D.J.; Elbaz, O.; El-Ghaffar, H.A.A.; Azmy, E.; Anber, N.; Mousa, S.A. Novel targeted nano-parthenolide molecule against NF-kB in acute myeloid leukemia. *Molecules* **2019**, *24*, 2103. [CrossRef]

223. Kim, S.L.; Liu, Y.C.; Park, Y.R.; Seo, S.Y.; Kim, S.H.; Kim, I.H.; Lee, S.O.; Lee, S.; Kim, D.G.; Kim, S.W. Parthenolide enhances sensitivity of colorectal cancer cells to TRAIL by inducing death receptor 5 and promotes TRAIL-induced apoptosis. *Int. J. Oncol.* **2015**, *46*, 1121–1130. [CrossRef] [PubMed]

224. Holcomb, B.K.; Yip-Schneider, M.T.; Waters, J.A.; Beane, J.D.; Crooks, P.A.; Schmidt, C.M. Dimethylamino parthenolide enhances the inhibitory effects of gemcitabine in human pancreatic cancer cells. *J. Gastrointest. Surg.* **2012**, *16*, 1333–1340. [CrossRef] [PubMed]

225. Yip-Schneider, M.T.; Wu, H.; Njoku, V.; Ralstin, M.; Holcomb, B.; Crooks, P.A.; Neelakantan, S.; Sweeney, C.J.; Schmidt, C.M. Effect of celecoxib and the novel anti-cancer agent, dimethylamino-parthenolide, in a developmental model of pancreatic cancer. *Pancreas* **2008**, *37*, 45–53. [CrossRef] [PubMed]

226. Yip-Schneider, M.T.; Wu, H.; Hruban, R.H.; Lowy, A.M.; Crooks, P.A.; Schmidt, C.M. Efficacy of dimethylaminoparthenolide and sulindac in combination with gemcitabine in a genetically engineered mouse model of pancreatic cancer. *Pancreas* **2013**, *42*, 160–167. [CrossRef]

227. Yip-Schneider, M.T.; Wu, H.; Stantz, K.; Agaram, N.; Crooks, P.A.; Schmidt, C.M. Dimethylaminoparthenolide and gemcitabine: A survival study using a genetically engineered mouse model of pancreatic cancer. *BMC Cancer* **2013**, *13*, 194. [CrossRef]

228. Carlisi, D.; Lauricella, M.; D'Anneo, A.; Buttitta, G.; Emanuele, S.; di Fiore, R.; Martinez, R.; Rolfo, C.; Vento, R.; Tesoriere, G. The synergistic effect of SAHA and parthenolide in MDA-MB231 breast cancer cells. *J. Cell Physiol.* **2015**, *230*, 1276–1289. [CrossRef]

229. Lamture, G.; Crooks, P.A.; Borrelli, M.J. Actinomycin-D and dimethylamino-parthenolide synergism in treating human pancreatic cancer cells. *Drug Dev. Res.* **2018**, *79*, 287–294. [CrossRef]

230. Rasmussen, U.; Broogger, C.S.; Sandberg, F. Thapsigargine and thapsigargicine, two new histamine liberators from *Thapsia garganica* L. *Acta Pharm. Suec.* **1978**, *15*, 133–140.

231. Andersen, T.B.; López, C.Q.; Manczak, T.; Martinez, K.; Simonsen, H.T. Thapsigargin - from *Thapsia* L. to mipsagargin. *Molecules* **2015**, *20*, 6113–6127. [CrossRef]

232. Doan, N.T.Q.; Paulsen, E.S.; Sehgal, P.; Møller, J.V.; Nissen, P.; Denmeade, S.R.; Isaacs, J.T.; Dionne, C.A.; Christensen, S.B. Targeting thapsigargin towards tumors. *Steroids* **2015**, *97*, 2–7. [CrossRef]

233. Mohamed Ibrahim, A.M.; Martinez-Swatson, K.A.; Benkaci-Ali, F.; Cozzi, F.; Zoulikha, F.; Simonsen, H.T. Effects of gamma irradiation and comparison of different extraction methods on sesquiterpene lactone yields from the medicinal plant *Thapsia garganica* L. (Apiaceae). *J. Appl. Res. Med. Aromat. Plants* **2018**, *8*, 26–32. [CrossRef]

234. Crestey, F.; Toma, M.; Christensen, S.B. Concise synthesis of thapsigargin from nortrilobolide. *Tetrahedron Lett.* **2015**, *56*, 5896–5898. [CrossRef]

235. Chen, D.; Evans, P.A. A concise, efficient and scalable total synthesis of thapsigargin and nortrilobolide from (R)-(-)-carvone. *J. Am. Chem. Soc.* **2017**, *139*, 6046–6049. [CrossRef] [PubMed]

236. Denmeade, S.R.; Isaacs, J.T. The SERCA pump as a therapeutic target: Making a "smart bomb" for prostate cancer. *Cancer Biol. Ther.* **2005**, *4*, 14–22. [CrossRef] [PubMed]

237. Wei, Y.; Meng, M.; Tian, Z.; Xie, F.; Yin, Q.; Dai, C.; Wang, J.; Zhang, Q.; Liu, Y.; Liu, C. Pharmacological preconditioning with the cellular stress inducer thapsigargin protects against experimental sepsis. *Pharmacol. Res.* **2019**, *141*, 114–122. [CrossRef] [PubMed]

238. Michelangeli, F.; East, J.M. A diversity of SERCA Ca^{2+} pump inhibitors. *Biochem. Soc. Transact.* **2011**, *39*, 789–797. [CrossRef]

239. Wootton, L.L.; Michelangeli, F. The effects of the phenylalanine to valine mutation on the sensitivity of sarcoplasmic/endoplasmic reticulum Ca^{2+} ATPase (SERCA) Ca^{2+} pump isoforms 1, 2, and 3 to thapsigargin and other inhibitors. *J. Biol. Chem.* **2006**, *281*, 6970–6976. [CrossRef]

240. Skytte, D.M.; Moller, J.V.; Liu, H.; Nielsen, H.O.; Svenningsen, L.E.; Jensen, C.M.; Olsen, C.E.; Christensen, S.B. Elucidation of the topography of the thapsigargin binding site in the sarco-endoplasmic calcium ATPase. *Bioorg. Med. Chem.* **2010**, *18*, 5634–5646. [CrossRef]

241. Shiraishi, M.; Hirasawa, N.; Kobayashi, Y.; Oikawa, S.; Murakami, A.; Ohuchi, K. Participation of mitogen-activated protein kinase in thapsigargin- and TPA-induced histamine production in murine macrophage RAW 264.7 cells. *Brit. J. Pharmacol.* **2000**, *129*, 515–524. [CrossRef]

242. Muramatsu, Y.; Maemoto, T.; Iwashita, A.; Matsuoka, N. Novel neuroprotective compound SCH-20148 rescues thymocytes and SH-SY5Y cells from thapsigargin-induced mitochondrial membrane potential reduction and cell death. *Eur. J. Pharmacol.* **2007**, *563*, 40–48. [CrossRef]

243. Wang, C.; Li, T.; Tang, S.; Zhao, D.; Zhang, C.; Zhang, S.; Deng, S.; Zhou, Y.; Xiao, X. Thapsigargin induces apoptosis when autophagy is inhibited in HepG2 cells and both processes are regulated by ROS-dependent pathway. *Environ. Toxicol. Pharmacol.* **2016**, *41*, 167–179. [CrossRef] [PubMed]

244. Silva, Z.; Verissimo, T.; Videira, P.A.; Novo, C. Protein disulfide isomerases: Impact of thapsigargin treatment on their expression in melanoma cell lines. *Int. J. Biol. Macromol.* **2015**, *79*, 44–48. [CrossRef] [PubMed]

245. Meldrum, D.R.; Cleveland, J.C.J.; Mitchell, M.B.; Rowland, R.T.; Banerjee, A.; Harken, A.H. Constractive priming of myocardium against ishemia-reperfusion injury. *Shock* **1996**, *6*, 238–242. [CrossRef] [PubMed]

246. Karunakaran, U.; Lee, J.E.; Elumalai, S.; Moon, J.S.; Won, K.C. Myricetin prevents thapsigargin-induced CDK5-P66Shc signalosome mediated pancreatic β-cell dysfunction. *Free Rad. Biol. Med.* **2019**, *141*, 59–66. [CrossRef] [PubMed]

247. Janyou, A.; Changtam, C.; Suksamrarn, A.; Tocharus, C.; Tocharus, J. Suppression effects of O-demethyldemethoxycurcumin on thapsigargin triggered on endoplasmic reticulum stress in SK-N-SH cells. *Neurotoxicology* **2015**, *50*, 92–100. [CrossRef]

248. Einbond, L.S.; Wu, H.A.; Sandu, C.; Ford, M.; Mighty, J.; Antonetti, V.; Redenti, S.; Ma, H. Digitoxin enhances the growth inhibitory effects of thapsigargin and simvastatin on ER negative human breast cancer cells. *Fitoterapia* **2016**, *109*, 146–154. [CrossRef]

249. Denmeade, S.R.; Jakobsen, C.M.; Janssen, S.; Khan, S.R.; Garrett, E.S.; Lilja, H.; Christensen, S.B.; Isaacs, J.T. Prostate-specific antigen-activated thapsigargin prodrug as targeted therapy for prostate cancer. *J. Nat. Cancer Inst.* **2003**, *95*, 990–1000. [CrossRef]

250. Zhong, W.; Chebolu, S.; Darmani, N.A. Thapsigargin-induced activation of Ca^{2+}-CaMKII-ERK in brainstem contributes to substance P release and induction of emesis in the least shrew. *Neuropharmacology* **2016**, *103*, 195–210. [CrossRef]

251. Kmonickova, E.; Harmatha, J.; Vokac, K.; Kostecka, P.; Farghali, H.; Zidek, Z. Sesquiterpene lactone trilobolide activates production of interferon-gamma and nitric oxide. *Fitoterapia* **2010**, *81*, 1213–1219. [CrossRef]

252. Mahalingam, D.; Wilding, G.; Denmeade, S.; Sarantopoulas, J.; Cosgrove, D.; Cetnar, J.; Azad, N.; Bruce, J.; Kurman, M.; Allgood, V.E.; et al. Mipsagargin, a novel thapsigargin-based PSMA-activated prodrug: Results of a first-in-man phase I clinical trial in patients with refractory, advanced or metastatic solid tumours. *Br. J. Cancer* **2016**, *114*, 986–994. [CrossRef]

253. Mahalingam, D.; Peguero, J.; Cen, P.; Arora, S.P.; Sarantopoulos, J.; Rowe, J.; Allgood, V.; Tubb, B.; Campos, L. A phase II, multicenter, single-arm study of mipsagargin (G-202) as a second-line therapy following sorafenib for adult patients with progressive advanced hepatocellular carcinoma. *Cancers* **2019**, *11*, 833. [CrossRef] [PubMed]

254. Park, H.H.; Kim, S.G.; Kim, M.J.; Lee, J.; Choi, B.K.; Jin, M.H.; Lee, E. Suppressive effect of tomentosin on the production of inflammatory mediators in RAW264.7 cells. *Biol. Pharm. Bull.* **2014**, *37*, 1177–1183. [CrossRef] [PubMed]

255. Merghoub, N.; El Btaouri, H.; Benbacer, L.; Gmouh, S.; Trentesaux, C.; Brassart, B.; Attaleb, M.; Madoulet, C.; Wenner, T.; Amzazi, S.; et al. Tomentosin induces telomere shortening and caspase-dependant apoptosis in cervical cancer cells. *J. Cell. Biochem.* **2017**, *118*, 1689–1698. [CrossRef]

256. Göpfert, J.C.; Heil, N.; Conrad, J.; Spring, O. Cytological development and sesquiterpene lactone secretion in capitate glandular trichomes of sunflower. *Plant Biol.* **2005**, *7*, 148–155. [CrossRef] [PubMed]

257. Lee, C.M.; Lee, J.; Nam, M.J.; Choi, Y.S.; Park, S.H. Tomentosin displays anti-carcinogenic effect in human osteosarcoma MG-63 cells via the induction of intracellular reactive oxygen species. *Int. J. Mol. Sci.* **2019**, *20*, 1508. [CrossRef] [PubMed]

258. De Laurentis, N.; Losacco, V.; Milillo, M.A.; Lai, O. Chemical investigations of volatile constituents of *Inula viscosa* (L.) Aiton (Asteraceae) from different areas of Apulia, Southern Italy. *Delpinoa* **2002**, *44*, 115–119.

259. Bar-Shalom, R.; Bergman, M.; Grossman, S.; Azzam, N.; Sharvit, L.; Fares, F. *Inula Viscosa* extract inhibits growth of colorectal cancer cells *in vitro* and *in vivo* through induction of apoptosis. *Front. Oncol.* **2019**, *9*, 1–14. [CrossRef]

260. Cohen, Y.; Wang, W.; Ben-Daniel, B.H.; Ben-Daniel, Y. Extracts of *Inula viscosa* control downy mildew of grapes caused by *Plasmopara viticola*. *Phytopathology* **2006**, *96*, 417–424. [CrossRef]

261. Ahern, J.R.; Whitney, K.D. Stereochemistry affects sesquiterpene lactone bioactivity against an herbivorous grasshopper. *Chemoecology* **2014**, *24*, 35–39. [CrossRef]

Review

Therapeutic Potential of Rosmarinic Acid: A Comprehensive Review

Muhammad Nadeem [1], Muhammad Imran [2], Tanweer Aslam Gondal [3], Ali Imran [4], Muhammad Shahbaz [5], Rai Muhammad Amir [6], Muhammad Wasim Sajid [7], Tahira Batool Qaisrani [8], Muhammad Atif [9], Ghulam Hussain [10], Bahare Salehi [11,*], Elise Adrian Ostrander [12], Miquel Martorell [13], Javad Sharifi-Rad [14,*], William C. Cho [15,*] and Natália Martins [16,17,*]

1 Department of Environmental Sciences, Comsats University Islamabad, Vehari Campus, Vehari 61100, Pakistan
2 University Institute of Diet and Nutritional Sciences, Faculty of Allied Health Sciences, The University of Lahore, Lahore 54000, Pakistan
3 School of Exercise and Nutrition, Deakin University, Victoria 3221, Australia
4 Department of Food Science, Nutrition & Home Economics, Institute of Home and Food Sciences, Government College University Faisalabad, Faisalabad 38000, Pakistan
5 Department of Food Science and Technology, MNS-University of Agriculture, Multan 66000, Pakistan
6 Institute of Food and Nutritional Sciences, PMAS-Arid Agriculture University, Rawalpindi 46300, Pakistan
7 Department of Biosciences, COMSATS Institute of Information Technology, Sahiwal 57000, Pakistan
8 Department of Agricultural Engineering and Technology, Ghazi University, Dera Ghazi Khan 32200, Pakistan
9 Department of Clinical Laboratory Sciences, College of Applied Medical Sciences, Jouf University, Sakaka 72341, Saudi Arabia
10 Department of Physiology, Faculty of Life Sciences, Government College University, Faisalabad 38000, Pakistan
11 Student Research Committee, School of Medicine, Bam University of Medical Sciences, Bam 44340847, Iran
12 Medical Illustration, Kendall College of Art and Design, Ferris State University, Grand Rapids, MI 49503, USA
13 Department of Nutrition and Dietetics, Faculty of Pharmacy, University of Concepcion, Concepcion 4070386, Chile
14 Zabol Medicinal Plants Research Center, Zabol University of Medical Sciences, Zabol 61615-585, Iran
15 Department of Clinical Oncology, Queen Elizabeth Hospital, 30 Gascoigne Road, Hong Kong, China
16 Faculty of Medicine, University of Porto, Alameda Prof. Hernâni Monteiro, 4200-319 Porto, Portugal
17 Institute for Research and Innovation in Health (i3S), University of Porto, 4200-135 Porto, Portugal
* Correspondence: bahar.saheli007@gmail.com (B.S.); javad.sharifirad@gmail.com (J.S.-R.); chocs@ha.org.hk (W.C.C.); ncmartins@med.up.pt (N.M.)

Received: 26 June 2019; Accepted: 29 July 2019; Published: 2 August 2019

Abstract: Naturally occurring food-derived active ingredients have received huge attention for their chemopreventive and chemotherapy capabilities in several diseases. Rosmarinic acid (RA) is a caffeic acid ester and a naturally-occurring phenolic compound in a number of plants belonging to the Lamiaceae family, such as *Rosmarinus officinalis* (rosemary) from which it was formerly isolated. RA intervenes in carcinogenesis through different ways, including in tumor cell proliferation, apoptosis, metastasis, and inflammation. On the other hand, it also exerts powerful antimicrobial, anti-inflammatory, antioxidant and even antidepressant, anti-aging effects. The present review aims to provide an overview on anticancer activities of RA and to deliberate its therapeutic potential against a wide variety of diseases. Given the current evidence, RA may be considered as part of the daily diet in the treatment of several diseases, with pre-determined doses avoiding cytotoxicity.

Keywords: rosemary; rosmarinic acid; anticancer; antidiabetic; cardioprotective; antioxidant; oxidative stress

---

1. Introduction

Rosmarinic acid (RA) is an ester of caffeic acid and 3,4-dihydroxyphenyl lactic acid that occurs in nature as phenolic compounds. Its molecular formula is $C_{18}H_{16}O_8$ and is formally known as (R)-α-[[3-(3,4-dihydroxyphenyl)-1-oxo-2 E-propenyl]oxy]-3,4-dihydroxy-enzenepropanoic acid (Figure 1). The main sources of RA are plants belonging to the Boraginaceae family, subfamily Nepetoideae. It was isolated for the first time in 1958 from the rosemary plant (*Rosmarinus officinalis* L.), and recently it has been reported in *Forsythia koreana* (Rehder) Nakai, *Hyptis pectinate* (L.) Poit., *Ocimum tenuiflorum* L., *Thymus mastichina* (L.) L., and plants belonging to Lamiaceae family [1].

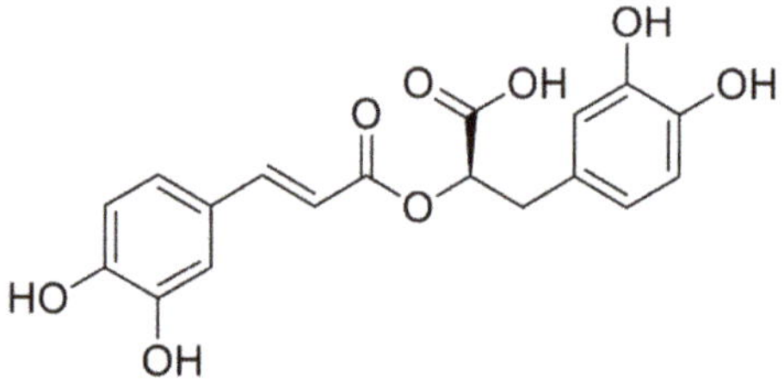

Figure 1. Chemical structure of rosmarinic acid.

RA has remarkable biological effects, including antiviral, antibacterial, anticancer, antioxidant, anti-aging, antidiabetic, cardioprotective, hepatoprotective, nephroprotective, antidepressant, antiallergic, and anti-inflammatory activities (Figure 2). RA and some rosemary extract-isolated compounds, like carnosic and ursolic acids and carnosol have also shown to be able to reduce the likelihood of tumor development in several body organs, such as stomach, colon, liver, breast, and leukemia cells [2–4]. Thus, here we review the various therapeutic potentials of RA in this article.

2. Bioavailability of Rosmarinic Acid and Its Metabolic Changes in the Human Body

RA is partially metabolized to the coumaric acid and caffeic acid in the body of the rat [5], and the hypolipidemic effect of RA may also be a consequence of the action of its metabolites. For example, caffeic acid inhibited the synthesis of hepatic fatty acid synthase, 3-hydroxy-3-methylglutaryl CoA reductase and acyl-CoA:cholesterol acyltransferase activities and increased fatty acid β-oxidation activity in high-fat diet-induced obese mice [6]. Caffeic acid and sinapic acid increased serum estradiol concentrations in rats with estrogen deficiency, which may have contributed to the observed metabolic effects [7]. In the rat ovary ovulation, external ovarian tissues such as fatty tissue, skin, bones and brain are the source of estradiol. In these sites, C19 cannot be synthetized. Steroids C19 (androgens) can be converted to estrogens by aromatase. Therefore, it seems possible that RA or its metabolites increase the activity of aromatase. Caffeic acid increased estradiol and reduced total cholesterol concentrations only in rats that were fed standard food containing soy, and these effects were not observed in rats fed without soy with reduced phenolic acid contents [8]. It is therefore possible that at least some of the RA effects reported depend on the diet. RA showed similar beneficial effects on some lipid parameters and insulin resistance (HOMA-IR) as that demonstrated for sinapic acid in a parallel study [7]. Moreover, RA had positive effects on expression of hepatic genes or proteins involved in signaling insulin and glucose and lipid metabolism, such as insulin receptor substrate-1 (IRS-1), 5′ AMP-activated protein kinase (AMPK), phosphoenolpyruvate carboxykinase (PEPCK), glucose transporter 2 (GLUT2), forkhead box protein O1 (FOXO1), sterol regulatory element-binding protein 1 (SREBP1), and carnitine palmitoyltransferase 1 (CPT1) in diabetic rats [9]. The possible mechanism of action of RA on glucose and lipid metabolism may be mediated by peroxisome proliferator-activated receptor (PPAR) peroxidation; RA has been shown to activate these receptors. It should be noted that the lower RA dose (10 mg/kg) was sufficient to reduce the HOMA-IR index and the concentration of fructosamine, while a higher dose (50 mg/kg) was required to reduce the total cholesterol and triglyceride levels in rats with estrogen deficiency. Moreover, RA and its metabolites can directly neutralize reactive oxygen species (ROS) [10] and thereby reduce the formation of oxidative damage products. The antioxidant activity of RA directly derives from its structure, namely the presence of 4 hydrogens in the phenolic system and two catecholic moieties, which give this compound polar character. Electrochemical studies have shown that RA oxidizes in two steps. In the first step, the rest of the caffeic acid is oxidized and in the second step the residue of 3,4-dihydroxyphenyl lactic acid.

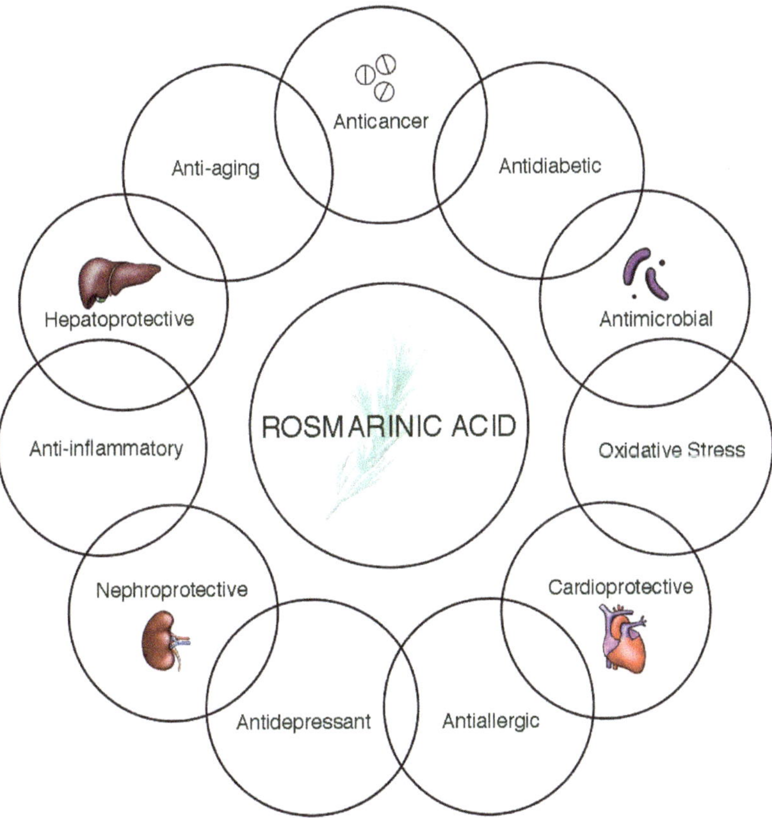

Figure 2. Rosmarinic acid and its potential functions.

RA is therefore considered to be the strongest antioxidant of all hydroxycinnamic acid derivatives [11]. Inhibition of the production of advanced glycation end products under the influence of RA was previously presented in vitro and in vivo [12]. The use of RA in doses of 10 and 50 mg/kg in rats with estrogen deficiency did not affect the body mass. RA administered at a dose of 10 mg/kg of ovariectomized rats did not affect estradiol and progesterone concentrations compared with ovariectomy control rats, whereas RA at a dose of 50 mg/kg of estradiol showed a trend of growth. Orchids containing RA are often used in self-healing and daily diets, so it is possible to consume 5–10 g of these plants daily in the form of infusions and spices [13]. RA is water-soluble, and according to literature data, the efficacy of secretion of this compound in infusions is about 90% [14]. Accordingly, it is possible to consume approximately 110 mg RA daily, i.e., approximately 1.6 mg/kg for adult men weighing 70 kg. Increasing the concentration of reduced glutathione (GSH) in plasma due to the use of RA was previously described in various models of diabetes [15]. RA has been shown to stimulate the regulation of the catalytic subunits of the glutamate cysteine ligase (the enzyme involved in the biosynthesis of GSH) in the hematopoietic stem cells [16]. It can be assumed that the increase in the concentration of GSH, after the administration of RA, was previously the result of the intense biosynthesis of GSH rather than its recovery from the oxidized form. Moreover, it should be noted that the RA appears to be absorbed into the rat mainly as its metabolites [5]. It is possible that these metabolites also play a role in the observed increase in GSH concentration. Furthermore, serum GSH/oxidized glutathione (GSSG) was calculated, as it is known to be an important indicator of redox cell status as well as for the state of redox at the tissue and whole body [17]. The adventitious effect of RA on redox homeostasis has been shown to increase the ratio of GSH/GSSG in serum rats.

3. Health Benefits of Rosmarinic Acid

3.1. Anticancer Potential

Several mechanisms have been proposed for RA anticancer activity (Figure 3). For instance, in rats with colon cancer, RA at the concentration of 5 mg/kg body weight (b.w.) impaired tumor formation and development, reduced lipid peroxidation by-products and pro-apoptotic proteins expression, modulated xenobiotic enzymes, and increased apoptotic proteins expression [18]. In human liver cancer cell line, HepG2, transfected with plasmid containing ARE-luciferin gene, RA predominantly enhances ARE-luciferin activity and promotes nuclear factor E2-related factor-2 (Nrf2) translocation from cytoplasm to the nucleus and also increases MRP2 and P-gp efflux activity along with intercellular ATP level [19]. A study conducted by Wu et al. [20] reported that RA inhibited CCRF-CEM and CEM/ADR5000 cells in a dose-dependent pattern but caused less cytotoxicity towards normal lymphocytes. RA concurrently induced necrosis and apoptosis and stimulated MMP dysfunction activated PARP-cleavage and caspase-independent apoptosis. RA also blocked the translocation of p65 from the cytosol to the nucleus [20]. Moreover, it inhibits transcription factor hypoxia-inducible factor-1α (HIF-1α) expression, which affects the glycolytic pathway; meanwhile, it also suppressed glucose consumption and lactate production in colorectal cells [21]. RA also inhibits micro RNAs and pro-inflammatory cytokines and thus may suppress the Warburg effects through an inflammatory pathway involving activator of transcription-3 (STAT3) and signal transducer of interleukin (IL)-6 [22]. Furthermore, RA inhibits HL-60 promyelocytic leukemia cells' growth and development and provides strong scavenging free radical effects, disturbing the balance of nuclear deoxyribonucleotide triphosphate (dNTP) levels without affecting protein levels of RR (R1, R2, p53R2) subunits, ultimately leading to apoptosis induction [23,24].

RA application, at a concentration of 5 mg/kg b.w. during 30 weeks in 1,2-dimethyldrazin stimulated colon carcinogenesis in the rat at 20 mg/kg b.w. and significantly stopped tumor formation and proliferation. RA supplementation also reduced tumor necrosis factor-α (TNF-α), cyclooxygenase-2 (COX-2) and IL-6 levels, and modulated p65 expression [25]. It is also able to inhibit the release of the highly mobile group box 1 (HMGB1) and to slow down HMGB1-dependent inflammatory responses in human endothelial cells, stopping HMBG1-mediated hyperpermeability and leukocytes migration in mice [26]. RA supplementation primarily decreases aberrant crypt foci (ACF) formation and multiplicity in rats [27].

RA inhibited APC10.1 cell growth that comes in Apc (Min) mouse model of colorectal carcinogenesis [28]. Through oral administration, RA totally prevented skin tumor cells formation in DMBA-induced mouse skin carcinogenesis and decreased lipid peroxidation byproducts levels [29]. It also inhibited human ovarian cancer A2780 cell line, disturbing the cell cycle at multiple phases and stimulating apoptosis by modifying multiple genes expression, involved in apoptosis regulation [30].

RA induced the cell cycle arrest and apoptosis in prostate cancer cell lines (PCa, PC-3, and DU145) [31]. These effects were mediated through modulation of histone deacetylases expression (HDACs), specifically HDAC2; the aberrant expression of theses enzymes is related with the onset of human cancer.

In rats with 1,2 dimethylhydrazine-induced colon carcinogenesis, RA administration at dose of 2.5, 5, and 10 mg/kg b.w. led to a decrease in the number of polyps (50%), reversed oxidative markers (21%), antioxidant status (38.6%), CYP450 contents (29.4%), and PNPH activities (21.9%) [32]. RA can also inhibit adhesion, invasion, and migration of Ls 174-T human colon carcinoma cells through enhancing GSH levels and decreasing ROS levels. Finally, RA may also inhibit colorectal carcinoma metastasis, by reducing the extracellular signal-regulated kinase pathway and the number and weight of lung tumors [33]. MDA-MB-231BO human bone homing breast cancer cells migrations are also inhibited by RA, whereas the number and size of mineralized nodules in ST-2 murine bone marrow stoma cells cultures raise [34].

RA also enhances chemosensitivity of human resistant gastric carcinoma SGC7901 cells [35]. The anticancer potential of RA analogues has also been tested. RA analogue-11 induces apoptosis of SGC7901 via the epidermal growth factor receptor (EGFR)/Akt/nuclear factor kappa B (NF-κB) pathway [36].

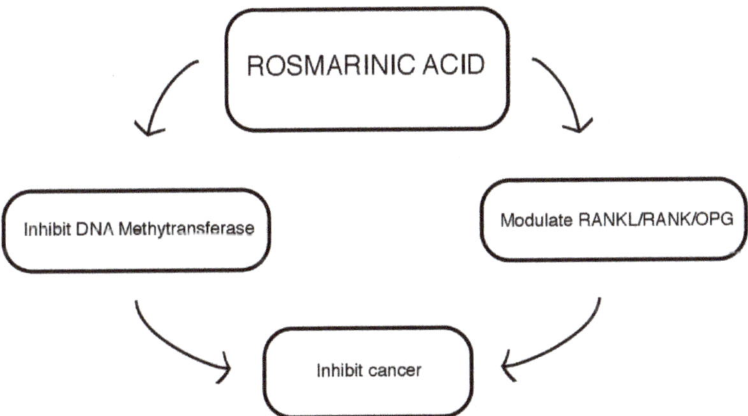

Figure 3. Mechanism of rosmarinic acid as an anticancer agent.

3.2. Antidiabetic Activity

RA supplementation increases the expression of mitochondrial biogenesis key genes, like sirtuin 1 (SIRT-1), peroxisome proliferator-activated receptor gamma coactivator 1-alpha (PGC-1α), and mitochondrial transcription factor A (TFAM) via activation of AMP-activated protein kinase (AMPK) in the skeletal muscle of insulin-resistant rats as well as in L6 myotubes. It also increased glucose faster and decreased serine IRS-1 phosphorylation, while increasing the glucose transporter type 4 (GLUT4) transfer [37].

In streptozotocin (STZ)-induced diabetic rats, RA exerted a noticeable hypoglycemic effect, whereas in high-fat diet (HFD) fed diabetic rats it increased glucose utilization and ameliorated insulin sensitivity. RA supplementation inverted the STZ- and HFD-induced increase in phosphoenolpyruvate carboxykinase (PEPCK) expression in the liver and the STZ- and HFD-induced decrease in GLUT4 expression in skeletal muscle. RA exerts hypoglycemic effects and improves insulin sensitivity, also increasing GLUT4 expression and decreasing PEPCK expression [38]. In addition, it also reverses memory and learning defects through improving cognition in healthy rats, inhibiting hyperglycemia, lipid peroxidation, and enhancing antioxidant defense system [39]. At a concentration of 10 mg/kg, RA decreased TBARS levels in kidney and liver of STZ-induced diabetic rats. This effect was mainly conferred by its ability to increase superoxide dismutase (SOD) and catalase (CAT) activity and to reverse the decrease in ascorbic acid and of non-protein thiol levels in diabetic rats [40]. RA administration also ameliorated oxidative stress markers in diabetic rats and water consumption and urination. Thus, it was proposed that RA mitigates STZ-induced diabetic manifestations by protecting rat's tissues against free radicals' damaging effects [15]. At 100 mg/kg, RA significantly increased insulin index sensitivity and reduced blood glucose, advanced glycation end-products, HbA1c, IL-1β, TNFα, IL-6, p-JNK, P38 mitogen-activated protein kinase (MAPK), and NF-κB levels. Moreover, it significantly reduced free fatty acids (FFA), triglycerides, serum cholesterol, AOPPs, lipid peroxides, and protein carbonyls levels in plasma and pancreas of diabetic rats. The reduced activities of CAT, SOD, glutathione *S*-transferases (GST), and glutathione peroxidase (GPx) and the reduced levels of vitamins C and E, ceruloplasmin, and GSH in plasma of diabetic rats were also significantly recovered by RA application. Furthermore, it protects pancreatic β-cells from oxidative stress in HFD-STZ-induced experimental diabetes [41]. The protective effects of RA (30 mg/kg) against hypoglycemia, hyperlipidemia, oxidative stress, and an imbalanced gut microbiota architecture was studied in diabetic rats. The treatment decreased the levels of fasting plasma glucose, total cholesterol, and triglyceride, exhibited an antioxidant and antiglycative effect, showed protective effects against tissue damage and inflammation in the abdominal aorta, increased the population of diabetes-resistant bacteria, and decreased the number of diabetes-sensitive bacteria [12].

RA also reduced diabetes occurrence and preserved normal insulin secretion, ROS, and reactive nitrogen species (RNS) by regulating antioxidant enzymes, and attenuating the pro-inflammatory T helper 2 and T regulatory cells levels [42]. At a concentration of 10 mg/kg, RA treatment significantly reduced lipid peroxidation levels in the hippocampus (28%), cortex (38%), and striatum (47%) of diabetic rats [43]. In Wistar rats, RA administrated orally at 50 mg/kg for 10 weeks in STZ-induced diabetes diminished endothelium-dependent relaxation accompanied by IL-1β, TNF-α, preproendothelin-1, and endothelin converting enzyme 1 overexpression. It also provided aortic endothelial function protection against diabetes-induced damage [44]. Finally, it significantly inhibited the carbohydrate-induced adaptive increase of sodium-dependent glucose cotransporter 1 (SGLT1) in the enterocyte brush border membrane [45].

3.3. Antimicrobial Activity

Regarding RA antimicrobial activity, it exerted antibacterial effects against *Staphylococcus aureus* strains, and the lowest blocking concentration was found to be 0.8 and 10 mg/mL against *S. aureus* and methicillin-resistant *S. aureus* (MRSA), respectively. Moreover, it displayed synergistic effects with amoxicillin, ofloxacin and vancomycin antibiotics against *S. aureus*, and only with vancomycin against MRSA. Time-kill analysis showed that using a combination of RA with antibiotics is more effective than using individual antibiotics. Microbial surface components recognizing adhesive matrix molecules (MSCRAMM) adhesion protein expression in MRSA and *S. aureus* was also significantly suppressed by using a combination of RA with vancomycin rather than RA alone [46].

On the other hand, RA administration reduced biofilm formation in a concentration- and time-dependent manner, suggesting that it could be used as an effective antimicrobial agent to kill the planktonic cells activity and to reduces the biofilm formation activity in early-stage development [47]. RA also exerts inhibitory effects against *Escherichia coli* K-12 and *Staphylococcus carnosus* LTH1502 growth, through decreasing cell counts and cell number [48]. Under acidic conditions, RA was reacted with nitrite ions to give 6,6-nitro and 6-dinitrorosmarinic acids. These compounds were active as HIV-1 integrase inhibitors at sub-molecular levels and inhibited viral replication in MT-4 cells. Without increasing cellular toxicity levels, RA nitration A strongly improved anti-integrase inhibition and antiviral effects [49]. Moreover, RA also exerted antimicrobial effects against *Enterobacteriaceae* spp., *Pseudomonas* spp., lactic acid bacteria, yeast and mold, and psychotropic counts, as well as fate *Listeria monocytogenes* inculcated in chicken meats [50]. Finally, RA also displayed inhibitory effects against *S. aureus* cocktail through inducing morphological changes and reducing viable cells counts and causing morphological changes in cheese and meat samples, such as cell shrinkage and appearance of blabbing-like structures in cell surfaces [51–53].

3.4. Cardioprotective Activity

RA at 25, 50, 10 mg/L had the capacity to maintain ATP levels in cells and inhibit the decrease in H/R-induced cell viability, lactate dehydrogenase (LDH) leakage, and excessive ROS. It also inhibited H/R-induced cardiomyocyte apoptosis and down-regulated p-AKt cleaved caspase expression [54].

The endothelial protein C receptor (EPCR) has a prominent role in inflammation and coagulation, whereas its activity is significantly changed by ectodomain cleavage and release as the soluble protein (sEPCR). RA has been found to be a strong anti-inflammatory agent. Monitoring RA effects in TNF-α, phorbol-12-myristate 13-acetate (PMA), IL-1β and in cecal ligation and puncture (CLP)-mediated EPCR shedding and underlying mechanisms, it was found that RA treatment led to a potent inhibition of PMA, TNF-α, and IL-induced EPCR shedding through TACE expression suppression. Furthermore, RA reduced extracellular regulated kinases (ERK) 1/2, PMA-stimulated p38, and c-Jun N-terminal kinase (JNK) phosphorylation. These results support the upcoming use of RA as an anti-sEPCR shedding reagent against IL-1β, TNF-α, PMA, and CLP-mediated EPCR shedding [55,56].

RA administration in fructose-fed rats (FFR) significantly enhanced insulin sensitivity, reduced lipid levels, oxidative damage, and a p22phox subunit of nicotinamide adenine dinucleotide phosphate reduced oxidase expression as well as prevented cardiac hypertrophy. RA lowered fructose-induced blood pressure through decreasing angiotensin-converting enzymes activity and endothelin-1 and increasing the level of nitric oxide (NO) [57]. RA also reduced fasting serum levels of vascular cell adhesion molecule 1 (VCAM-1), inter-cellular adhesion molecule 1 (ICAM-1), plasminogen-activator-inhibitor-1 (PAI-1), and increased GPX and SOD levels [58]. RA also intervenes in many important steps of angiogenesis, including adhesion, migration, proliferation, and tube formation of human umbilical vein endothelial cells (HUVEC) in a dose-dependent pattern. It also decreased IL-8 release from endothelial cells, H_2O_2-dependent vascular endothelial growth factor (VEGF) expression, and intracellular ROS levels [59]. Finally, in H9c2 cardiac muscle cells, RA inhibited apoptosis by decreasing intracellular ROS generation and recovering mitochondria membrane potential [60,61].

3.5. Antioxidant Activity

RA exhibited free radicals scavenging activity in hepatic stellate cells (HSCs) as a result of its antioxidant effects by boosting GSH synthesis and participating in NF-κB-dependent inhibition of MMP-2 activity. It also has the ability to reverse activated HSCs to quiescent cells and ultimately inhibits MMP-2 activity. RNA interference-imposed knockdown of NF-κB abolished MMP-2 down-regulation by RA. NF-κB inactivation mediated by RA could be blocked by the diphenyleneiodonium chloride, a potent inhibitor of NADH/NADPH oxidase. Moreover, transfection of dominant-negative (DN) mutant JNK1, p38α kinase, or extracellular signal-regulated kinases 2 (ERK2) had no such effect. At once, RA suppresses lipid peroxidation (LPO) and ROS generation, whereas in HSC-T6 cells it increases cellular GSH. Additionally, it significantly increases Nrf2 translocation and catalytic subunits from glutamate cysteine ligase (GCLc) expression but was not able to modulate GCL (GCLm) subunits and antioxidant response element (ARE)-mediated luciferase activity. GClc up-regulation mediated by RA is inhibited by shRNA-induced Nrf2 knockdown. The knocking down of Nrf2 abolished RA-mediated inhibition of ROS [16,62].

On the other hand, lycopene and RA administration reduced elevated blood urea nitrogen, renal malondialdehyde (MDA), proapoptotic protein (Bax) immuno-expression, serum creatinine, inducible nitric oxide synthase (iNOS), and autophagic marker protein (LC3/B) levels induced by gentamicin. This combination also increased the reduced SOD, an antiapoptotic protein (Bcl2) immuno-expression, GPx, and GSH levels and ameliorated gentamicin-induced histopathological changes. Moreover, it also evidenced a greater protective effect than corresponding monotherapy [63].

The in vivo antioxidant defense system consists of antioxidant enzymes including CAT, GPx, SOD, and nutritional antioxidants. Any disturbance in normal antioxidant defense system triggers several diseases, including diabetes, cancer, atherosclerosis, and degenerative diseases [64]. In experimental animals, carbon tetrachloride (CCl_4) induced neurotoxicity, whereas CCl_4 skin absorption, inhalation, and ingestion increased lipid peroxidation and reduced protein and antioxidant enzymes contents. This molecule produces free radicals in the lungs, heart, blood cells, and kidney. Under aerobic conditions, CCl_4 is converted into highly reactive trichloromethyl radicals through the action of the cytochrome P450 system [65].

Brain amyloid-β (Aβ) accumulation is a hallmark of Alzheimer's disease (AD) and has an important role in cognitive dysfunction [66]. At a dose of 0.25 mg/kg, RA significantly enhanced cognitive function and object discrimination and recognition test. Furthermore, RA decreased the time to reach the platform and increased the number of crossings over the removed platform, when compared with Aβ25-35-induced group in Morris water maze test; moreover, it reduced NO and MDA levels in kidney, brain, and liver [67]. RA also suppresses AD development by reducing amyloid β aggregation by increasing monoamine secretion in mice [68].

RA also prevented oxidative stress in C6 glial cells by increasing cell viability and inhibiting lipid peroxidation. It also decreased H_2O_2-induced COX-2 and iNOS expression at the transcriptional level and down-regulated COX-2 protein expression and iNOS in C6 glial cells treated with RA [67]. It significantly reduced oxidative stress and increased antioxidant status in Wistar rats post-spinal cord injury (SCI). RA also facilitated the inflammatory process through pro-inflammatory post-SCI and down-regulated NF-κB [69]. It also exerted a significant cytoprotective effect through covering the intercellular ROS in HaCat keratinocytes. RA also increased CAT, SOD, heme oxygenase-1 (HO-1), and transcription factor Nrf2 expression and activity, markedly reduced by UVB radiation [70].

On the other hand, in G93A-SOD1 transgenic mice with amyotrophic lateral sclerosis (ALS), RA application at a daily dose of 400 mg/kg significantly increased their survival by reliving function motor neurons deficits. These types of systematic changes were closely correlated with a decrease in neuronal loss and oxidative stress in ventral horns of G93A-SOD1 mice [71]. RA also increased SOD, GSH and GPx activities and decreased MDA levels in kidney and liver of sepsis-induced rats [72].

RA also reduced threshold shift, attenuated noise-induced hearing loss, and promoted hair cells survival. Moreover, it enhanced the endogenous antioxidant defense system by decreasing SOD production and up-regulation and decreased 4-HNE expression [73,74]. SOD, CAT, and GPx activities were increased through RA application at a dose of 50, 100, and 200 mg/kg. Furthermore, it induced structural changes in the kidney and liver at a dose of 200 mg/kg [75,76].

3.6. Hepatoprotective Activity

Sepsis, shock, and renal artery stenosis are the major clinical problems of acute renal failure, usually associated with high morbimortality rates. Ischemia-reperfusion (I-R) injury provokes cell damage, cell death, tissue necrosis, multiorgan dysfunction and increases vascular permeability. These types of physiopathological processes include RNS, ROS, neutrophils, cytokines, platelets, coagulation system, endothelium, and xanthine oxidoreductase enzyme system activation. During I-R injury, cell death occurs as a result of both apoptosis and necrosis [77–79]. RA modulates lipid peroxidation, production of ROS, peroxynitrite formation, complement factors and proinflammatory mediators, such as cytokines and chemokines. These processes are involved in hepatic diseases [80].

On the other hand, RA at dose of 150 mg/kg was able to treat rats with I-R, lowered lipid peroxidation and nitro tyrosine levels, enhanced GSH contents, and reduced neutrophil infiltration, hepatocellular damage, and all oxidative or nitrative stress markers. It also exerted anti-inflammatory and antioxidant effects in the ischemic liver, protecting hepatocytes from ischemic injury [81,82]. Moreover, it also conferred marked protection against oxidative stress through increasing CAT and GPx contents, thus preventing hepatic steatosis. Different RA derivatives have also been reported as having anti-secretory and antiulcer effects in epithelial tissues, being able to heal gastric ulceration [83]. Indeed, RA led to a significant reduction in hepatic toxicity, including alanine aminotransferase (ALT), aspartate aminotransferase (AST), lipid peroxidation, and oxidized glutathione levels and improved antioxidant effects of GPx, CAT, and SOD enzymes [26]. A marked improvement in liver serum markers and histology and inflammatory process decrease were also stated after RA administration at a dose of 10, 25, and 50 mg/kg by gavage once daily for two consecutive days against CCl_4-induced hepatic necrosis. Furthermore, RA prevented α-smooth muscle actin (α-SMA) and transforming growth factor β1 (TGF-β1) expression, suggesting profibrotic response suppression [84].

Peroxisome proliferator-activated receptor γ (PPARγ) is required for HSCs differentiation, and its epigenetic repression carries on the HSCs activation. RA inhibits HSCs signaling and expression by canonical Wnts as well as suppresses liver fibrosis progression and activation [85–87]. In CCl_4-induced rat liver fibrosis model, RA inhibited HSCs proliferation and TGF-β1, connective transforming growth factor (CTGF), and α-SMA expression. Much evidence has shown that RA can decrease fibrosis grade and ameliorate biochemical and histopathological morphology in CCl_4-induced liver fibrosis [88].

In lipopolysaccharide (LPS)-activated RAW164.7 cells, RA concentration-dependently down-regulated IL-6, TNF-α, and high mobility group box1 protein levels. RA also inhibited I kappa B kinase pathway and modulated NF-κB. RA intravenous injection also decreased puncture-induced lethality and cecal ligation in rats. Additionally, RA down-regulated serum IL-6 and TNF-α levels, triggering receptor expressed on myeloid cells, high mobility group box1 protein, and endotoxin whilst up-regulating the serum IL-6 level. Moreover, post-RA injection, a marked decrease in serum enzyme activities was observed, along with amelioration of liver, lungs, and small intestine hemodynamics; this anti-inflammatory mechanism may be explained through inhibition of NF-κB pathway activation by inhibiting I kappa B kinase activity [89]. RA also decreased ROS production and protein and DNA synthesis inhibition in a dose-dependent manner [90].

In extrahepatic cholestasis rat model by bile-duct ligation, RA showed hepatoprotective effect via mechanisms involving resolution of oxidative burden and down-regulation of HMGB1/TLR4, NF-κB, AP 1, and TGF-β1/Smad signaling [91].

3.7. Antidepressant Potential

RA effects at a dose of 0, 3, 10, or 30 mg/kg were investigated in female C57BL/6 mice for 60 min before pilocarpine (300 mg/kg) or pentylenetetrazol (PTZ, 60 mg/kg) injection. Generalized seizure duration and myoclonic generalized tonic-clonic seizure latencies were analyzed by electroencephalographic (EEG) and behavioral methods. The effect of an acute RA dose on mice was also evaluated in behavior in the open field, rotarod, novel object recognition, and forced swim tests. In the PTZ model, RA dose-dependently increased generalized seizures and latency to myoclonic jerks and improved pilocarpine-induced myoclonic jerks' latency. Additionally, RA (30 mg/kg) improved the time at the center of the open field, the crossings number and the immobility time in forced swim test [92].

Worldwide mood disorders are the most spreading forms of mental illness and, in main cases, morbidity. According to the World Health Organization (WHO), depression is one of the top causes of morbimortality throughout the world. Depression is a recurrent, potentially life-threatening, and chronic mood disorder that has been estimated to affect about 21% of the world population [93]. RA administration (10 mg/kg, daily) in chronic stress Sprague Dawley rats changed depressive behaviors in rats exposed to impulsive stress modal and restored hippocampal brain-derived neurotrophic factor (BDNF) and pERK1/2 protein expression. Thus, RA could be conceived as a great molecule in the treatment of depression and in triggering changes in BDNF levels and in ERK11/2 signaling in pharmacological science [94]. In PC 12 cells, RA showed significant neurotrophic effects and improved cholinergic functions in correlation with the ERK1/2 signaling pathway and MAPK. RA also caused an extensive up-regulation of pyruvate carboxylase (PC) and tyrosine hydroxylase (TH), involved in serotonergic, GABAergic, and dopaminergic pathways regulation, whereas against corticosterone-induced toxicity it provides neuronal cells protection [95].

3.8. Nephroprotective Activity

Currently, the most often used antimicrobial agents provoke an acute renal injury in 60% of acquired infections in hospitals with noticeable morbimortality rates [96]. Among them, aminoglycosides have been frequently used to treat bacterial infections, and gentamicin is the most commonly used aminoglycoside, given its lower costs and lower rate of antibiotic-acquired resistance; nevertheless, its therapeutic use at 80 mg/kg/day for more than 7 days induces nephrotoxicity in about 30% of patients [97]. Acute renal toxicity characterization is assessed by the sudden decrease of kidney function due to the accumulation of urea, creatinine, and other waste products. Gentamicin-induced nephrotoxicity is related to enhanced oxidative stress levels, which may be a major contributing factor for renal injury [98].

RA led to a decrease in serum blood urea nitrogen and creatinine levels in Sprague Dawley rats and ultimately decreased myeloperoxidase and MDA levels. In fact, the level of incident injury decreased in the RA-treated group. In addition, RA extensively reduced Bowman's capsules dilatation, glomerular necrosis, tubular epithelium degeneration, tubular epithelium necrosis and dilatation, and focal glomerular necrosis [99].

RA at doses of 1, 2, and 5 mg/kg for 2 days significantly increased serum creatinine and blood urea nitrogen levels and reduced cisplatin (CP)-induced histopathological changes. RA also reduced CP-produced oxidative stress and amplified cytochrome P450 2E1 (CYP2E1), HO-1, and renal-4-hydroxynonenal expression. Additionally, RA repressed TNF-α and NF-κB expression, as stated through inflammation inhibition. Moreover, RA reduced p53, phosphorylated p53, and active caspase-3-expression in kidney through exerting antiapoptotic activity [100]. RA also notably reduced MDA, tubular necrosis, urea, and creatinine levels, and increased renal GSH, SOD, CAT, GPS, volume density creatinine clearance, and PCT [101]. Finally, RA could provide protective effects against 6-hydroxydopamine-induced neurotoxicity via its antioxidant activity [102].

3.9. Anti-Aging Activity

RA administration could effectively reverse chaperones- and Pin1-induced abnormal changes and suppress P-tau and insoluble P-tau formation, induced by chronic restraint stress (CRS), particularly in middle-aged mice [103].

AD is a progressive neurodegenerative disorder that causes dementia in older people. Disease indicators include the appearance of plaques and tangles in brain tissues, which progressively kill neurons from brain cortex, amygdala, hippocampus, and other non-regeneratable brain regions. Consequently, acetylcholine (ACh) levels decline, the widely known cholinergic deficit hypothesis for AD. ACh has a significant role in brain functions, such as thinking, reasoning, remembering, and behavioral abilities [104,105].

In AD, stress is an important risk factor, since it induces tau phosphorylation and enhances tau insolubility in the brain. RA application is able to dominantly suppress the increase in tau phosphorylation levels and insoluble P-tau formation, facilitated by chronic resistant stress, and overturn the abnormal changes in middle-aged mice [103]. At 1.6, 16 and 32 mg/kg, RA exerted markedly useful effects on memory and learning and also reduced the levels of protein carbonyls in the hippocampus [106]. In ALS, a neurodegenerative disease, RA significantly delays motor neuron dysfunction in paw grip endurance tests, through attenuating motor neurons degeneration and extending the life span of ALS mice model, detected at later stages, and about 2% patients present an associated mutation in the gene encoding Cu/Zn-SOD [107].

RA also exerted protective effects against 6-hydroxydopamine-facilitated neurotoxicity and prevented 1-methyl-4-phenypyridinium effects in MES23.5 dopaminergic cells. In fact, 1-methyl-4-phenylpyridinium treatment reduces cell viability and dopamine contents, as well as causes apoptotic morphological changes. Additionally, 1-methyl-4-phenylpyridinium precedes mitochondrial dysfunction, easily detected through inhibiting mitochondrial respiratory chain complex 1-associated activity, suggesting mitochondrial transmembrane collapse and ROS generation. Thus, RA pretreatment was able to restore mitochondrial respiratory chain complex 1 activity and to reverse the other MPP positive damaging effects [91,92] partially. In mice, RA improves oxidative stress parameters and mitochondrial respiratory chain activity [108]. RA also proved to be effective in preventing in vitro amyloid peptide aggregation and in delaying disease progression in animal models [109]. It also provided neuroprotective effects against Aβ-induced toxicity by lowering lipid peroxidation, DNA and ROS formation, and inhibiting phosphorylated p38 MAPK levels [110,111]. RA improved antioxidant properties and healthspan via the IIS and MAPK pathways in *Caenorhabditis elegans* [112].

N2A cells' H_2O_2-induced cytotoxicity is also positively affected by RA; in fact, it is able to attenuate LDH, intercellular ROS, and mitochondrial membrane potential disruption. RA also promoted TH and BDNF genes up-regulation and prevented genotoxicity [113,114]. Clovamide in combination with RA, at the rate of 10–100 μM, in SH-SY5Y cells significantly enhanced PPARγ expression and inhibited NF-κB translocation, respectively [115].

3.10. Anti-Allergic Activity

RA significantly decreased murine double minute (MDM) 2 and thymic stromal lymphopoietin (TSLP) expression in induced mast cells proliferation. It also significantly decreased the levels of phosphorylated signal transducer, IL-13, and transcription-6 activation in TSLP-stimulated HMC-1 cells. Moreover, RA triggered an increment of p53 levels, poly-ADP-ribose polymerase cleavage, caspase-3 activation, and a reduction in Bcl2 and procaspase-3 levels. Furthermore, it significantly reduced TNF-α, IL-6, and IL-1β production in TSLP-stimulated HMC-1 cells. It also reduced IL-4, immunoglobulin E (IgE), and TSLP levels in short ragweed pollen-induced allergic conjunctivitis mouse model [116].

In murine model of respiratory allergy caused by Bloma tropicalis (Bt) mite, RA led to a considerable decrease in leukocytes or eosinophils numbers in bronchoalveolar lavage (BAL), of mucus presence in the respiratory tract, reduced lung histopathological changes, and eosinophil peroxidase activity and IL-4 changes [117,118]. At a dose of 1 or 5 μM in NC/Nga mice under specific pathogenic free conditions, RA was shown to be the most effective treatment against 2,4-dinitrofluorobenzene (DNFB)-induced AD-like skin lesion. Moreover, it suppressed IL-4 and interferon (INF) production by activated CD4$^+$ cells. RA also inhibited skin lesions and ears thickness development and increased total serum IgE levels in DNFB-treated NC/Nga mice [119]. Furthermore, it inhibited IgE levels increase in spleen, nasal mucosa, and serum, inhibited the increase in rubs number, and reduced histamine levels in ovalbumin unsensitized (OVA) mice. RA also inhibited protein levels and mRNA expressions of IL-6, IL-1β, and TNF-α in nasal mucosa or spleen tissues in OVA-sensitized mice [120,121]. In a murine model of allergic asthma triggered by house dust mites (HDMs), RA inhibited the boost up of mononuclear, eosinophils, and neutrophils cells levels around airways and in BAL fluid. Moreover, it also significantly inhibited IL-13 expression increased by HDM allergen [122].

RA (200 mg or 50 mg for 21 days) also reduced the number of eosinophils and neutrophils considerably in nasal lavage fluid. Up-regulation of VCAM-1, COX-2, MIP-2, and ICAM-1 by 2-tetradecanoylphorbol 13-acetate (TPA) were significantly reduced with RA pretreatment. ROS production detected as LPO, 8-hydroxy-2'deoxyguanosine (8OH-dG) and thiobarbituric acid reactive substance (TBARS) and TPA levels were markedly reduced by RA pretreatment. Furthermore, RA is conceived as a potential agent against seasonal allergic rhino conjunctivitis (SAR), through mediated polymorphonuclear leukocytes (PMNL) infiltration inhibition [123].

RA also inhibited significantly the increases in eosinophils levels in BAL fluid along with murine airways. In the lungs of sensitized mice, RA inhibited the increase in protein expression of IL-5, IL-4, and eotaxin. Thus, RA seems to be a successful intervention for allergic asthma, given its ability to increase chemokines, cytokines, and allergen-specific antibody levels [124]. Additionally, RA reduced inflammation and allergic immunoglobulin responses occurring in mice PMNL. It also noticeably increased the response rates for itchy eyes, watery eyes, and itchy nose and reduced eosinophils and neutrophils levels in nasal lavage fluid of SAR [125].

3.11. Anti-Inflammatory Activity

RA administration (40 mg/kg) in a rat model of sciatic nerve chronic constriction injury (CCI)-induced neuropathic pain reduced spinal inflammatory markers, such as matrix metallopeptidase 2 (MMP2), prostaglandin E2 (PGE-2), IL-1β, and COX-2 [126]. In addition, RA administration (75, 150, and 300 mg/kg) in hepatocellular carcinoma (HCC) for 10 days reduced inflammatory and angiogenic factors levels, including TNF-α, IL-6, IL-1β, TGF-β, and VEGF. Moreover, it also decreased NF-κB and p65 expression in the xenograft microenvironment [127].

On the other hand, RA (IC_{50} = 14.25 μM) inhibited LPS-induced NO production in RAW 264.7 cells. It also repressed LPS-induced pro-inflammatory cytokines expression, including INF-β, MCP-1, iNOS, IL-1β, IL-6, IL-10, and NF-κB activation. In dependent and independent pathways, down-regulation of iNOS by RA was due to myeloid differentiation primary response gene 88 (MyD88). Additionally, RA triggered HO-1 expression through inducing Nrf2 activity [92,128].

RA (5, 10, and 20 mg/kg) also inhibited Th2 cytokines in BAL fluid, increased inflammatory cells, ameliorated hyper airway responsiveness (AHR), and reduced total IgE and Ova-specific IgE concentrations in a murine model of asthma (female BALB/c mice). In upper airways, RA reduced the number of mucus hypersecretion and inflammatory cells. It seems that RA protective effects might be mediated by p38 phosphorylation, JNK, and ERK suppression. Additionally, RA pretreatment provoked the reduction of Ym2, CC chemokine receptor 3 (CCR3), CCL11 (eotaxin), AMCase, and E-selectin mRNA expression in lung tissues [129]. In SPI, RA noticeably increased antioxidant status and reduced oxidative stress levels in Wistar rats post-SCI. It also improved inflammatory mechanisms through pro-inflammatory cytokines reduction and NF-κB down regulation [69]. RA (10, 25, and 50 mg/kg) also remarkably decreased the serum transaminases (ALT and AST) and LDH concentration in liver ischemia-reperfusion rats. Furthermore, it reduced multiorgan dysfunction markers (lung, liver, and kidney) through metalloproteinase-9 and NF-κB modulating [130]. RA also displayed both peripheral and central antinociceptive effects and anti-inflammatory activity against chronic and acute inflammation [83,131–133]. In Freund's complete adjuvant (FCA)-induced arthritic rats, RA attenuates inflammation [134].

RA also decreased blocked TNF-α-induced NF-κB activation and cytotoxicity, oxygen-glucose deprivation (OGD)-induced apoptosis, and high-mobility group box1 (HMGB1) expression in SH-SY5Y cells. It also decreased brain edema, reduced NF-κB activation and HMGB1 expression, and attenuated histopathological damage at a dose of 50 mg/kg [135]. RA also considerably lowered allergic asthma through a significant decrease in the number of leukocytes/eosinophils, of mucus present in the respiratory tract, eosinophil peroxidase activity, and IL-4 and histopathological changes in lung BAL fluid [117,136]. RA-derived water and ethanol extract also markedly inhibited LPS-stimulated PGE-2 and NO production in a dose-dependent manner in RAW 264.7 mouse macrophages [137].

4. Conclusions

Current evidence supports the deepened exploration of RA as a promising therapeutic agent against a wide variety of modern lifestyle disorders (Table 1). However, mechanisms underlying RA's therapeutic activity need further investigation. Initial studies indicate that RA may act through various mechanisms, such as exerting anti-inflammatory and antioxidant effects as well as inhibiting cell proliferation, migration, and selectively inducing cancer cells apoptosis. In addition, RA's anti-angiogenic effects, as demonstrated through human umbilical vein endothelial cells proliferation, migration, adhesion, and tube formation inhibition, suggest that it can be beneficial in preventing tumor growth and metastasis. Thus, given the above-highlighted aspects, rosemary extract can be conceived as a rich source of potential candidates to be included in the diet with promising effects at pre-determined doses, avoiding toxicity.

Table 1. Bioactive effects of rosmarinic acid.

Bioactive Effects	Mechanisms	References
Anticancer	Prevent tumor formation development, reduce lipid peroxidation byproducts and proapoptotic proteins expression	[18]
	Cause cell cycle arrest and stimulate MMP dysfunction-activated PARP-cleavage Block p65 translocation from cytosol to the nucleus	[20]
	Inhibit HL-60 promyelocytic leukemia cells' growth and development Induce apoptosis	[23,24]
	Inhibit transcription factor HIF-1α expression	[21]
	Promote Nrf2 translocation from cytoplasm to the nucleus Increase MRP2 activity efflux	[19]
	Stop tumor formation and proliferation Reduce TNF-α, COX-2, IL-6 levels and modulates p65 expression	[25]
	Modulate histone deacetylases expression	[31]
Antidiabetic	Increase key genes expression involved in mitochondrial biogenesis like PGC-1α, SIRT-1, and TFAM via AMPK activation Decrease serine IRS-1phosphorylation and enhance GLUT4 translocation	[37]
	Increase GLUT4 expression and decrease PEPCK expression	[38]
	Enhance antioxidant defense system	[39]
	Reduce blood glucose, advanced glycation end-products, HbA1c, IL,1β, TNFα, IL6, p-JNK, P38 MAPK, and NF-kB Reduce FFA, triglycerides, serum cholesterol, AOPPs, lipid peroxides, and protein carbonyls levels	[41]
	Preserve normal insulin secretion Attenuate pro-inflammatory T helper 2 and T regulatory cells	[42]
	Increase the population of diabetes-resistant bacteria and decrease the number of diabetes-sensitive bacteria	[12]
Antimicrobial	Exhibit antibacterial activity against *S. aureus* Suppress MSCRAMM's protein expression in *S. aureus*	[46]
	Exert antimicrobial activity against *Enterobacteriaceae*, lactic acid bacteria, *Pseudomonas* spp., psychotropic, yeast, and mold	[50]
	Inhibit *S. carnosus* LTH1502 and *E. coli* K-12 LTH4263 growth	[48]
Cardioprotective	Inhibit H/R-induced cardiomyocyte apoptosis and down-regulate the expression of cleaved caspase of p-AKt	[54]
	Improve insulin sensitivity, reduce lipid levels and p22phox subunit of nicotinamide adenine dinucleotide phosphate reduced oxidase expression	[57]
	Inhibit PMA, TNF-α, IL-induced EPCR shedding by TACE expression suppression Reduce ERK1/2, PMA-stimulated p38 and JNK phosphorylation	[55,56]

Table 1. *Cont.*

Bioactive Effects	Mechanisms	References
Oxidative stress	Enhance cognitive function Reduce nitric oxide and MDA levels	[67]
	Inhibit cellular lipid peroxidation and decrease H_2O_2-induced COX-2 expression	[67]
	Enhance defense system of endogenous antioxidant Decrease 4-HNE expression	[73,74]
	Down-regulate NF-kB	[69]
	Increase CAT, HO-1, SOD activity and expression Reduce factor Nrf2 transcription	[70]
	Inhibit liver fibrosis progression and activation	[85,86]
	Prevent α-SMA expression and TGF-β1	[84]
Antidepressant	Restore hippocampal BDNF and pERK1/2 protein expression	[94]
	Inhibit monoamine oxidase and monoamine transporters	[138]
	Up-regulate PC and TH	[95]
Nephroprotective	Decrease serum levels of blood urea nitrogen and creatinine Decrease myeloperoxidase and MDA levels	[99]
	Repress TNF-α and NF-κB expression, demonstrating inhibition of inflammation Reduce p53, phosphorylated p53, and active caspase-3-expression	[100]
Anti-aging	Reduce protein carbonyls in the hippocampus	[106]
	Attenuate disruption of LDH, intercellular ROS, and mitochondrial membrane potential	[113]
	Improves oxidative stress parameters and mitochondrial respiratory chain activity	[108]
	Inhibit phosphorylated p38 MAPK	[110,111]
	Exert antioxidant effects against 6-hydroxydopamine facilitate neurotoxicity Restore activity of complex 1 of mitochondrial respiratory chain	[139,140]
	Prevent amyloid peptide aggregation	[68,109]
Anti-allergy	Decrease eosinophils number	[118]
	Inhibit IgE, protein levels and mRNA expressions of IL-6, IL-1β, and TNF-α and reduce histamine levels	[120,121]
	Suppress IL-4 and INF production Inhibit skin lesions development and ears thickness	[119]

Table 1. *Cont.*

Bioactive Effects	Mechanisms	References
Anti-inflammatory	Inhibit Th2 cytokines, ameliorate AHR Reduce total IgE and Ova-specific IgE concentrations Reduce Ym2, CCR3, CCL11, AMCase, and E-selectin mRNA expression	[129]
	Inhibit LPS-induced NO production Repress LPS-induced pro-inflammatory cytokines expression including INF-β, monocyte chemo attractant protein-1, iNOS, IL-1β, IL-6, IL-10, and activation of NF-κB	[92,128]
	Decrease serum transaminases (ALT and AST) and LDH levels Inhibit NF-κB	[130]
	Reduce oxidative stress levels and down-regulate NF-κB	[69]

Author Contributions: All authors contributed to the manuscript. Conceptualization: M.I., M.N. and J.S.-R.; Validation investigation, resources, data curation, writing: all authors; review and editing: B.S., J.S.-R., W.C.C. and N.M. All the authors read and approved the final manuscript.

Funding: None.

Acknowledgments: N. Martins thank to Portuguese Foundation for Science and Technology (FCT–Portugal) for the Strategic project ref. UID/BIM/04293/2013 and "NORTE2020-Programa Operacional Regional do Norte" (NORTE-01-0145-FEDER-000012). Also, this work was supported by CONICYT PIA/APOYO CCTE AFB170007.

Conflicts of Interest: The authors declare no conflict of interest.

References

1. Gordo, J.; Maximo, P.; Cabrita, E.; Lourenco, A.; Oliva, A.; Almeida, J.; Filipe, M.; Cruz, P.; Barcia, R.; Santos, M.; et al. *Thymus mastichina*: Chemical constituents and their anti-cancer activity. *Nat. Prod. Commun.* **2012**, *7*, 1491–1494. [CrossRef] [PubMed]

2. Baliga, M.S.; Jimmy, R.; Thilakchand, K.R.; Sunitha, V.; Bhat, N.R.; Saldanha, E.; Rao, S.; Rao, P.; Arora, R.; Palatty, P.L. *Ocimum sanctum* L. (Holy Basil or Tulsi) and its phytochemicals in the prevention and treatment of cancer. *Nutr. Cancer* **2013**, *65* (Suppl. 1), 26–35. [CrossRef] [PubMed]

3. Andrade, J.M.; Faustino, C.; Garcia, C.; Ladeiras, D.; Reis, C.P.; Rijo, P. *Rosmarinus officinalis* L.: An update review of its phytochemistry and biological activity. *Future Sci. OA* **2018**, *4*, FSO283. [CrossRef] [PubMed]

4. Ngo, S.N.; Williams, D.B.; Head, R.J. Rosemary and cancer prevention: Preclinical perspectives. *Crit. Rev. Food Sci. Nutr.* **2011**, *51*, 946–954. [CrossRef] [PubMed]

5. Baba, S.; Osakabe, N.; Natsume, M.; Terao, J. Orally administered rosmarinic acid is present as the conjugated and/or methylated forms in plasma, and is degraded and metabolized to conjugated forms of caffeic acid, ferulic acid and m-coumaric acid. *Life Sci.* **2004**, *75*, 165–178. [CrossRef] [PubMed]

6. Cho, A.S.; Jeon, S.M.; Kim, M.J.; Yeo, J.; Seo, K.I.; Choi, M.S.; Lee, M.K. Chlorogenic acid exhibits anti-obesity property and improves lipid metabolism in high-fat diet-induced-obese mice. *Food Chem. Toxicol. Int. J. Publ. Br. Ind. Biol. Res. Assoc.* **2010**, *48*, 937–943. [CrossRef] [PubMed]

7. Zych, M.; Kaczmarczyk-Sedlak, I.; Wojnar, W.; Folwarczna, J. The Effects of Sinapic Acid on the Development of Metabolic Disorders Induced by Estrogen Deficiency in Rats. *Oxid. Med. Cell. Longev.* **2018**, *2018*, 9274246. [CrossRef] [PubMed]

8. Folwarczna, J.; Pytlik, M.; Zych, M.; Cegiela, U.; Nowinska, B.; Kaczmarczyk-Sedlak, I.; Sliwinski, L.; Trzeciak, H.; Trzeciak, H.I. Effects of caffeic and chlorogenic acids on the rat skeletal system. *Eur. Rev. Med Pharmacol. Sci.* **2015**, *19*, 682–693. [PubMed]

9. Jayanthy, G.; Subramanian, S. RA abrogates hepatic gluconeogenesis and insulin resistance by enhancing IRS-1 and AMPK signalling in experimental type 2 diabetes. *RSC Adv.* **2015**, *5*, 44053–44067. [CrossRef]

10. Han, J.; Wang, D.; Ye, L.; Li, P.; Hao, W.; Chen, X.; Ma, J.; Wang, B.; Shang, J.; Li, D.; et al. Rosmarinic Acid Protects against Inflammation and Cardiomyocyte Apoptosis during Myocardial Ischemia/Reperfusion Injury by Activating Peroxisome Proliferator-Activated Receptor Gamma. *Front. Pharmacol.* **2017**, *8*, 456. [CrossRef] [PubMed]

11. Amoah, S.K.; Sandjo, L.P.; Kratz, J.M.; Biavatti, M.W. Rosmarinic Acid—Pharmaceutical and Clinical Aspects. *Planta Med.* **2016**, *82*, 388–406. [CrossRef] [PubMed]

12. Ou, J.; Huang, J.; Zhao, D.; Du, B.; Wang, M. Protective effect of rosmarinic acid and carnosic acid against streptozotocin-induced oxidation, glycation, inflammation and microbiota imbalance in diabetic rats. *Food Funct.* **2018**, *9*, 851–860. [CrossRef] [PubMed]

13. European Medicines Agency. *Assessment Report on Melissa officinalis L., Folium*; EMA/HMPC/196746/2012; European Medicines Agency: London, UK, 2013; pp. 1–20.

14. Fecka, I.; Turek, S. Determination of water-soluble polyphenolic compounds in commercial herbal teas from *Lamiaceae*: Peppermint, melissa, and sage. *J. Agric. Food Chem.* **2007**, *55*, 10908–10917. [CrossRef]

15. Sotnikova, R.; Kaprinay, B.; Navarova, J. Rosmarinic acid mitigates signs of systemic oxidative stress in streptozotocin-induced diabetes in rats. *Gen. Physiol. Biophys.* **2015**, *34*, 449–452. [CrossRef] [PubMed]

16. Lu, C.; Zou, Y.; Liu, Y.; Niu, Y. Rosmarinic acid counteracts activation of hepatic stellate cells via inhibiting the ROS-dependent MMP-2 activity: Involvement of Nrf2 antioxidant system. *Toxicol. Appl. Pharmacol.* **2017**, *318*, 69–78. [CrossRef] [PubMed]

17. Enns, G.M.; Cowan, T.M. Glutathione as a Redox Biomarker in Mitochondrial Disease-Implications for Therapy. *J. Clin. Med.* **2017**, *6*, 50. [CrossRef]

18. Venkatachalam, K.; Gunasekaran, S.; Namasivayam, N. Biochemical and molecular mechanisms underlying the chemopreventive efficacy of rosmarinic acid in a rat colon cancer. *Eur. J. Pharmacol.* **2016**, *791*, 37–50. [CrossRef]

19. Wu, J.; Zhu, Y.; Li, F.; Zhang, G.; Shi, J.; Ou, R.; Tong, Y.; Liu, Y.; Liu, L.; Lu, L.; et al. *Spica prunellae* and its marker compound rosmarinic acid induced the expression of efflux transporters through activation of Nrf2-mediated signaling pathway in HepG2 cells. *J. Ethnopharmacol.* **2016**, *193*, 1–11. [CrossRef]

20. Wu, C.F.; Hong, C.; Klauck, S.M.; Lin, Y.L.; Efferth, T. Molecular mechanisms of rosmarinic acid from *Salvia miltiorrhiza* in acute lymphoblastic leukemia cells. *J. Ethnopharmacol.* **2015**, *176*, 55–68. [CrossRef]

21. Xu, Y.; Han, S.; Lei, K.; Chang, X.; Wang, K.; Li, Z.; Liu, J. Anti-Warburg effect of rosmarinic acid via miR-155 in colorectal carcinoma cells. *Eur. J. Cancer Prev. Off. J. Eur. Cancer Prev. Organ.* **2016**, *25*, 481–489. [CrossRef]

22. Han, S.; Yang, S.; Cai, Z.; Pan, D.; Li, Z.; Huang, Z.; Zhang, P.; Zhu, H.; Lei, L.; Wang, W. Anti-Warburg effect of rosmarinic acid via miR-155 in gastric cancer cells. *Drug Des. Dev. Ther.* **2015**, *9*, 2695–2703. [CrossRef]

23. Saiko, P.; Steinmann, M.T.; Schuster, H.; Graser, G.; Bressler, S.; Giessrigl, B.; Lackner, A.; Grusch, M.; Krupitza, G.; Bago-Horvath, Z.; et al. Epigallocatechin gallate, ellagic acid, and rosmarinic acid perturb dNTP pools and inhibit de novo DNA synthesis and proliferation of human HL-60 promyelocytic leukemia cells: Synergism with arabinofuranosylcytosine. *Phytomed. Int. J. Phytother. Phytopharm.* **2015**, *22*, 213–222. [CrossRef] [PubMed]

24. Heo, S.K.; Noh, E.K.; Yoon, D.J.; Jo, J.C.; Koh, S.; Baek, J.H.; Park, J.H.; Min, Y.J.; Kim, H. Rosmarinic acid potentiates ATRA-induced macrophage differentiation in acute promyelocytic leukemia NB4 cells. *Eur. J. Pharmacol.* **2015**, *747*, 36–44. [CrossRef] [PubMed]

25. Karthikkumar, V.; Sivagami, G.; Viswanathan, P.; Nalini, N. Rosmarinic acid inhibits DMH-induced cell proliferation in experimental rats. *J. Basic Clin. Physiol. Pharmacol.* **2015**, *26*, 185–200. [CrossRef] [PubMed]

26. Yang, S.Y.; Hong, C.O.; Lee, G.P.; Kim, C.T.; Lee, K.W. The hepatoprotection of caffeic acid and rosmarinic acid, major compounds of *Perilla frutescens*, against t-BHP-induced oxidative liver damage. *Food Chem. Toxicol. Int. J. Publ. Br. Ind. Biol. Res. Assoc.* **2013**, *55*, 92–99. [CrossRef] [PubMed]

27. Karthikkumar, V.; Sivagami, G.; Vinothkumar, R.; Rajkumar, D.; Nalini, N. Modulatory efficacy of rosmarinic acid on premalignant lesions and antioxidant status in 1.2-dimethylhydrazine induced rat colon carcinogenesis. *Environ. Toxicol. Pharmacol.* **2012**, *34*, 949–958. [CrossRef] [PubMed]

28. Karmokar, A.; Marczylo, T.H.; Cai, H.; Steward, W.P.; Gescher, A.J.; Brown, K. Dietary intake of rosmarinic acid by Apc(Min) mice, a model of colorectal carcinogenesis: Levels of parent agent in the target tissue and effect on adenoma development. *Mol. Nutr. Food Res.* **2012**, *56*, 775–783. [CrossRef] [PubMed]

29. Sharmila, R.; Manoharan, S. Anti-tumor activity of rosmarinic acid in 7.12-dimethylbenz(a)anthracene (DMBA) induced skin carcinogenesis in Swiss albino mice. *Indian J. Exp. Biol.* **2012**, *50*, 187–194.

30. Tai, J.; Cheung, S.; Wu, M.; Hasman, D. Antiproliferation effect of Rosemary (*Rosmarinus officinalis*) on human ovarian cancer cells in vitro. *Phytomed. Int. J. Phytother. Phytopharm.* **2012**, *19*, 436–443. [CrossRef]

31. Jang, Y.G.; Hwang, K.A.; Choi, K.C. Rosmarinic Acid, a Component of Rosemary Tea, Induced the Cell Cycle Arrest and Apoptosis through Modulation of HDAC2 Expression in Prostate Cancer Cell Lines. *Nutrients* **2018**, *10*, 1784. [CrossRef]

32. Venkatachalam, K.; Gunasekaran, S.; Jesudoss, V.A.; Namasivayam, N. The effect of rosmarinic acid on 1.2-dimethylhydrazine induced colon carcinogenesis. *Exp. Toxicol. Pathol. Off. J. Ges. Fur Toxikol. Pathol.* **2013**, *65*, 409–418. [CrossRef] [PubMed]

33. Xu, Y.; Xu, G.; Liu, L.; Xu, D.; Liu, J. Anti-invasion effect of rosmarinic acid via the extracellular signal-regulated kinase and oxidation-reduction pathway in Ls174-T cells. *J. Cell. Biochem.* **2010**, *111*, 370–379. [CrossRef] [PubMed]

34. Xu, Y.; Jiang, Z.; Ji, G.; Liu, J. Inhibition of bone metastasis from breast carcinoma by rosmarinic acid. *Planta Med.* **2010**, *76*, 956–962. [CrossRef] [PubMed]

35. Yu, C.; Chen, D.Q.; Liu, H.X.; Li, W.B.; Lu, J.W.; Feng, J.F. Rosmarinic acid reduces the resistance of gastric carcinoma cells to 5-fluorouracil by downregulating FOXO4-targeting miR-6785-5p. *Biomed. Pharmacother.* **2019**, *109*, 2327–2334. [CrossRef] [PubMed]

36. Li, W.; Li, Q.; Wei, L.; Pan, X.; Huang, D.; Gan, J.; Tang, S. Rosmarinic Acid Analogue-11 Induces Apoptosis of Human Gastric Cancer SGC-7901 Cells via the Epidermal Growth Factor Receptor (EGFR)/Akt/Nuclear Factor kappa B (NF-κB) Pathway. *Med. Sci. Monit. Basic Res.* **2019**, *25*, 63–75. [CrossRef] [PubMed]

37. Jayanthy, G.; Roshana Devi, V.; Ilango, K.; Subramanian, S.P. Rosmarinic Acid Mediates Mitochondrial Biogenesis in Insulin Resistant Skeletal Muscle Through Activation of AMPK. *J. Cell. Biochem.* **2017**, *118*, 1839–1848. [CrossRef]

38. Runtuwene, J.; Cheng, K.C.; Asakawa, A.; Amitani, H.; Amitani, M.; Morinaga, A.; Takimoto, Y.; Kairupan, B.H.; Inui, A. Rosmarinic acid ameliorates hyperglycemia and insulin sensitivity in diabetic rats, potentially by modulating the expression of PEPCK and GLUT4. *Drug Des. Dev. Ther.* **2016**, *10*, 2193–2202. [CrossRef]

39. Hasanein, P.; Felehgari, Z.; Emamjomeh, A. Preventive effects of *Salvia officinalis* L. against learning and memory deficit induced by diabetes in rats: Possible hypoglycaemic and antioxidant mechanisms. *Neurosci. Lett.* **2016**, *622*, 72–77. [CrossRef]

40. Mushtaq, N.; Schmatz, R.; Ahmed, M.; Pereira, L.B.; da Costa, P.; Reichert, K.P.; Dalenogare, D.; Pelinson, L.P.; Vieira, J.M.; Stefanello, N.; et al. Protective effect of rosmarinic acid against oxidative stress biomarkers in liver and kidney of strepotozotocin-induced diabetic rats. *J. Physiol. Biochem.* **2015**, *71*, 743–751. [CrossRef]

41. Govindaraj, J.; Sorimuthu Pillai, S. Rosmarinic acid modulates the antioxidant status and protects pancreatic tissues from glucolipotoxicity mediated oxidative stress in high-fat diet: Streptozotocin-induced diabetic rats. *Mol. Cell. Biochem.* **2015**, *404*, 143–159. [CrossRef]

42. Vujicic, M.; Nikolic, I.; Kontogianni, V.G.; Saksida, T.; Charisiadis, P.; Orescanin-Dusic, Z.; Blagojevic, D.; Stosic-Grujicic, S.; Tzakos, A.G.; Stojanovic, I. Methanolic extract of *Origanum vulgare* ameliorates type 1 diabetes through antioxidant, anti-inflammatory and anti-apoptotic activity. *Br. J. Nutr.* **2015**, *113*, 770–782. [CrossRef]

43. Mushtaq, N.; Schmatz, R.; Pereira, L.B.; Ahmad, M.; Stefanello, N.; Vieira, J.M.; Abdalla, F.; Rodrigues, M.V.; Baldissarelli, J.; Pelinson, L.P.; et al. Rosmarinic acid prevents lipid peroxidation and increase in acetylcholinesterase activity in brain of streptozotocin-induced diabetic rats. *Cell Biochem. Funct.* **2014**, *32*, 287–293. [CrossRef]

44. Sotnikova, R.; Okruhlicova, L.; Vlkovicova, J.; Navarova, J.; Gajdacova, B.; Pivackova, L.; Fialova, S.; Krenek, P. Rosmarinic acid administration attenuates diabetes-induced vascular dysfunction of the rat aorta. *J. Pharm. Pharmacol.* **2013**, *65*, 713–723. [CrossRef]

45. Azevedo, M.F.; Lima, C.F.; Fernandes-Ferreira, M.; Almeida, M.J.; Wilson, J.M.; Pereira-Wilson, C. Rosmarinic acid, major phenolic constituent of Greek sage herbal tea, modulates rat intestinal SGLT1 levels with effects on blood glucose. *Mol. Nutr. Food Res.* **2011**, *55* (Suppl. 1), S15–S25. [CrossRef] [PubMed]

46. Ekambaram, S.P.; Perumal, S.S.; Balakrishnan, A.; Marappan, N.; Gajendran, S.S.; Viswanathan, V. Antibacterial synergy between rosmarinic acid and antibiotics against methicillin-resistant *Staphylococcus aureus*. *J. Intercult. Ethnopharmacol.* **2016**, *5*, 358–363. [CrossRef]

47. Slobodnikova, L.; Fialova, S.; Hupkova, H.; Grancai, D. Rosmarinic acid interaction with planktonic and biofilm *Staphylococcus aureus*. *Nat. Prod. Commun.* **2013**, *8*, 1747–1750. [CrossRef]

48. Suriyarak, S.; Bayrasy, C.; Schmidt, H.; Villeneuve, P.; Weiss, J. Impact of fatty acid chain length of rosmarinate esters on their antimicrobial activity against *Staphylococcus carnosus* LTH1502 and *Escherichia coli* K-12 LTH4263. *J. Food Prot.* **2013**, *76*, 1539–1548. [CrossRef]

49. Dubois, M.; Bailly, F.; Mbemba, G.; Mouscadet, J.F.; Debyser, Z.; Witvrouw, M.; Cotelle, P. Reaction of rosmarinic acid with nitrite ions in acidic conditions: Discovery of nitro- and dinitrorosmarinic acids as new anti-HIV-1 agents. *J. Med. Chem.* **2008**, *51*, 2575–2579. [CrossRef]

50. Raeisi, M.; Tabaraei, A.; Hashemi, M.; Behnampour, N. Effect of sodium alginate coating incorporated with nisin, *Cinnamomum zeylanicum*, and rosemary essential oils on microbial quality of chicken meat and fate of *Listeria monocytogenes* during refrigeration. *Int. J. Food Microbiol.* **2016**, *238*, 139–145. [CrossRef]

51. Honorio, V.G.; Bezerra, J.; Souza, G.T.; Carvalho, R.J.; Gomes-Neto, N.J.; Figueiredo, R.C.; Melo, J.V.; Souza, E.L.; Magnani, M. Inhibition of *Staphylococcus aureus* cocktail using the synergies of oregano and rosemary essential oils or carvacrol and 1.8-cineole. *Front. Microbiol.* **2015**, *6*, 1223. [CrossRef]

52. Zhuang, Y.; Jiang, J.; Bi, H.; Yin, H.; Liu, S.; Liu, T. Synthesis of rosmarinic acid analogues in *Escherichia coli*. *Biotechnol. Lett.* **2016**, *38*, 619–627. [CrossRef] [PubMed]

53. Suriyarak, S.; Gibis, M.; Schmidt, H.; Villeneuve, P.; Weiss, J. Antimicrobial mechanism and activity of dodecyl rosmarinate against *Staphylococcus carnosus* LTH1502 as influenced by addition of salt and change in pH. *J. Food Prot.* **2014**, *77*, 444–452. [CrossRef] [PubMed]

54. Li, X.L.; Liu, J.X.; Li, P.; Zheng, Y.Q. Protective effect of rosmarinic acid on hypoxia/reoxygenation injury in cardiomyocytes. *J. Chin. Mater. Med.* **2014**, *39*, 1897–1901.

55. Ku, S.K.; Yang, E.J.; Song, K.S.; Bae, J.S. Rosmarinic acid down-regulates endothelial protein C receptor shedding in vitro and in vivo. *Food Chem. Toxicol. Int. J. Publ. Br. Ind. Biol. Res. Assoc.* **2013**, *59*, 311–315. [CrossRef] [PubMed]

56. Ferreira, L.G.; Celotto, A.C.; Capellini, V.K.; Albuquerque, A.A.; Nadai, T.R.; Carvalho, M.T.; Evora, P.R. Is rosmarinic acid underestimated as an experimental cardiovascular drug? *Acta Cir. Bras.* **2013**, *28* (Suppl. 1), 83–87. [CrossRef]

57. Karthik, D.; Viswanathan, P.; Anuradha, C.V. Administration of rosmarinic acid reduces cardiopathology and blood pressure through inhibition of p22phox NADPH oxidase in fructose-fed hypertensive rats. *J. Cardiovasc. Pharmacol.* **2011**, *58*, 514–521. [CrossRef]

58. Sinkovic, A.; Suran, D.; Lokar, L.; Fliser, E.; Skerget, M.; Novak, Z.; Knez, Z. Rosemary extracts improve flow-mediated dilatation of the brachial artery and plasma PAI-1 activity in healthy young volunteers. *Phytother. Res.* **2011**, *25*, 402–407. [CrossRef]

59. Huang, S.S.; Zheng, R.L. Rosmarinic acid inhibits angiogenesis and its mechanism of action in vitro. *Cancer Lett.* **2006**, *239*, 271–280. [CrossRef]

60. Kim, D.S.; Kim, H.R.; Woo, E.R.; Hong, S.T.; Chae, H.J.; Chae, S.W. Inhibitory effects of rosmarinic acid on adriamycin-induced apoptosis in H9c2 cardiac muscle cells by inhibiting reactive oxygen species and the activations of c-Jun N-terminal kinase and extracellular signal-regulated kinase. *Biochem. Pharmacol.* **2005**, *70*, 1066–1078. [CrossRef]

61. Chlopcikova, S.; Psotova, J.; Miketova, P.; Sousek, J.; Lichnovsky, V.; Simanek, V. Chemoprotective effect of plant phenolics against anthracycline-induced toxicity on rat cardiomyocytes. Part II. caffeic, chlorogenic and rosmarinic acids. *Phytother. Res.* **2004**, *18*, 408–413. [CrossRef]

62. Adomako-Bonsu, A.G.; Chan, S.L.; Pratten, M.; Fry, J.R. Antioxidant activity of rosmarinic acid and its principal metabolites in chemical and cellular systems: Importance of physico-chemical characteristics. *Toxicol. Vitr. Int. J. Publ. Assoc.* **2017**, *40*, 248–255. [CrossRef] [PubMed]

63. Bayomy, N.A.; Elbakary, R.H.; Ibrahim, M.A.A.; Abdelaziz, E. Effect of Lycopene and Rosmarinic Acid on Gentamicin Induced Renal Cortical Oxidative Stress, Apoptosis, and Autophagy in Adult Male Albino Rat. *Anat. Rec.* **2017**, *300*, 1137–1149. [CrossRef] [PubMed]

64. Govindarajan, R.; Vijayakumar, M.; Pushpangadan, P. Antioxidant approach to disease management and the role of 'RasayanA' herbs of Ayurveda. *J. Ethnopharmacol.* **2005**, *99*, 165–178. [CrossRef] [PubMed]

65. Khan, M.R.; Rizvi, W.; Khan, G.N.; Khan, R.A.; Shaheen, S. Carbon tetrachloride-induced nephrotoxicity in rats: Protective role of *Digera muricata*. *J. Ethnopharmacol.* **2009**, *122*, 91–99. [CrossRef] [PubMed]

66. Gandy, S. The role of cerebral amyloid beta accumulation in common forms of Alzheimer disease. *J. Clin. Investig.* **2005**, *115*, 1121–1129. [CrossRef]

67. Lee, A.Y.; Wu, T.T.; Hwang, B.R.; Lee, J.; Lee, M.H.; Lee, S.; Cho, E.J. The Neuro-Protective Effect of the Methanolic Extract of *Perilla frutescens* var. japonica and Rosmarinic Acid against $H_{(2)}O_{(2)}$-Induced Oxidative Stress in C_6 Glial Cells. *Biomol. Ther.* **2016**, *24*, 338–345. [CrossRef]

68. Hase, T.; Shishido, S.; Yamamoto, S.; Yamashita, R.; Nukima, H.; Taira, S.; Toyoda, T.; Abe, K.; Hamaguchi, T.; Ono, K.; et al. Rosmarinic acid suppresses Alzheimer's disease development by reducing amyloid beta aggregation by increasing monoamine secretion. *Sci. Rep.* **2019**, *9*, 8711. [CrossRef]

69. Shang, A.J.; Yang, Y.; Wang, H.Y.; Tao, B.Z.; Wang, J.; Wang, Z.F.; Zhou, D.B. Spinal cord injury effectively ameliorated by neuroprotective effects of rosmarinic acid. *Nutr. Neurosci.* **2017**, *20*, 172–179. [CrossRef]

70. Fernando, P.M.; Piao, M.J.; Kang, K.A.; Ryu, Y.S.; Hewage, S.R.; Chae, S.W.; Hyun, J.W. Rosmarinic Acid Attenuates Cell Damage against UVB Radiation-Induced Oxidative Stress via Enhancing Antioxidant Effects in Human HaCaT Cells. *Biomol. Ther.* **2016**, *24*, 75–84. [CrossRef]

71. Seo, J.S.; Choi, J.; Leem, Y.H.; Han, P.L. Rosmarinic Acid Alleviates Neurological Symptoms in the G93A-SOD1 Transgenic Mouse Model of Amyotrophic Lateral Sclerosis. *Exp. Neurobiol.* **2015**, *24*, 341–350. [CrossRef]

72. Bacanli, M.; Aydin, S.; Taner, G.; Goktas, H.G.; Sahin, T.; Basaran, A.A.; Basaran, N. Does rosmarinic acid treatment have protective role against sepsis-induced oxidative damage in Wistar Albino rats? *Hum. Exp. Toxicol.* **2016**, *35*, 877–886. [CrossRef] [PubMed]

73. Fetoni, A.R.; Paciello, F.; Rolesi, R.; Eramo, S.L.; Mancuso, C.; Troiani, D.; Paludetti, G. Rosmarinic acid up-regulates the noise-activated Nrf2/HO-1 pathway and protects against noise-induced injury in rat cochlea. *Free Radic. Biol. Med.* **2015**, *85*, 269–281. [CrossRef] [PubMed]

74. Khamse, S.; Sadr, S.S.; Roghani, M.; Hasanzadeh, G.; Mohammadian, M. Rosmarinic acid exerts a neuroprotective effect in the kainate rat model of temporal lobe epilepsy: Underlying mechanisms. *Pharm. Biol.* **2015**, *53*, 1818–1825. [CrossRef] [PubMed]

75. Zhang, Y.; Chen, X.; Yang, L.; Zu, Y.; Lu, Q. Effects of rosmarinic acid on liver and kidney antioxidant enzymes, lipid peroxidation and tissue ultrastructure in aging mice. *Food Funct.* **2015**, *6*, 927–931. [CrossRef] [PubMed]

76. Nabavi, S.F.; Tenore, G.C.; Daglia, M.; Tundis, R.; Loizzo, M.R.; Nabavi, S.M. The cellular protective effects of rosmarinic acid: From bench to bedside. *Curr. Neurovascular Res.* **2015**, *12*, 98–105. [CrossRef]

77. Yazihan, N.; Ataoglu, H.; Kavas, G.O.; Akyurek, N.; Yener, B.; Aydm, C. The effect of K-ATP channel blockage during erythropoietin treatment in renal ischemia-reperfusion injury. *J. Investig. Surg. Off. J. Acad. Surg. Res.* **2008**, *21*, 340–347. [CrossRef] [PubMed]

78. Fan, L.H.; He, L.; Cao, Z.Q.; Xiang, B.; Liu, L. Effect of ischemia preconditioning on renal ischemia/reperfusion injury in rats. *Int. Braz. J. Urol. Off. J. Braz. Soc. Urol.* **2012**, *38*, 842–854. [CrossRef] [PubMed]

79. Sedaghat, Z.; Kadkhodaee, M.; Seifi, B.; Salehi, E.; Najafi, A.; Dargahi, L. Remote preconditioning reduces oxidative stress, downregulates cyclo-oxygenase-2 expression and attenuates ischaemia-reperfusion-induced acute kidney injury. *Clin. Exp. Pharmacol. Physiol.* **2013**, *40*, 97–103. [CrossRef] [PubMed]

80. Elufioye, T.O.; Habtemariam, S. Hepatoprotective effects of rosmarinic acid: Insight into its mechanisms of action. *Biomed. Pharmacother.* **2019**, *112*, 108600. [CrossRef] [PubMed]

81. Ferlemi, A.V.; Katsikoudi, A.; Kontogianni, V.G.; Kellici, T.F.; Iatrou, G.; Lamari, F.N.; Tzakos, A.G.; Margarity, M. Rosemary tea consumption results to anxiolytic- and anti-depressant-like behavior of adult male mice and inhibits all cerebral area and liver cholinesterase activity; phytochemical investigation and *in silico* studies. *Chem. Biol. Interact.* **2015**, *237*, 47–57. [CrossRef]

82. Ramalho, L.N.; Pasta, A.A.; Terra, V.A.; Augusto, M.; Sanches, S.C.; Souza-Neto, F.P.; Cecchini, R.; Gulin, F.; Ramalho, F.S. Rosmarinic acid attenuates hepatic ischemia and reperfusion injury in rats. *Food Chem. Toxicol. Int. J. Publ. Br. Ind. Biol. Res. Assoc.* **2014**, *74*, 270–278. [CrossRef] [PubMed]

83. Kamyab, A.A.; Eshraghian, A. Anti-Inflammatory, gastrointestinal and hepatoprotective effects of *Ocimum sanctum* Linn: An ancient remedy with new application. *Inflamm. Allergy Drug Targets* **2013**, *12*, 378–384. [CrossRef] [PubMed]

84. Domitrovic, R.; Skoda, M.; Vasiljev Marchesi, V.; Cvijanovic, O.; Pernjak Pugel, E.; Stefan, M.B. Rosmarinic acid ameliorates acute liver damage and fibrogenesis in carbon tetrachloride-intoxicated mice. *Food Chem. Toxicol. Int. J. Publ. Br. Ind. Biol. Res. Assoc.* **2013**, *51*, 370–378. [CrossRef] [PubMed]

85. Yang, M.D.; Chiang, Y.M.; Higashiyama, R.; Asahina, K.; Mann, D.A.; Mann, J.; Wang, C.C.; Tsukamoto, H. Rosmarinic acid and baicalin epigenetically derepress peroxisomal proliferator-activated receptor gamma in hepatic stellate cells for their antifibrotic effect. *Hepatology* **2012**, *55*, 1271–1281. [CrossRef] [PubMed]

86. Tandogan, B.; Kuruuzum-Uz, A.; Sengezer, C.; Guvenalp, Z.; Demirezer, L.O.; Ulusu, N.N. In vitro effects of rosmarinic acid on glutathione reductase and glucose 6-phosphate dehydrogenase. *Pharm. Biol.* **2011**, *49*, 587–594. [CrossRef]

87. Zhang, J.J.; Wang, Y.L.; Feng, X.B.; Song, X.D.; Liu, W.B. Rosmarinic acid inhibits proliferation and induces apoptosis of hepatic stellate cells. *Biol. Pharm. Bull.* **2011**, *34*, 343–348. [CrossRef]

88. Li, G.S.; Jiang, W.L.; Tian, J.W.; Qu, G.W.; Zhu, H.B.; Fu, F.H. In vitro and in vivo antifibrotic effects of rosmarinic acid on experimental liver fibrosis. *Phytomed. Int. J. Phytother. Phytopharm.* **2010**, *17*, 282–288. [CrossRef]

89. Jiang, W.L.; Chen, X.G.; Qu, G.W.; Yue, X.D.; Zhu, H.B.; Tian, J.W.; Fu, F.H. Rosmarinic acid protects against experimental sepsis by inhibiting proinflammatory factor release and ameliorating hemodynamics. *Shock* **2009**, *32*, 608–613. [CrossRef]

90. Renzulli, C.; Galvano, F.; Pierdomenico, L.; Speroni, E.; Guerra, M.C. Effects of rosmarinic acid against aflatoxin B1 and ochratoxin-A-induced cell damage in a human hepatoma cell line (Hep G2). *J. Appl. Toxicol.* **2004**, *24*, 289–296. [CrossRef]

91. Lin, S.Y.; Wang, Y.Y.; Chen, W.Y.; Liao, S.L.; Chou, S.T.; Yang, C.P.; Chen, C.J. Hepatoprotective activities of rosmarinic acid against extrahepatic cholestasis in rats. *Food Chem. Toxicol. Int. J. Publ. Br. Ind. Biol. Res. Assoc.* **2017**, *108*, 214–223. [CrossRef]

92. Grigoletto, J.; Oliveira, C.V.; Grauncke, A.C.; Souza, T.L.; Souto, N.S.; Freitas, M.L.; Furian, A.F.; Santos, A.R.; Oliveira, M.S. Rosmarinic acid is anticonvulsant against seizures induced by pentylenetetrazol and pilocarpine in mice. *Epilepsy Behav.* **2016**, *62*, 27–34. [CrossRef] [PubMed]

93. Berton, O.; Nestler, E.J. New approaches to antidepressant drug discovery: Beyond monoamines. *Nat. Rev. Neurosci.* **2006**, *7*, 137–151. [CrossRef] [PubMed]

94. Jin, X.; Liu, P.; Yang, F.; Zhang, Y.H.; Miao, D. Rosmarinic acid ameliorates depressive-like behaviors in a rat model of CUS and Up-regulates BDNF levels in the hippocampus and hippocampal-derived astrocytes. *Neurochem. Res.* **2013**, *38*, 1828–1837. [CrossRef] [PubMed]

95. Sasaki, K.; El Omri, A.; Kondo, S.; Han, J.; Isoda, H. *Rosmarinus officinalis* polyphenols produce anti-depressant like effect through monoaminergic and cholinergic functions modulation. *Behav. Brain Res.* **2013**, *238*, 86–94. [CrossRef] [PubMed]

96. Dashti-Khavidaki, S.; Moghaddas, A.; Heydari, B.; Khalili, H.; Lessan-Pezeshki, M.; Lessan-Pezeshki, M. Statins against drug-induced nephrotoxicity. *J. Pharm. Pharm. Sci.* **2013**, *16*, 588–608. [CrossRef] [PubMed]

97. Balakumar, P.; Rohilla, A.; Thangathirupathi, A. Gentamicin-induced nephrotoxicity: Do we have a promising therapeutic approach to blunt it? *Pharmacol. Res.* **2010**, *62*, 179–186. [CrossRef] [PubMed]

98. Kang, C.; Lee, H.; Hah, D.Y.; Heo, J.H.; Kim, C.H.; Kim, E.; Kim, J.S. Protective Effects of *Houttuynia cordata* Thunb. on Gentamicin-induced Oxidative Stress and Nephrotoxicity in Rats. *Toxicol. Res.* **2013**, *29*, 61–67. [CrossRef]

99. Ozturk, H.; Ozturk, H.; Terzi, E.H.; Ozgen, U.; Duran, A.; Uygun, I. Protective effects of rosmarinic acid against renal ischaemia/reperfusion injury in rats. *J. Pak. Med Assoc.* **2014**, *64*, 260–265.

100. Domitrovic, R.; Potocnjak, I.; Crncevic-Orlic, Z.; Skoda, M. Nephroprotective activities of rosmarinic acid against cisplatin-induced kidney injury in mice. *Food Chem. Toxicol.* **2014**, *66*, 321–328. [CrossRef]

101. Tavafi, M.; Ahmadvand, H. Effect of rosmarinic acid on inhibition of gentamicin induced nephrotoxicity in rats. *Tissue Cell* **2011**, *43*, 392–397. [CrossRef]

102. Ren, P.; Jiang, H.; Li, R.; Wang, J.; Song, N.; Xu, H.M.; Xie, J.X. Rosmarinic acid inhibits 6-OHDA-induced neurotoxicity by anti-oxidation in MES23.5 cells. *J. Mol. Neurosci.* **2009**, *39*, 220–225. [CrossRef]

103. Shan, Y.; Wang, D.D.; Xu, Y.X.; Wang, C.; Cao, L.; Liu, Y.S.; Zhu, C.Q. Aging as a Precipitating Factor in Chronic Restraint Stress-Induced Tau Aggregation Pathology, and the Protective Effects of Rosmarinic Acid. *J. Alzheimer's Dis.* **2016**, *49*, 829–844. [CrossRef] [PubMed]

104. Kim, J.; Lee, H.J.; Lee, K.W. Naturally occurring phytochemicals for the prevention of Alzheimer's disease. *J. Neurochem.* **2010**, *112*, 1415–1430. [CrossRef]

105. Holland, D.; Desikan, R.S.; Dale, A.M.; McEvoy, L.K. Rates of decline in Alzheimer disease decrease with age. *PLoS ONE* **2012**, *7*, e42325. [CrossRef]

106. Farr, S.A.; Niehoff, M.L.; Ceddia, M.A.; Herrlinger, K.A.; Lewis, B.J.; Feng, S.; Welleford, A.; Butterfield, D.A.; Morley, J.E. Effect of botanical extracts containing carnosic acid or rosmarinic acid on learning and memory in SAMP8 mice. *Physiol. Behav.* **2016**, *165*, 328–338. [CrossRef]

107. Shimojo, Y.; Kosaka, K.; Noda, Y.; Shimizu, T.; Shirasawa, T. Effect of rosmarinic acid in motor dysfunction and life span in a mouse model of familial amyotrophic lateral sclerosis. *J. Neurosci. Res.* **2010**, *88*, 896–904. [CrossRef] [PubMed]

108. Luft, J.G.; Steffens, L.; Moras, A.M.; da Rosa, M.S.; Leipnitz, G.; Regner, G.G.; Pfluger, P.F.; Goncalves, D.; Moura, D.J.; Pereira, P. Rosmarinic acid improves oxidative stress parameters and mitochondrial respiratory chain activity following 4-aminopyridine and picrotoxin-induced seizure in mice. *Naunyn-Schmiedeberg's Arch. Pharmacol.* **2019**. [CrossRef]

109. Airoldi, C.; Sironi, E.; Dias, C.; Marcelo, F.; Martins, A.; Rauter, A.P.; Nicotra, F.; Jimenez-Barbero, J. Natural compounds against Alzheimer's disease: Molecular recognition of Abeta1-42 peptide by *Salvia sclareoides* extract and its major component, rosmarinic acid, as investigated by NMR. *Chem. Asian J.* **2013**, *8*, 596–602. [CrossRef]

110. Iuvone, T.; De Filippis, D.; Esposito, G.; D'Amico, A.; Izzo, A.A. The spice sage and its active ingredient rosmarinic acid protect PC12 cells from amyloid-beta peptide-induced neurotoxicity. *J. Pharmacol. Exp. Ther.* **2006**, *317*, 1143–1149. [CrossRef]

111. Dashti, A.; Soodi, M.; Amani, N. Cr (VI) induced oxidative stress and toxicity in cultured cerebellar granule neurons at different stages of development and protective effect of Rosmarinic acid. *Environ. Toxicol.* **2016**, *31*, 269–277. [CrossRef]

112. Lin, C.; Xiao, J.; Xi, Y.; Zhang, X.; Zhong, Q.; Zheng, H.; Cao, Y.; Chen, Y. Rosmarinic acid improved antioxidant properties and healthspan via the IIS and MAPK pathways in Caenorhabditis elegans. *BioFactors* **2019**. [CrossRef] [PubMed]

113. Ghaffari, H.; Venkataramana, M.; Jalali Ghassam, B.; Chandra Nayaka, S.; Nataraju, A.; Geetha, N.P.; Prakash, H.S. Rosmarinic acid mediated neuroprotective effects against H_2O_2-induced neuronal cell damage in N2A cells. *Life Sci.* **2014**, *113*, 7–13. [CrossRef] [PubMed]

114. Braidy, N.; Matin, A.; Rossi, F.; Chinain, M.; Laurent, D.; Guillemin, G.J. Neuroprotective effects of rosmarinic acid on ciguatoxin in primary human neurons. *Neurotox. Res.* **2014**, *25*, 226–234. [CrossRef] [PubMed]

115. Fallarini, S.; Miglio, G.; Paoletti, T.; Minassi, A.; Amoruso, A.; Bardelli, C.; Brunelleschi, S.; Lombardi, G. Clovamide and rosmarinic acid induce neuroprotective effects in in vitro models of neuronal death. *Br. J. Pharmacol.* **2009**, *157*, 1072–1084. [CrossRef] [PubMed]

116. Yoou, M.S.; Park, C.L.; Kim, M.H.; Kim, H.M.; Jeong, H.J. Inhibition of MDM2 expression by rosmarinic acid in TSLP-stimulated mast cell. *Eur. J. Pharmacol.* **2016**, *771*, 191–198. [CrossRef] [PubMed]

117. Costa, R.S.; Carneiro, T.C.; Cerqueira-Lima, A.T.; Queiroz, N.V.; Alcantara-Neves, N.M.; Pontes-de-Carvalho, L.C.; Velozo Eda, S.; Oliveira, E.J.; Figueiredo, C.A. *Ocimum gratissimum* Linn. and rosmarinic acid, attenuate eosinophilic airway inflammation in an experimental model of respiratory allergy to *Blomia tropicalis*. *Int. Immunopharmacol.* **2012**, *13*, 126–134. [CrossRef] [PubMed]

118. Zhu, F.; Asada, T.; Sato, A.; Koi, Y.; Nishiwaki, H.; Tamura, H. Rosmarinic acid extract for antioxidant, antiallergic, and alpha-glucosidase inhibitory activities, isolated by supramolecular technique and solvent extraction from *Perilla* leaves. *J. Agric. Food Chem.* **2014**, *62*, 885–892. [CrossRef]

119. Jang, A.H.; Kim, T.H.; Kim, G.D.; Kim, J.E.; Kim, H.J.; Kim, S.S.; Jin, Y.H.; Park, Y.S.; Park, C.S. Rosmarinic acid attenuates 2.4-dinitrofluorobenzene-induced atopic dermatitis in NC/Nga mice. *Int. Immunopharmacol.* **2011**, *11*, 1271–1277. [CrossRef]

120. Oh, H.A.; Park, C.S.; Ahn, H.J.; Park, Y.S.; Kim, H.M. Effect of *Perilla frutescens* var. acuta Kudo and rosmarinic acid on allergic inflammatory reactions. *Exp. Biol. Med.* **2011**, *236*, 99–106. [CrossRef]

121. Lee, J.; Jung, E.; Koh, J.; Kim, Y.S.; Park, D. Effect of rosmarinic acid on atopic dermatitis. *J. Dermatol.* **2008**, *35*, 768–771. [CrossRef]

122. Inoue, K.; Takano, H.; Shiga, A.; Fujita, Y.; Makino, H.; Yanagisawa, R.; Ichinose, T.; Kato, Y.; Yamada, T.; Yoshikawa, T. Effects of volatile constituents of a rosemary extract on allergic airway inflammation related to house dust mite allergen in mice. *Int. J. Mol. Med.* **2005**, *16*, 315–319. [CrossRef] [PubMed]

123. Osakabe, N.; Takano, H.; Sanbongi, C.; Yasuda, A.; Yanagisawa, R.; Inoue, K.; Yoshikawa, T. Anti-inflammatory and anti-allergic effect of rosmarinic acid (RA); inhibition of seasonal allergic rhinoconjunctivitis (SAR) and its mechanism. *BioFactors* **2004**, *21*, 127–131. [CrossRef] [PubMed]

124. Sanbongi, C.; Takano, H.; Osakabe, N.; Sasa, N.; Natsume, M.; Yanagisawa, R.; Inoue, K.I.; Sadakane, K.; Ichinose, T.; Yoshikawa, T. Rosmarinic acid in *Perilla* extract inhibits allergic inflammation induced by mite allergen, in a mouse model. *Clin. Exp. Allergy J. Br. Soc. Allergy Clin. Immunol.* **2004**, *34*, 971–977. [CrossRef] [PubMed]

125. Takano, H.; Osakabe, N.; Sanbongi, C.; Yanagisawa, R.; Inoue, K.; Yasuda, A.; Natsume, M.; Baba, S.; Ichiishi, E.; Yoshikawa, T. Extract of *Perilla frutescens* enriched for rosmarinic acid, a polyphenolic phytochemical, inhibits seasonal allergic rhinoconjunctivitis in humans. *Exp. Biol. Med.* **2004**, *229*, 247–254. [CrossRef] [PubMed]

126. Ghasemzadeh Rahbardar, M.; Amin, B.; Mehri, S.; Mirnajafi-Zadeh, S.J.; Hosseinzadeh, H. Anti-inflammatory effects of ethanolic extract of *Rosmarinus officinalis* L. and rosmarinic acid in a rat model of neuropathic pain. *Biomed. Pharmacother.* **2017**, *86*, 441–449. [CrossRef] [PubMed]

127. Cao, W.; Hu, C.; Wu, L.; Xu, L.; Jiang, W. Rosmarinic acid inhibits inflammation and angiogenesis of hepatocellular carcinoma by suppression of NF-kappaB signaling in H22 tumor-bearing mice. *J. Pharmacol. Sci.* **2016**, *132*, 131–137. [CrossRef] [PubMed]

128. So, Y.; Lee, S.Y.; Han, A.R.; Kim, J.B.; Jeong, H.G.; Jin, C.H. Rosmarinic Acid Methyl Ester Inhibits LPS-Induced NO Production via Suppression of MyD88-Dependent and -Independent Pathways and Induction of HO-1 in RAW 264.7 Cells. *Molecules* **2016**, *21*, 1083. [CrossRef]

129. Liang, Z.; Xu, Y.; Wen, X.; Nie, H.; Hu, T.; Yang, X.; Chu, X.; Yang, J.; Deng, X.; He, J. Rosmarinic Acid Attenuates Airway Inflammation and Hyperresponsiveness in a Murine Model of Asthma. *Molecules* **2016**, *21*, 769. [CrossRef]

130. Rocha, J.; Eduardo-Figueira, M.; Barateiro, A.; Fernandes, A.; Brites, D.; Bronze, R.; Duarte, C.M.; Serra, A.T.; Pinto, R.; Freitas, M.; et al. Anti-inflammatory effect of rosmarinic acid and an extract of *Rosmarinus officinalis* in rat models of local and systemic inflammation. *Basic Clin. Pharmacol. Toxicol.* **2015**, *116*, 398–413. [CrossRef]

131. Boonyarikpunchai, W.; Sukrong, S.; Towiwat, P. Antinociceptive and anti-inflammatory effects of rosmarinic acid isolated from *Thunbergia laurifolia* Lindl. *Pharmacol. Biochem. Behav.* **2014**, *124*, 67–73. [CrossRef]

132. Usha, T.; Middha, S.K.; Bhattacharya, M.; Lokesh, P.; Goyal, A.K. Rosmarinic Acid, a New Polyphenol from *Baccaurea ramiflora* Lour. Leaf: A Probable Compound for Its Anti-Inflammatory Activity. *Antioxidants* **2014**, *3*, 830–842. [CrossRef] [PubMed]

133. Lucarini, R.; Bernardes, W.A.; Ferreira, D.S.; Tozatti, M.G.; Furtado, R.; Bastos, J.K.; Pauletti, P.M.; Januario, A.H.; Silva, M.L.; Cunha, W.R. In vivo analgesic and anti-inflammatory activities of *Rosmarinus officinalis* aqueous extracts, rosmarinic acid and its acetyl ester derivative. *Pharm. Biol.* **2013**, *51*, 1087–1090. [CrossRef] [PubMed]

134. Gautam, R.K.; Gupta, G.; Sharma, S.; Hatware, K.; Patil, K.; Sharma, K.; Goyal, S.; Chellappan, D.K.; Dua, K. Rosmarinic acid attenuates inflammation in experimentally induced arthritis in Wistar rats, using Freund's complete adjuvant. *Int. J. Rheum. Dis.* **2019**, *22*, 1247–1254. [CrossRef] [PubMed]

135. Luan, H.; Kan, Z.; Xu, Y.; Lv, C.; Jiang, W. Rosmarinic acid protects against experimental diabetes with cerebral ischemia: Relation to inflammation response. *J. Neuroinflammation* **2013**, *10*, 28. [CrossRef] [PubMed]

136. Pearson, W.; Fletcher, R.S.; Kott, L.S. Oral rosmarinic acid-enhanced *Mentha spicata* modulates synovial fluid biomarkers of inflammation in horses challenged with intra-articular LPS. *J. Vet. Pharmacol. Ther.* **2012**, *35*, 495–502. [CrossRef] [PubMed]

137. Huang, N.; Hauck, C.; Yum, M.Y.; Rizshsky, L.; Widrlechner, M.P.; McCoy, J.A.; Murphy, P.A.; Dixon, P.M.; Nikolau, B.J.; Birt, D.F. Rosmarinic acid in *Prunella vulgaris* ethanol extract inhibits lipopolysaccharide-induced prostaglandin E2 and nitric oxide in RAW 264.7 mouse macrophages. *J. Agric. Food Chem.* **2009**, *57*, 10579–10589. [CrossRef] [PubMed]

138. Abdelhalim, A.; Karim, N.; Chebib, M.; Aburjai, T.; Khan, I.; Johnston, G.A.; Hanrahan, J. Antidepressant, Anxiolytic and Antinociceptive Activities of Constituents from *Rosmarinus officinalis*. *J. Pharm. Pharm. Sci.* **2015**, *18*, 448–459. [CrossRef] [PubMed]

139. Du, T.; Li, L.; Song, N.; Xie, J.; Jiang, H. Rosmarinic acid antagonized 1-methyl-4-phenylpyridinium (MPP+)-induced neurotoxicity in MES23.5 dopaminergic cells. *Int. J. Toxicol.* **2010**, *29*, 625–633. [CrossRef] [PubMed]

140. Demirezer, L.O.; Gurbuz, P.; Kelicen Ugur, E.P.; Bodur, M.; Ozenver, N.; Uz, A.; Guvenalp, Z. Molecular docking and ex vivo and in vitro anticholinesterase activity studies of *Salvia* sp. and highlighted rosmarinic acid. *Turk. J. Med Sci.* **2015**, *45*, 1141–1148. [CrossRef] [PubMed]

MDPI

St. Alban-Anlage 66

4052 Basel

Switzerland

Tel. +41 61 683 77 34

Fax +41 61 302 89 18

www.mdpi.com

Applied Sciences Editorial Office

E-mail: applsci@mdpi.com

www.mdpi.com/journal/applsci